RESISTANCE '91:

ACHIEVEMENTS AND DEVELOPMENTS IN COMBATING PESTICIDE RESISTANCE

Proceedings of the SCI Symposium 'Resistance '91: Achievements and Developments in Combating Pesticide Resistance', held at Rothamsted Experimental Station, Harpenden, UK, 15–17 July 1991.

ORGANISING COMMITTEE

Dr I. Denholm, AFRC Institute of Arable Crops Research, Rothamsted Experimental Station

Dr A. L. Devonshire, AFRC Institute of Arable Crops Research, Rothamsted Experimental Station

Dr M. G. Ford, Portsmouth Polytechnic

Dr M. B. Green, Independent Consultant

Dr D. W. Hollomon, AFRC Institute of Arable Crops Research, Long Ashton Research Station

Dr B. P. S. Khambay, AFRC Institute of Arable Crops Research, Rothamsted Experimental Station

Dr R. Greenwood (*Treasurer*), Portsmouth Polytechnic

RESISTANCE '91:

ACHIEVEMENTS AND DEVELOPMENTS IN COMBATING PESTICIDE RESISTANCE

Edited by

IAN DENHOLM

ALAN L. DEVONSHIRE

Rothamsted Experimental Station, Harpenden, Herts, UK

and

DEREK W. HOLLOMON

Long Ashton Experimental Station, Bristol, UK

Published for SCI
by
ELSEVIER APPLIED SCIENCE
LONDON and NEW YORK

ELSEVIER SCIENCE PUBLISHERS LTD
Crown House, Linton Road, Barking, Essex IG11 8JU, England

WITH 35 TABLES AND 55 ILLUSTRATIONS

© 1992 SCI

British Library Cataloguing in Publication Data

Denholm, Ian
  Resistance '91: Achievements and
  Developments in Combating Pesticide
  Resistance.—1991: Proceedings of the
  SCI Symposium Held at Rothamsted
  Experimental Station, Harpenden, UK,
  15–17 July 1991
  I. Title
  632

  ISBN 1-85166-886-1

Library of Congress CIP data applied for

Preface

The development of pesticide resistance in arthropod pests, plant pathogens and weeds can be viewed and studied from two contrasting perspectives. At a fundamental level, resistance provides an almost ideal example of adaptation to withstand severe environmental stress. Population geneticists, biochemists and, most recently, molecular biologists have cast considerable light on the nature of this adaptation in diverse taxonomic groups, and on factors determining its selection and spread within and between populations.

Unlike most evolutionary phenomena, however, resistance is also of immediate practical and economic significance. Not only has the number of resistant species continued to increase inexorably, but there has been an alarming increase in the severity and extent of some resistance problems. Cases of organisms resisting virtually all available pesticides are by no means uncommon, and pose a formidable challenge in view of present difficulties in discovering and developing novel chemicals. Although most occurrences of resistance were initially monofactorial, resistance now frequently involves a suite of coexisting mechanisms that protect organisms against the same or different pesticide groups, and may even predispose them to resist new, as yet unused chemicals.

Against this backdrop, the SCI organised the Resistance '91 International Symposium at Rothamsted Experimental Station in July 1991, to review progress in reconciling fundamental and practical perspectives to comprehend and overcome the challenge of resistance. The symposium also provided a timely opportunity to examine how rapid advances in molecular biology can complement conventional chemical and biological approaches to pest and disease management, and thereby help to reduce resistance risks. By aspiring to these objectives, Resistance '91 followed in the tradition of an earlier SCI Meeting in July 1986 at Southampton University, and of other conferences held in the intervening period—most notably the American Chemical Society Symposium 'Managing Resistance to Agrochemicals' at Los Angeles in September 1988. Like both earlier meetings, Resistance '91 brought together specialists from several disciplines, from public and private sectors, and sought to integrate work on the three pesticide classes—insecticides, fungicides and herbicides—most threatened by the development of resistance.

In publishing the Symposium Proceedings the editors have retained the original programme structure, grouping papers under five interrelated themes. The first covers techniques used or advocated to monitor resistance, and aspects of the analysis and interpretation of field monitoring data. The second presents examples of control strategies already implemented to manage resistance, or at an advanced stage of planning. It emphasises the essential involvement of academics, extension specialists and industrialists for the success of these measures, and ends with an outline of a major new collaborative venture for combating resistance on a global scale. The third group of papers investigates factors determining the dynamics of resistance genes, and in particular the prospects for integrating ecological and genetic data for a more accurate prediction of the build-up and spread of resistance. The fourth section demonstrates how current research in genetics, biochemistry and molecular biology is not only resolving the nature of resistance mechanisms, but is breaking exciting new ground in fundamental areas such as nerve function and the regulation of gene expression. Papers in the final section deal with advances in biotechnology and novel approaches to pest and disease management; they provide enticing glimpses of the likely make-up of future control strategies. New toxophores, biopesticides, semiochemicals and genetically transformed crops expressing insecticidal proteins all have potential to relieve pressure on existing chemicals, but none of these is likely to be entirely free of resistance risks. This being so, we can be certain that current research to underpin resistance management for conventional pesticides will remain highly relevant amid the rapidly changing face of crop protection through the 1990s and beyond.

IAN DENHOLM
ALAN L. DEVONSHIRE
DEREK W. HOLLOMON

Acknowledgements

The organisers extend their sincere thanks to the following companies for financial contributions which were used to sponsor the attendance of some speakers and provide bursaries for the attendance of student delegates from colleges and universities:

Bayer AG
BASF AG
CIBA-Geigy AG
DuPont Agricultural Products
ICI Agrochemicals plc
ISK Biotech Europe Ltd
Rhône-Poulenc Agriculture
Sandoz Agrochemicals Ltd
Schering Agrochemicals Ltd
Shell Research Ltd
Sumitomo Chemicals (UK) Ltd
Takeda Chemical Industries Ltd

Roman Sawicki: An Appreciation

Roman M. Sawicki FRS

With the untimely death of Roman Sawicki on 22 July 1990, the pesticide resistance community lost one of its most influential spokesmen and innovative experimentalists. In a career spanning 35 years based entirely at Rothamsted, Roman not only contributed substantially to all aspects of resistance research, but pioneered many scientific, technical and conceptual advances that underpin insect pest management today. For example, many of the widely occurring resistance mechanisms now being subjected to sophisticated biochemical and/or molecular analyses were first identified in houseflies or aphids in Roman's laboratory, and isolated using painstaking genetic and toxicological procedures. Related work with houseflies showed how the continuous exposure of field populations to a succession of insecticides had caused the sequential selection and build-up of several resistance genes—an early and unequivocal demonstration of the dangers of the 'pesticide treadmill' syndrome. For most of the last ten years, he focused his attention on the scientific and practical aspects of insecticide resistance management (IRM), travelling tirelessly on behalf of governments, overseas researchers and the agrochemical industry to devise, champion and support the introduction of IRM strategies on cotton and other crops. Roman was always eager to share his knowledge gained from studying insecticide resistance with those struggling to combat resistance to other pesticide classes.

Tragically, Roman's death occurred only two months after his official retirement from Rothamsted, and denied him the eagerly awaited opportunity to devote more time to his family and outside interests. Nor did he survive to participate in the Resistance '91 symposium, with which he was involved from its inception and through its early planning. This being so, it is entirely appropriate that the current volume, with contributions from many of Roman's closest colleagues, be dedicated to his life and distinguished scientific career. From comments received before and during the symposium, it was abundantly clear that Roman's qualities and friendship were valued just as highly outside Rothamsted as by those of us privileged to work alongside him on a day-to-day basis. Individuals combining Roman's scientific abilities and dedication with utmost professional and personal integrity come along very rarely; we all miss him greatly.

IAN DENHOLM

Contents

Simulation and Prediction

Mechanisms of Resistance

Future Trends

MONITORING FUNGICIDE RESISTANCE: PURPOSES, PROCEDURES AND PROGRESS

KEITH BRENT
Department of Agricultural Sciences,
University of Bristol,
AFRC Institute of Arable Crops Research,
Long Ashton Research Station, Bristol, BS18 9AF, UK.

ABSTRACT

The effectiveness of monitoring as an early warning system is discussed in relation to both discrete and continuous modes of resistance build-up. Monitoring can also be done to check that avoidance strategies or limitations are working, to investigate cases of suspected loss of performance, to track the spread or intensification of resistance with time, to guide fungicide selection at local level, to gain under-standing of the behaviour of resistant forms in field populations, and to validate predictive models. The choice of methods of sampling and bioassay is considerable, and will depend upon the pathogen, the host, the degree of estimated risk, availability of labour and other factors. Agreement between methods is often good, but some anomalies and pitfalls are described as is the great advantage of having 'base-line data' and a collection of early isolates. The need to monitor practical performance as well as isolate responses is also emphasised. DNA probes or other biochemical detectors of fungicide resistance are not yet available, but would be valuable especially for epidemiological studies of resistance. The value of some reported monitoring exercises is appraised, and the requirement is stressed for more sustained studies, for full, prompt publication of results, and for increased public-sector and industry interaction in planning and interpretation.

INTRODUCTION

The term 'monitor' derives from the Latin *monere* 'to warn'. In the context of resistance research, monitoring may indeed be done primarily to give warning of potential resistance problems. However, there are several other important aims which will be discussed in the next section, and in a broader and generally used sense 'monitoring for fungicide resistance' simply means testing samples of target pathogen populations for their degree of sensitivity to one or more fungicides.

World-wide results of monitoring obtained over the years form the basis of our now substantial knowledge and understanding of fungicide resistance as a practical problem in crop protection. Such results were first published in the early to mid 1960s and provided clear evidence for the first detected cases of failures of crop disease control caused by fungicide resistance:- resistance to biphenyl fumigation and orthophenylphenate dipping in *Penicillium* spp. causing post-harvest rots in lemons in California (1), resistance to hexachlorobenzene seed treatment in *Tilletia foetida* causing bunt disease in wheat in Australia (2), and resistance to organo-mercurial seed treatment in the leaf-stripe pathogen (*Pyrenophora avenae*) in oats in Scotland (3). These early cases, which were remarkably well investigated and reported, occurred many years after initial introduction of the fungicide treatments concerned. Whilst causing considerable difficulties in disease control in some situations, they have proved to be of relatively limited significance in comparison with the more widespread and immediate problems encountered by most of the systemic fungicides which were introduced from the late 1960s onwards.

A large amount of monitoring for resistance to the systemic fungicides has been done over the past twenty years, especially in Europe, USA and Japan. This contribution discusses first the several aims of these studies, then the methodology that has been developed, and finally the progress which has been made so far. It extends and updates earlier reviews of the detection and monitoring of pesticide and fungicide resistance (4,5).

PURPOSES

Early warning

This is often considered to be the main intention of resistance monitoring, in the same way that monitoring for radioactivity in a nuclear power station is done to give warning of impending danger. The obvious question arises as to whether reliable and early indications of a potential resistance problem can in fact be obtained through monitoring. It is not so easy to give a simple answer. High-level resistance can arise by single-step mutation as is the case for

resistance to benzimidazole and phenylamide fungicides (6). It is
difficult to detect the resistant mutants in a field population of the
target fungus until relatively high frequencies are obtained (> 1%) and
failure of disease control is either happening or imminent. To have a
95% probability of detecting even a 1 in 100 frequency of resistant
forms, one would need to test 300 samples. Such a frequency is very
much higher than the initial 10^{-7} or 10^{-8} frequency of mutation. Thus
the selection and increase in frequency of resistant mutants in
successive pathogen generations is likely to occur undetected until the
point is reached when the next, or next-but-one, fungicide application
will fail to give normal levels of disease control in the field. Thus
monitoring generally will not give adequate warning of resistance
build-up unless very large and generally impracticable numbers of
samples are tested.

Nevertheless a useful early warning role can still exist in some
situations. If resistance is polygenic or multistep, as can occur with
the DMI (sterol demethylase inhibitor) fungicides (6), then resistant
strains may exist in high frequency before appreciable loss of
practical disease control is observed. Detection of these resistant
strains is both feasible and worthwhile as a warning of greater, more
practically significant resistance. Even with the single-step type of
resistance some useful warning can be obtained of the likelihood of
resistance occurring in other farms, regions or countries where the
fungicide was introduced later, where use was less intensive, or where
there were other reasons for build-up occurring at a later time. If
pathogen reproduction is slow, as for example in the cereal eyespot
pathogen (*Pseudocercosporella herpotrichoides*) which generally sporulates only
once or twice per year, some degree of warning may be obtained. If the
resistant mutants are less fit, build-up of resistance may be slow or
interrupted and useful warning may be obtained. For example, the
resistance of *Botrytis cinerea* to the dicarboximide fungicides was detected
through monitoring before widespread problems occurred (7); in this
case the resistant variants survived less well in the absence of the
fungicide than wild-type fungus so that build-up was not strictly
proportional to their degree of selection. Again, if a strategy of
fungicide use that minimises the risk of resistance has been adopted,
as is increasingly and rightly the case, the rate of build-up of

resistance may be brought down to a level at which monitoring can provide a warning to modify the strategy or change to another treatment. For example, it has been possible to gain early warning of the build-up of resistance to phenylamide fungicides in *Phytophthora infestans* in potato crops in England that have been treated from the start with formulated mixtures of phenylamides with mancozeb (8), whereas in Holland or Eire, where initial treatments were with a phenylamide alone, monitoring only revealed resistance after a major breakdown of performance had occurred (9).

Checking that avoidance strategies are working

This is a valuable and achievable aim of resistance monitoring. However, it generally requires monitoring to be done extensively, over large regions, and this can be too expensive an exercise to support year by year except in certain special situations of known high risk coupled with major fungicide use. The alternative, cheaper, and generally adequate approach, is to monitor only for practical performance of the fungicide when applied to crops either commercially or in field experiments, by assessing amounts of disease in treated crops. If signs of deterioration of effect are observed, then resistance monitoring can be done to check that resistance is the cause. With some diseases, such as apple scab (caused by *Venturia inaequalis*), that give obvious symptoms and are normally controlled almost completely by fungicides, any loss of performance is usually easy to observe. However, some other diseases are less easy to assess, especially if the loss of effectiveness is gradual, and careful examination backed wherever possible by reference to standard treatments, is required.

However, the difficulty of assessing the field performance of phenylamide fungicides when applied in the presence of mancozeb is a further complicating factor. It is necessary in this context to monitor at sites of both good and poor control, and at treated and untreated sites, and to seek correlation of results of sensitivity tests with the treatments applied (or not applied) and with degree of disease suppression. Good correlation between results of performance monitoring and resistance monitoring often has provided convincing evidence of the occurrence of resistance problems. On occasions, however, such correlations are not found and underlying reasons have to be sought. For example, in paired treated and untreated fields in the UK the sensitivity of barley powdery mildew samples to ethirimol did not correlate with efficacy of fungicide applications in the field. This was partly because the degree of resistance fluctuated during the season according to selection pressure, and partly because a highly resistant residual population resulted directly from a period of very effective practical control and hence high selection pressure, and actually indicated sustained good performance rather than loss of effectiveness (10). In surveys of phenylamide resistance, correlation between sensitivity scores and degree of field control by phenylamide-mancozeb mixtures has also been poor (Cooke, this volume). In this case, the method of monitoring, which involves multiple spore inocula and fails to distinguish between minority and majority resistance within spore populations, has been cited as an underlying reason for the poor correlation.

Complaint monitoring

This is done at specific sites, as a detective exercise to investigate the cause of an apparent lack of performance of a fungicide observed and reported by the grower. Before this is done, other possible reasons, such as faulty application technique, mistaken dose rate, misidentification of the pathogen or unusually high disease pressure should be checked and eliminated. If the loss of effectiveness is severe, and monitoring experience has already been acquired, relatively few samples should suffice to determine whether or not there is a resistance problem. A 'first-time' complaint may require more attention, involve comparison with samples taken from neighbouring farms, and warrant reference to standard sensitive isolates and records of base-line data if these are available.

To advance knowledge of the resistance phenomenon and to validate models

Monitoring is sometimes done in well-defined field experiments or other situations in which different fungicide or fungicide mixtures, different rates or timing of application, and other determining factors can be examined for their effects on the rate of build-up of resistance. This is difficult work, generally involving intensive and frequent sampling, large plots, and accurate well-replicated assays for fungicide sensitivity. A long term approach, in which the same plots and treatments are maintained over several years, is often required in order to gain conclusive results. Ingress of spores (sensitive or resistant) of the pathogen from sources external to the experimental area, and movement of spores between plots, are difficult to detect and quantify, and can give rise to confusing results. Possibly the use of experimental pathogen populations identifiable by particular RFLP, isozyme, or other taxonomic markers could allow assessment of, and correction for such spore movements.

Insufficient studies of this type have been done so far. Some examples have been discussed by Skylakakis (11). These included field experiments on resistance of *Cercospora beticola* on sugar beet to benomyl (12), *Monilinia fructicola* on nectarines to benomyl (13), *Botrytis cinerea* on grapes to dicarboximides (14), and *Botrytis cinerea* on strawberries to dicarboximides (15). Further field studies have been reported for resistance of *Botrytis cinerea* on grapes (16,17) and strawberries (18) to dicarboximides, and *Blumeria graminis* (syn. *Erysiphe graminis* f.sp. *hordei*) on barley to ethirimol and triadimenol (19,20). Despite the above work, the relationship of the speed of development of resistance to factors such as fungicide dose and frequency of application, mixture or rotation with companion compounds, and use in the eradicant or protectant mode, are still poorly understood, and much further work is justified.

Guidance to fungicide use at local level

If monitoring methods give rapid results, and if pathogen populations fluctuate in their response to fungicide treatment, then growers could usefully determine on-the-spot which particular fungicides are most likely to be effective at any given time. Du Pont have pioneered this approach for the control of Sigatoka disease in bananas caused by *Mycosphaerella fijiensis* (21). Spores are collected from sporulating lesions and tested for fungicide response *in vitro* as a guide to the choice of fungicide for the next spray. This approach has also been followed to some extent in citrus packing houses in the USA (Eckert, J.W., personal communication, 1986). Generally, however, it is not useful; shifts in fungicide response often occur over lengthy periods and are fairly stable, and so there is little point in making frequent checks. Moreover, few growers have the facilities or expertise reliably to obtain and interpret results of bioassays.

PROCEDURES

Having discussed the 'why' of monitoring, I will now consider 'how' this is, and should be, done. The overall framework of a monitoring programme comprises several successive phases: establishment of methods, base-line determination, checking performance, and checking sensitivity. The need to establish methods of sampling and testing requires no explanation. According to situation these may have to be developed de novo, or there may be well-tried methods which can be adopted either as they stand or after minor modifications. Some further comments on methodology are given later.

It is important to obtain base-line data, i.e. data on the initial variation between samples of the target pathogen in their response to a particular fungicide before it has been used widely in the field. This gives an assessment of 'background noise', against which later results of monitoring operations can be compared, so that any substantial change in response can be detected clearly. Base-line monitoring also serves to test out methods of sampling and bioassay on a sizeable scale. If there are no initial data on variation in response, it can be very difficult to interpret subsequent monitoring results. It is useful to preserve some specimen samples obtained during base-line monitoring so that any subsequent change made in the bioassay method can be tested on these 'original' isolates. If base-line monitoring has not been done, and regrettably this is often the case, then it may be possible to retrieve the situation to some extent by obtaining and testing specimens preserved in culture collections, or specimens obtained from other regions or countries where the fungicide has never been used.

After base-line studies are completed, checking for sustained field performance can be a relatively cheap and easy way to determine whether resistance problems are arising. Sharp observation by growers and advisors should reveal any significant changes in response to a fungicide by those pathogens whose symptoms are easy to see and where a

high degree of resistance arises from the wild-type level. An example
referred to earlier is the appearance of resistant strains of the apple
scab fungus (*V. inaequalis*) to benzimidazole fungicides, which causes
obvious incidents of scab where little or none would be expected. In
many other situations however it is difficult to know whether
resistance is building up by visual observation alone. Resistance can
build up gradually and insidiously, in cases where polygenic regulation
of resistance occurs, and where increased disease severity is not
obvious unless field experiments with control and standard plots are
done. For example apparent increases in cereal diseases such as rust
and mildews may well result from pathogen adaptation to crop varieties,
or from changes in the weather. Also, if use strategies such as
rotation or mixing of different fungicides are in operation, a slow or
patchy build-up of resistance could occur and be rather difficult to
detect visually. In such cases, it may be advantageous (although more
expensive) to check performance by re-submitting the chemical to field
testing against standard treatments.

If evidence for a possible build-up of resistance is produced by
such performance checks, and if other possible causes of performance
loss such as poor application, unusual disease pressure or adverse
weather conditions have been ruled out or shown to be unlikely, then
some form of monitoring of the target pathogen population to enable its
sensitivity to be compared with the baseline position probably will be
justified.

Sampling methods
It is extremely important to plan very carefully the number and the
nature of samples to be taken since this can affect greatly the cost
and the effectiveness of any monitoring exercise. Initial monitoring
for a new fungicide, or in a new region of use, will generally be done
on an extensive basis, a relatively small number of representative
samples being taken from a sizeable number of sites throughout the
region. It will often be adequate to take one, or a few bulk samples
of spores from a field. At the extreme, the technique of transporting
a mobile spore trap, containing treated bait plants, on a vehicle along
roads through relevant agricultural areas has been adopted for
monitoring barley powdery mildew (22). Within single fields, it may be

possible to collect at different sampling points leaves or fruit
bearing lesions, and to obtain viable spores from these in the
laboratory. Alternatively, fungicide-treated bait plants can be placed
in the field for exposure to the spore population for a particular
period and then incubated in a glasshouse or growth room for subsequent
disease development and assessment (20). This latter method has the
advantage of testing a completely fresh spore population; however,
inoculum density in the field can vary greatly according to the weather
and microclimate conditions and therefore much replication in terms of
sites and exposure events is often necessary in order to give
conclusive results.

Intensive monitoring, involving the bioassay of many samples from
a single site is necessary when there is a need to determine the
proportion of resistant forms, and any differences in degree of
sensitivity, in populations where heterogeneity or fluctuation in
responses may exist. In this case small samples, such as single spores
or sporophores, or single-pustule samples will be required. It is
sometimes possible to 'split' larger samples for testing, for example
by observing colony formation from individual spores spread on
fungicide-containing agar, or by spraying a spore suspension onto
plants so as to produce many discrete infections.

It is important to keep any necessary multiplication of samples
for bioassay purposes as limited as possible, since sensitivity could
change during repeated sub-culture in the absence or presence of the
fungicide. On the other hand it is useful to re-test samples after
sub-culturing *in vivo* or *in vitro* to determine for the stability of any
response.

Sensitivity assays

Many different procedures for testing fungicide sensitivity have been
used over the years; these have been reviewed by Georgopoulos (23) and
Ogawa *et al.* (24). In general these different methods have given
similar overall results when used to investigate the same problems, but
it is clearly advantageous for international standardised methods to be
used as far as possible in order to permit direct comparison of
results, and some recommended methods have been published in detail by
FAO (25), and by the Fungicide Resistance Action Committee, FRAC (26).

Obligate pathogens are normally tested *in vivo*, either on seedling plants treated by spray, seed or soil application or on leaf discs or segments floating on fungicide solutions at different concentrations. The amount of inoculum should be standardised as far as possible, and it may be necessary to screen or separate plants to prevent cross-contamination. Facultative pathogens have mainly been tested *in vitro*, generally by measuring their growth on agar plates incorporating standard amounts of fungicide. Both spore and mycelial inocula have been used; generally these have given similar results. Tests for spore germination or germ-tube elongation, generally in liquid media, have also been used, and give particularly rapid results. When *in vitro* methods have been used, it is wise to make some *in vivo* comparison, to check that any differences detected in response are also expressed in disease situations on treated plants. Whatever method is used, it should not be changed or modified during subsequent surveys unless it is totally unsatisfactory. The production of a series of strictly comparable results over the years through the use of an unchanged method can be very valuable; results obtained by ICI for sensitivity of barley powdery mildew (*B. graminis*) to ethirimol between 1973 and 1990 provide an excellent example (27).

In practical agriculture, pathogens encounter a wide range of concentrations, and for this reason as well as for increased accuracy it is best to use a range of concentrations in bioassays, rather than a single, arbitrary, discriminating dose. However, for large-scale initial surveys of single-step resistance use of a single dose can give conclusive results, and this method has often been used. When a range of doses has been included in tests, different authors have used as a response score the highest concentration that allows some degree of pathogen development, the lowest concentration to give complete control, the estimated ED50 or a summation of degrees of control obtained at each of the concentrations used. The ED50 value has been the commonest form of scoring, and in the author's view this should be used wherever possible. It is sometimes argued that the ED95 value comes closer to the practical degree of disease control that is expected, but this is generally more difficult to measure accurately. However, the merits and limitations of different scoring methods will

vary according to individual circumstances, and it is unwise to lay
down rigid rules.

Biochemical, immunological and molecular biological methods have
been developed for monitoring insecticide resistance (28,29). These
are very convenient, and some of the methods can indicate degrees of
resistance without the need to test over a range of doses. In certain
situations they can detect resistance at lower frequencies than do
normal bioassays. There are dangers, however, in relying too heavily
on this approach. Biochemical, immunochemical and gene-probe tests
normally will only detect one mechanism of resistance, and other
mechanisms may well exist. An example where an insecticide-resistant
biotype reverted to a sensitive form which still contained the
amplified esterase gene that caused resistance (30), but did not
apparently express it, also illustrates the danger of total reliance on
gene probes. Nevertheless, some resistance problems may prove to
depend in practice on the presence or absence of a clear-cut resistance
mechanism and this approach may find useful practical value. The
ability to track particular resistance genes in field populations is,
of course, be an important tool for epidemiological research.

Efforts are being made to find biochemical or immunochemical
methods for detecting fungicide resistance, and prospects for this have
been discussed by Hollomon in this volume. The most advanced work to
date is probably that done on the apple scab fungus (*V. inaequalis*) in
relation to resistance to benzimidazole fungicides (31). Allele-
specific oligonucleotide probes have been produced, which allow
detection of resistance within three days in mycelial isolates.
However, several different point mutations causing resistance have been
identified, and it remains to be seen whether one or a few probes will
suffice to give reliable results for practical monitoring against field
resistant forms.

Interpretation of results

Caution must be the watchword when results of glasshouse or laboratory
tests on fungicide sensitivity are used to describe or predict the
performance of the fungicide in the field. For example, one cannot
assume that a spray concentration which is effective on small plants in
greenhouse tests will be equally effective in the field. Often

complete disease control may be obtained in the glasshouse from sprays at 10 μg/ml or lower concentrations, whereas in the field 50 μg/ml or more is necessary. Also one person's 'resistant' may be another person's 'moderate' or even 'sensitive', and a precise definition of what is meant by these terms should be made clearly in any report.

The successful isolation of resistant forms from field samples does not necessarily imply that there is a major resistance problem in practice. Before this deduction can be made, it is necessary to make sufficient tests to establish that a high proportion of the population is sufficiently resistant and pathogenic to cause trouble, and to correlate the incidence of such populations with failures of disease control.

It is necessary to stress that the results of monitoring studies, whether done to establish baseline data or undertaken at a later stage, are fully and promptly reported in scientific journals. Sufficient experimental detail should be given to permit critical appraisal of the work and repetition if appropriate. A great deal of monitoring work, especially that done by industrial companies, has gone unreported or has only been reported very briefly in meeting abstracts, newsletters or reviews, and this is a pity.

PROGRESS

Overall an impressive amount of monitoring data for fungicide resistance has been built up over the years, although a lot more has gone unreported. It is impossible to review comprehensively all the published work, but a few points about particular cases are made below.

Following the early work mentioned in the Introduction, national or regional programmes of monitoring began in the early 1970s. Surveys of possible resistance of cucumber powdery mildew (*Sphaerotheca fuliginea*) to dimethirimol were initiated in Holland in 1970, following many reports of failures of disease control by this relatively new and initially highly effective systemic fungicide. Clear-cut results were obtained, from floating-leaf-disc tests with a good correlation between incidence of resistance and loss of performance in many glasshouses (32,10). We now know that the prolonged periods of repeated application, the highly specific mode of action, and the frequent and abundant reproduction of the pathogen under enclosed conditions all added up to a very high risk situation. The product was withdrawn; subsequently the resistance gradually declined but soon returned when efforts were made to reintroduce the product.

In contrast, a very similar fungicide, ethirimol has generally given good performance over many years against barley powdery mildew (*B. graminis*) in several European countries. Monitoring in the UK over many years by the manufacturers, ICI, revealed that sensitivity was always somewhat lower in treated fields than in untreated fields (10,33). Sensitivity also fluctuated within seasons and the range of sensitivity gradually narrowed to an intermediate level in the mid-1970s. The detection of variation in response prompted the withdrawal of a recommendation for use of ethirimol as a seed treatment on winter barley; this may have helped to limit the development of ethirimol resistance at a time when the fungicide was being used widely as a sole treatment. More recently, as the intensity of ethirimol use has declined and other products have become increasingly used, ethirimol sensitivity has increased (27,34). These gradual and reversible changes suggest that polygenic (directional) resistance was involved. In these studies it was not possible to obtain a clear-cut correlation between resistance scores and build-up of mildew populations in treated and untreated fields.

Resistance of barley powdery mildew to DMI fungicides has been monitored extensively in recent years in the UK and other European countries (22,34,35,36). In general, behaviour has been rather similar to that of ethirimol resistance, with gradual shifts that correlated with usage of these fungicides. These were associated with a clear loss of performance in some regions, but again evidence of reversal has been obtained when changes in fungicide use led to decreased selection pressure. Together with observations of loss of effectiveness in areas of intense DMI usage, monitoring results indicating the occurrence of DMI-resistant forms of barley mildew have stimulated manufacturers to adopt modified, less intensive recommendations for use.

Monitoring for dicarboximide resistance in *Botrytis cinerea* in grape vines has been done in France, Germany and Italy during the 1980s (37). In this case also many different degrees of resistance were encountered in isolates. Those with a moderate degree of resistance were sufficiently fit to cause major problems of disease control in some

areas. However, even these appeared to be somewhat less fit than
sensitive forms, tending to decline in proportion during the winter
months (38,14). In response to these results of monitoring, use
strategies which limited the number of applications of dicarboximides
were formulated and promoted by FRAC; these appear to have delayed,
but not to have stopped, a gradual deterioration of effectiveness in
many areas.

Resistance to phenylamide fungicides in *Phytophthora infestans* in
potatoes (reviewed by Cooke in this volume) and in *Plasmopara viticola* in
grape vine (39) has been the subject of much monitoring in Europe. Use
of metalaxyl and related fungicides alone was associated with a rapid
appearance of pathogen populations highly resistant to phenylamides.
Sudden failures of control occurred, and these were clearly associated
with the incidence of resistance as indicated by monitoring. The
policy in certain countries of using this group of fungicides only when
mixed with mancozeb appears to have been associated with a slower
build-up of resistant forms, although eventually the incidence of these
forms has become very high. However, the question of loss of
performance in this situation is not clear, and mixed formulations are
still recommended by manufacturers and used in regions where the
apparent incidence of resistant isolates is high. Unfortunately the
bioassay method most widely adopted involves use of a multiple spore
inoculum, so that a positive indication of resistance can result from a
sample where resistance is present only as a minor component (40).

Probably the most dramatic and pervading fungicide resistance
problems to date are those that resulted from the use of the
benzimidazole fungicides. Many monitoring studies, large and small,
have been done over the years and have been reviewed (41,42). Simple
agar plate tests or germ-tube growth tests have given clear results,
and generally these have correlated well with the presence or absence
of control in the field. Sometimes the development of resistance has
been very rapid, as in cucurbit powdery mildew (43) and sugar-beet
leaf-spot caused by *Cercospora beticola* (44), but in other cases it has
taken many years, as in *Rhynchosporium secalis* in barley (45; D.W. Hollomon,
personal communication, 1991).

An enormous amount has been learnt from these and many other studies, regarding the factors that determine the build-up of resistance, and the risks attached to different strategies of disease management. However, sustained studies of resistance status in particular regions and sites over a number of years have been relatively few and should be encouraged. As new uses for existing products continue to emerge, and as new products appear, the need for continuing monitoring efforts is obvious. FRAC has done much to encourage sound monitoring procedures and collaboration between companies. The more recently founded IORPM (International Organisation for Resistant Pest Management) should provide some opportunities for increasing collaboration between agrochemical companies and the public sector, which is at present rather variable, and for a wider range of monitoring and resistance management programmes particularly in developing countries.

REFERENCES

1. Harding, P.R., Jr., Differential sensitivity to sodium orthophenylphenate by biphenyl-sensitive and biphenyl-resistant strains of *Penicillium digitatum. Plant Dis. Rep.*, 1962, **46**, 100-104.

2. Kuiper, J., Failure of hexachlorobenzene to control common bunt of wheat. *Nature*, 1965, **206**, 1219-1220.

3. Noble, M., Maggarvie, Q.D., Hams, A.F. and Leafe, L.L., Resistance to mercury of *Pyrenophora avenae* in Scottish seed oats. *Plant Pathol.*, 1966, **15**, 23-28.

4. Brent, K.J., Detection and monitoring of resistant forms: An overview. In *Pesticide Resistance-Strategies and Tactics for Management.* NRC Board on Agriculture, National Academy Press, Washington, DC, 1986, pp. 298-312.

5. Brent, K.J., Monitoring for fungicide resistance. In *Fungicide Resistance in North America,* ed. C.J. Delp, American Phytopathological Society Press, St. Paul, Minnesota, 1988, 9-11.

6. Georgopoulos, S.G. and Skylakakis, G., Genetic variability in the fungi and the problem of fungicide resistance. *Crop Prot.*, 1986, **5**, 299-305.

7. Beever, R.E. and Byrde, R.J.W., Resistance to the dicarboximide fungicides. In *Fungicide Resistance in Crop Protection,* eds J. Dekker and S.G. Georgopoulos, Pudoc, Wageningen, Netherlands, 1982, 101-117.

8. Carter, G.A., Smith, R.M. and Brent, K.J., Sensitivity to metalaxyl of *Phytophthora infestans* populations in potato crops in south-west England in 1980 & 1981. *Ann.Appl.Biol.*, 1982, **100**, 433-441.

9. Davidse, L.C., Looijen, D., Turkensteen, L.J. and Van der Wal, D., Occurrence of metalaxyl-resistant strains of *Phytophthora infestans* in Dutch potato fields. *Neth. J. Plant Pathol.*, 1981, **87**, 65–68.

10. Brent, K.J., Case study 4. Powdery mildew of barley and cucumber. In *Fungicide Resistance in Crop Protection*, eds J. Dekker and S.G. Georgopoulos, Pudoc, Wageningen, Netherlands, 1982, 219–230.

11. Skylakakis, G., Quantitative evaluation of strategies to delay fungicide resistance. *Proc. Br. Crop Prot. Conf. - Pests and Diseases*, 1984, **2**, 565–572.

12. Dovas, C., Skylakakis, G. and Georgopoulos, S.G., The adaptability of the benomyl-resistant population of *Cercospora beticola* in Northern Greece. *Phytopathology*, 1976, **66**, 1452–1456.

13. Sonada, R.M., Ogawa, J.M., Manji, B.T., Shabi, E. and Rough, D., Factors affecting control of blossom blight in a peach orchard with low level benomyl-resistant *Monilinia fructicola*. *Plant Dis.*, 1983, **67**, 681–684.

14. Löcher, F., Lorenz, G. and Beetz, K.J., Influence of vinclozolin mixtures on the development of resistance and on disease control in *Botrytis cinerea* Pers. of grapes. *Proc. 10th Int. Congr. Pl. Prot.*, 1983, **2**, 627–628.

15. Hunter, T. and Brent, K.J., Effects of different spray regimes on dicarboximide resistance on *Botrytis cinerea* in strawberries. *Proc. 10th Int. Congr. Pl. Prot.*, 1983, **2**, 631.

16. Lorenz, G. and Löcher, F., Strategies to control dicarboximide-resistant strains of *Botrytis cinerea*. *Proc. Brighton Crop Prot. Conf. - Pests and Diseases*, 1988, 1107–1115.

17. Beever, R.E., Pak, A. and Laracy, E.P., An hypothesis to account for the behaviour of dicarboximide-resistant strains of *Botrytis cinerea* in vineyards. *Pl. Pathol.*, 1991, **40**, 342–346.

18. Hunter, T., Brent, K.J., Carter, G.A. and Hutcheon, J.A., Effects of fungicide spray regimes on incidence of dicarboximide resistance in grey mould (*Botrytis cinerea*) on strawberry plants. *Ann. Appl. Biol.*, 1987, **110**, 515–525.

19. Hunter, T., Brent, K.J. and Carter, G.A., Effects of fungicide regimes on sensitivity and control of barley mildew. *Proc. 1984 Br. Crop. Prot. Conf. - Pests and Diseases*, 1984, **2**, 471–476.

20. Brent, K.J., Carter, G.A., Hollomon, D.W., Hunter, T., Locke T. and Proven, M., Factors affecting build-up of fungicide resistance in powdery mildew in spring barley. *Neth. J. Pl. Path.*, 1989, **95**(1), 31–41.

21. Anon., Black and Yellow Sigatoka, Improved identification and management techniques. Du Pont Latin America, Coral Gables, Florida, 1982.

22. Fletcher, J.T. and Wolfe, M.S., Insensitivity of *Erysiphe graminis* f.sp. *hordei* to triadimefon, triadimenol and other fungicides. *Proc. Br. Crop Prot. Conf. - Pests and Diseases,* 1981, **2**, 633–640.

23. Georgopoulos, S.G., Detection and measurement of fungicide resistance. In *Fungicide Resistance in Crop Protection,* eds. J. Dekker and S.G. Georgopoulos, Pudoc, Wageningen, Netherlands, 1982, 24–31.

24. Ogawa, J.M., Manji, B.T., Heaton, C.R., Petrie, J. and Sonada, R.M. Methods for detecting and monitoring the resistance of plant pathogens to chemicals. In Pest Resistance to Pesticides, eds. G.P. Georghiou and T. Saito, Plenum, New York, 1983, 117–162.

25. Anon., Recommended methods for for the detection and measurement of resistance of agricultural pests to pesticides. *Plant Prot. Bull.,* **30**, 36–71, 141–143.

26. Anon., FRAC methods for monitoring fungicide resistance. EPPO Bulletin, 1991, **2**, 292–354.

27. Stott, I.P.H., Noon, R.A. and Heaney, S.P., Flutriafol, ethirimol and thiabendazole seed treatment – an update on field performance and resistance monitoring. *Proc. Br. Crop Prot. Conf. - Pests and Diseases,* 1990, 1169–1174.

28. Brown, T.M. and Brogdon, W.G., Improved detection of insecticide resistance through conventional and molecular techniques. *Ann. Rev. Entomol.,* 1987, **32**, 145–162.

29. Devonshire, A.L., Biochemical and genetic analysis of insect populations resistant to insecticides. *Proc. Brighton Crop Prot. Conf. - Pests and Diseases,* 1990, **3**, 889–896.

30. Field,L.M., Devonshire, A.L., ffrench–Constant, R.H. and Forde, B.G., Changes in DNA methylation are associated with loss of insecticide resistance in the peach–potato aphid, *Myzus persicae* (Sulz.). *FEBS Letters.* 1989, **243**, 323–327.

31. Koenraadt, H., Jones, A.L. and Somerville, S.C., Use of allele-specific oligonucleotide probes to characterise resistance to benomyl in field strains of *Venturia inaequalis. Phytopathology,* 1991, **81**, Abstr.

32. Bent, K.J., Cole, A.M. Turner, J.A.W. and Woolner, M., Resistance of cucumber powdery mildew to dimethirimol. *Proc. Br. Insectic. Fungic. Conf.* 1971, **1**, 274–282.

33. Shephard, M.C., Brent, K.J., Woolner, M. and Cole, A.M., Sensitivity to ethirimol of powdery mildew from UK barley crops. *Proc. Br. Insectic. Fungic. Conf.*, 1975, **1**, 59-60.

34. Heaney, S.P., Hutt, R.T. and Miles, V.G., Sensitivity to fungicides of barley powdery mildew populations in England and Scotland; status and implications for fungicide use. *Proc. 1986 Br. Crop Prot. Conf. - Pests and Diseases*, 1986, **2**, 793-800.

35. Fletcher, J.T., Cooper, S.T. and Prestidge, A.L.H., An investigation of the sensitivity of *Erysiphe graminis* f.sp. *tritici* to various ergosterol inhibiting fungicides. In *Integrated Control of Cereal Mildews: Monitoring the Pathogen*, eds. M.S. Wolfe and E. Limpert, Martinus Nijhoff Publishers, Dordrecht, Netherlands, 1987, 129-135.

36. Limpert, E., Frequencies of virulence and fungicide resistance in the European barley mildew population in 1985. *J. Phytopathol.*, 1987, **119**, 298-311.

37. Lorenz, G., Dicarboximide fungicides: history of resistance development and monitoring methods. In *Fungicide Resistance in North America*, American Phytopathol. Soc. Press, St. Paul, Minnesota, 1988, 45-51.

38. Leroux, P. and Gredt, M., Phénomènes de résistance de *Botrytis cinerea* aux fongicides. *La Défense des Végétaux*, 1982, **217**, 3-17.

39. Staub, T. and Diriwaechter, G., Status and handling of fungicide resistance in pathogens of grapevine. *Proc. 1986 Br. Crop Prot. Conf. - Pests and Diseases*, 1986, **2**, 771-780.

40. King-Watson, Eileen D., Sensitivity monitoring methods for phenyl-amide fungicides. In *Fungicide Resistance in North America*, American Phytopathol. Soc. Press, St. Paul, Minnesota, 1988, 61-62.

41. Smith, Constance M., History of benzimidazole use and resistance. In *Fungicide Resistance in North America*, American Phytopathol. Soc. Press, St. Paul, Minnesota, 1988, 23-24.

42. Trivellas, Alice E., Benzimidazole resistance monitoring techniques and the use of monitoring studies to guide benomyl marketing. In *Fungicide Resistance in North America*, American Phytopathol. Soc. Press, St. Paul, Minnesota, 1988, 28-30.

43. Schroeder, W.T. and Provvidenti, R., Resistance to benomyl in powdery mildew of curcurbits. *Plant Dis. Rep.*, 1969, **53**, 271-275.

44. Georgopoulos, S.G. and Dovas, C., A serious outbreak of strains of *Cercospora beticola* resistant to benzimidazole fungicides in northern Greece. *Plant Dis. Rep.*, 1973, **62**, 205-208.

45. Jones, D.R., Sensitivity of *Rhynchosporium secalis* to DMI fungicides. *Proc. Brighton Crop Prot. Conf. - Pests and Diseases*, 1990, 1135-1140.

ROLE OF MUTATION AND MIGRATION IN THE EVOLUTION OF INSECTICIDE RESISTANCE IN THE MOSQUITO *CULEX PIPIENS*

MICHEL RAYMOND, MAITE MARQUINE and NICOLE PASTEUR
Laboratoire Génétique et Environnement,
URA 327, Case courrier 64,
U.S.T.L., place E. Bataillon, 34095 Montpellier, France

ABSTRACT

The well documented history of organophosphorous (OP) insecticide resistance in two heavily controlled areas, southern France and Corsica, indicates that mutation is a rare event and a limiting step for generating new resistance alleles in a given area. In contrast, the role of migration in the evolution of insecticide resistance has been underestimated, as illustrated by resistant overproduced esterases A1 or A2-B2 which have a unique origin and are, respectively, distributed over the whole Mediterranean area or Africa, North America and Asia.

INTRODUCTION

The evolution of insecticide resistance in insect populations is dependent on the existence and the incidence of resistance genes. Two forces -mutation and migration- are responsible for the occurrence and distribution of resistance genes within and between pest populations. Mutation, mediated through events such as nucleotide substitution, insertion, deletion, gene duplication or amplification etc., is ultimately responsible for generating resistant variants within a species. In any particular population, however, the presence of resistance genes can result not only from mutation *in situ*, but also from the immigration of individuals possessing these genes, either through active dispersal or by the passive transport of insects, usually by human agency.

In this paper we explore the relative contributions of mutation and migration to the evolution of organophosphorus (OP) resistance in the *Culex pipiens* complex. On the basis of monitoring data from many parts of the world we argue that mutations causing resistance are very rare events that can severely limit the capacity of populations to respond to insecticide selection. In contrast, there is now evidence that migration, in some cases at least, has been a powerful force promoting the geographical spread of resistance genes.

MUTATION IS A RARE EVENT

Evidence for mutation being a rate-limiting step in the evolution of resistance comes from two cases with a carefully documented history of insecticide treatment and a continuous record of resistance monitoring based on toxicological and biochemical assays.

Resistance to chlorpyrifos in southern France
In 1969, a control program involving only chlorpyrifos was initiated along the French Mediterranean coast with an organisation capable of treating up to 20,000 larval breeding sites of *C. pipiens*. Resistance to chlorpyrifos was first detected by routine monitoring three years later in the Montpellier area (1). This *ca.* 10-fold resistance was initially restricted to a single village, but subsequently spread (2, 3) and was present throughout the treated area in 1978 (4). This general and low level of resistance did not greatly reduce control efficacy, and no increase in dose or application frequency was apparently needed (1).

In 1974, the putative biochemical mechanism responsible for this resistance was identified as an overproduced esterase first termed A' (5, 6) and then A1 (7), coded by the *Est-3^A* allele. Susceptible mosquitoes possessed this esterase or another allozyme in an amount undetectable by starch gel electrophoresis, while resistant insects overproduced A1 at a level of up to 3% of the total body protein (4, 8, 9). The strict correlation between resistance and the presence of A1 in various natural populations in 1974 and 1975 implied that this esterase was the sole significant resistance mechanism in France during this period (2, 4, 10).

In 1978, chlorpyrifos resistance increased to a level of 100-fold, an event that was subsequently attributed to the appearance of a second resistance mechanism: target-site insensitivity mediated through altered acetylcholinesterase (11-13). The allele responsible (termed *Ace^R*) was initially localised on the coast of Hérault department in 1978 (4) but subsequently increased in frequency and geographical range, impairing the efficacy of chlorpyrifos to the extent that other insecticides were needed to regain adequate control. Consequently, temephos and fenitrothion were introduced into the control program from 1981 onwards.

Since Ace^R confers considerably higher resistance to chlorpyrifos than A1 (100 fold for Ace^R vs 3-10 fold for A1, (12)), its presence prior to 1978 should have been detectable through resistance monitoring or from localised control difficulties. The obvious conclusion is that the appearance of Ace^R was delayed well beyond the period in which it could potentially have been selected in field populations. Whether its eventual appearance resulted from mutation within the treated area or by migration of insects from other localities is not known, but in either case it seems that the rarity of the mutation needed to generate Ace^R from its wildtype homologue (Ace^S) was a primary constraint on the selection of this more powerful mechanism of chlorpyrifos resistance.

In contrast, the rapid selection of A1 resistance, within three years of the start of chlorpyrifos treatments, implies either that this mechanism was already present at low frequencies or arose through mutation or migration very early during the control program. Again, there is evidence that the mutation leading to overproduction of A1 was a rare, possibly unique, event and that its subsequent spread was a consequence of gene flow rather than several homologous mutations. Wherever it has been studied, the A1 mechanism is in strong linkage disequilibrium with an unamplified allele ($Est-2^{0.64}$) at another esterase locus ($Est-2$, later named esterase B) that is seemingly unconnected with OP resistance (14). The close association of A1 and a particular allele at the highly polymorphic $Est-2$ locus is only readily explicable if the mutation causing A1 overproduction occurred in a single locality (possibly in the Montpellier area, see above), the association A1 - $Est-2^{0.64}$ being subsequently enhanced by a hitch-hiking effect (15).

Resistance to temephos in Corsica

A control program was initiated in Corsica in 1973, involving weekly applications of temephos during the mosquito breeding season. Yet despite this selection pressure being sustained over 15 years, the maximum level of resistance documented by a survey in 1988 was only ca. 30 fold, and this did not significantly impair the effectiveness of the control treatments.

The relatively low resistance to temephos, apparently involving the increased production of several esterases (including A1, see below), contrasts with that documented elsewhere in the world, particularly in California where a very powerful mechanism conferring ca. 800-fold resistance to temephos occurred in 1974 (16). If present in Corsica, the Californian mechanism (the overproduction of a different esterase termed B1 (8, 9)), would be expected to have given a substantial selective advantage in populations intensively exposed to temephos. Once again, the non-appearance of B1 resistance in Corsica between 1973 and 1988 indicates that neither mutation nor migration occurred to generate this mechanism *de novo* or to transport it into Corsican populations.

MIGRATION CAN BE A FREQUENT EVENT

Discussions of the impact of migration on resistance development have focussed on the role of immigration of susceptible insects as a factor retarding selection of resistance in treated populations (*e.g.* 17, 18). In comparison, the importance of gene flow for transferring alleles for resistance into susceptible populations has received little attention (but see 19). For *C. pipiens*, however, there appear to be at least two good examples of migration not merely promoting the evolution of resistance but being responsible for the spread of particular alleles between adjacent countries and even between continents.

Esterase A1 in the Mediterranean

Following 15 years of temephos usage in Corsica, several putative resistance genes have been detected at high frequencies in *C. pipiens* populations, including the overproduced esterase A1 (20). Two lines of evidence suggest that the presence in Corsica of A1, at least, is a consequence of one or more migration events. Firstly, overproduction of A1 was initially recognised as a resistance mechanism in the Montpellier area in 1972, and was already spreading through southern France when temephos treatment began in Corsica. Secondly, A1 resistance in Corsica, as in southern France, is in strong linkage disequilibrium with the $Est\text{-}2^{0.64}$ allele (20). Both these observations support a hypothesis that A1 resistance arose only once and was subsequently introduced into Corsica by mosquitoes migrating from mainland France.

A1 resistance is now widely distributed in a contiguous region bordering the Mediterranean (Fig. 1), having been reported from Spain, France, Italy, Greece, Egypt and Tunisia (4,5, 21-23, and unpublished data). It is still unknown outside this region (e.g. tropical Africa, Asia, North America) despite extensive monitoring and widespread use of chemicals capable of selecting this mechanism to detectable frequencies. The apparent absence of A1 elsewhere in the world reinforces our previous conclusions regarding the rarity of mutations leading to overproduction of esterases. It is far more likely that the migration events already implicated to explain the introduction of A1 resistance into Corsica have also caused its appearance in other Mediterranean countries and will, if unchecked, promote its spread outside this restricted geographical area.

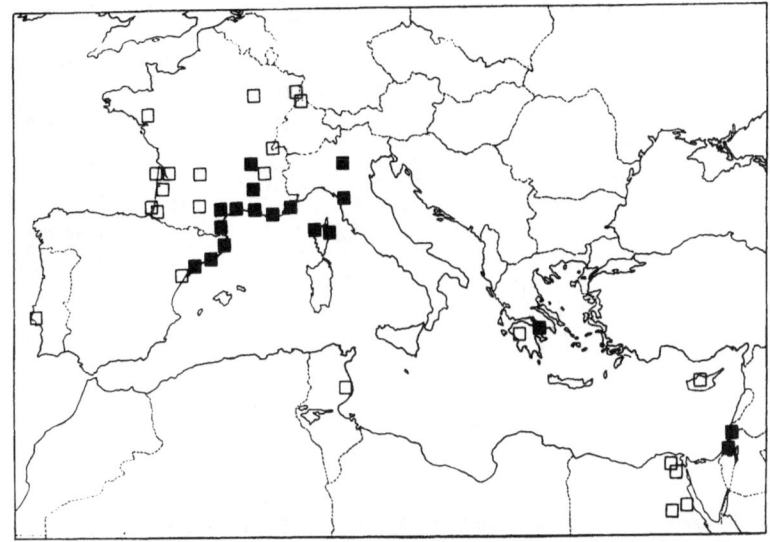

Figure 1. Distribution of A1 esterase in Mediterranean. Open squares indicate the absence of A1, and black ones indicate its presence. Data are from (4, 21, 23, 32, 33).

Esterases A2-B2 throughout the world

Proof that migration can disperse resistance genes over large geographical areas comes from work on two overproduced esterases (A2 and B2), generally associated, and implicated in OP resistance. This A2-B2 association has now been reported from at least four continents (Fig. 2), and may have appeared only within the last ten years in at least three localities: California, France and Italy (7, 21, 23).

Despite this wide distribution, recent mapping of the usually highly polymorphic esterase B region for six A2-B2 strains collected in Asia (Pakistan), Africa (Egypt, Congo, Ivory-coast) and North America (California, Texas), using 13 restriction enzymes, produced identical banding patterns (24). This finding indicates strongly that B2 (and A2) overproduction had a single origin and has subsequently been transported throughout the world. The source of the original mutation has been tentatively assigned to Africa or Asia, where the A2-B2 mechanism was first detected (25). This worldwide migration, almost certainly mediated through passive transport of mosquitoes on planes and boats (e.g. 26), has probably occurred within the last 30 years, i.e. subsequent to the widespread use of OP insecticides.

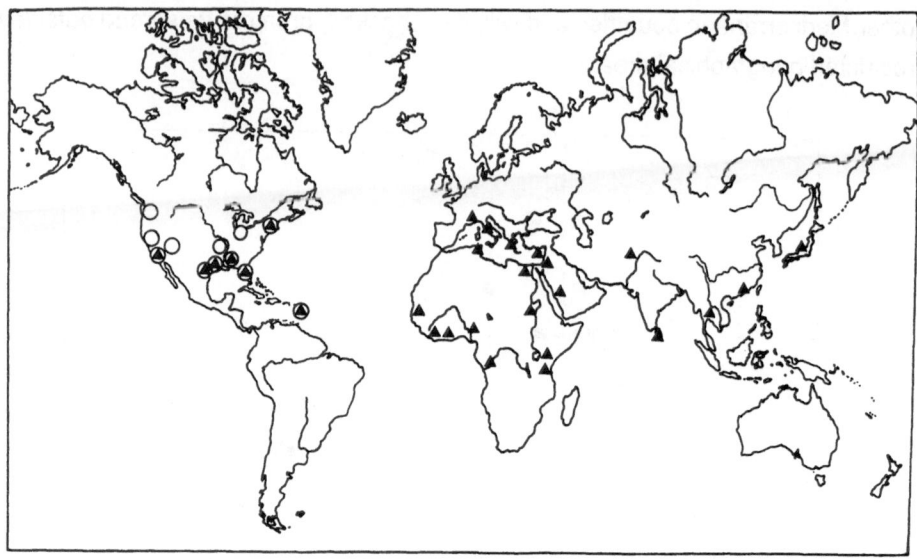

Figure 2. Known geographic distribution of overproduced esterase B1 ( ) and A2-B2 ( ) in the world, from (24).

CONCLUSIONS

The fact that mutation is a rare event is of no surprise, but that only a handful of efficient resistance genes have occurred in *Culex pipiens* since OP were used (*ca.* 30 years) indicates a limitation in the ability of this species to find adequate answers to the sudden presence of OP in its environment. This might be due to the nature of mutations leading to overproduce esterases through constitutive gene amplification (27- 29). If several independent steps, each with a low probability of occurrence, are required to generate an amplification, the combined probability, or overall mutation rate, could be vanishingly low.

Evidence of extensive migration within or between continents has been popularized among population geneticists with the now classical example of *Drosophila melanogaster* with transposable elements (*e.g.* P or I) which have invaded several continents from a single origin within the last few decades (*e.g.* 30), or with the repetitive north-south cline of the Adh^F gene (*e.g.* 31). It seems that OP insecticide resistance genes in *Culex pipiens* mosquitoes is another dramatic example of such extensive migration.

The occurrence of a new resistance gene in a treated population through a migration event has long been neglected as a significant phenomenon by people working on insecticide resistance. This was probably due to 1) an overestimate of mutation rate, leading to the conclusion that, as treated population sizes are generally large, resistant mutants will be produced within a few generations, so that migration has no dramatic effect on the evolution of resistance, and 2) an underestimate of migration rate, particularly long range passive migration by human activity.

In *Culex pipiens*, the evolution of OP resistance is more likely to be directed by migration events rather than by *de novo* mutations. It now remains to establish if this is the case for other pest species.

ACKNOWLEDGMENTS

This work was in part supported by grant N°90.76 from the Ministère de l'Education Nationale and by the CNRS/Programme Environnement. We thank I. Denholm for helpful suggestions.

REFERENCES

1. Sinègre, G., Jullien, J.L. and Crespo, O., Résistances de certaines populations de *Culex pipiens* L. au chlorpyrifos Dursban en Languedoc Roussillon France . *Cah. ORSTOM, sér. Ent. méd. Parasitol.*, 1976, **14**, 49-59.

2. Pasteur, N. and Sinègre, G., Esterase polymorphism and sensitivity to Dursban organophosphorous insecticide in *Culex pipiens pipiens* populations. *Biochem. Genet.*, 1975, **13**, 789-803.

3. Sinègre, G., Jullien, J.L. and Gaven, G., Acquisition progressive de la résistance au chlorpyrifos chez les larves de *Culex pipiens* (L.) dans le midi de la France. *Parassitologia*, 1977, **19**, 79-94.

4. Pasteur, N., Sinègre, G. and Gabinaud, A., *Est-2* and *Est-3* polymorphism in *Culex pipiens* L. from southern France in relation to organophosphate resistance. *Biochem. Genet.*, 1981, **19**, 499-508.

5. Pasteur, N., Recherches de génétique chez *Culex pipiens pipiens* L. Polymorphisme enzymatique, autogénèse et résistance aux insecticides organophosphorés. Thèse de Doctorat d'Etat, Université des Sciences et Techniques du Languedoc, Montpellier, 1977, 170 pp.

6. Pasteur, N., Iseki, A. and Georghiou, G.P., Genetic and biochemical studies on the highly active esterases A' and B associated with organophosphate resistance

in the *Culex pipiens* complex. *Biochem. Genet.*, 1981, **19**, 909-919.

7. Raymond, M., Pasteur, N., Georghiou, G.P., Mellon, R.B., Wirth, M. C. and Hawley, M., Detoxification esterases new to California in organophosphate resistant *Culex quinquefasciatus* (Diptera: Culicidae). *J. Med. Entomol.*, 1987, **24**, 24-27.

8. Fournier, D., Bride, J.M., Mouchès, C., Raymond, M., Magnin, M., Bergé, J.B., Pasteur, N. and Georghiou, G.P., Biochemical characterization of the esterases A1 and B1 associated with organophosphate resistance in the *Culex pipiens* complex. *Pest. Biochem. Physiol.*, 1987, **27**, 211-217.

9. Mouchès, C., Magnin, M., Bergé, J.B., de Silvestri M., Beyssat, V., Pasteur, N. and Georghiou, G.P., Overproduction of detoxifying esterases in organophosphate-resistant *Culex* mosquitoes and their presence in other insects. *Proc. Natl. Acad. Sci. USA*, 1987, **84**, 2113-2116.

10. Pasteur, N. and Sinègre, G., Autogenesis versus esterase polymorphism and chlorpyrifos resistance in *Culex pipiens pipiens* L. *Biochem. Genet.*, 1978, **16**, 941-943.

11. Raymond, M., Gaven, B., Pasteur, N. and Sinègre, G., Etude de la résistance au chlorpyrifos à partir de quelques souches du moustique Culex pipiens L. du sud de la France. *Génét. Sél. Evol.*, 1985, **17**, 73-88.

12. Raymond, M., Fournier, D., Bride, J.M., Cuany, A., Bergé, J.B., Magnin, M. and Pasteur, N., Identification of resistance mechanisms in *Culex pipiens* (Diptera: Culicidae) from southern France: Insensitive acetylcholinesterase and detoxifying oxidases. *J. Econ. Entomol.*, 1986, **79**, 1452-1458.

13. Raymond, M., Pasteur, N. and Georghiou, G.P., Inheritance of chlorpyrifos resistance in *Culex pipiens* L. (Diptera: Culicidae) and estimation of the number of genes involved. *Heredity*, 1987, **58**, 351-356.

14. Pasteur, N. and Sinègre, G., Chlorpyrifos Dursban[R] resistance in *Culex pipiens pipiens* L. from southern France, inheritance and linkage. *Experientia*, 1978, **34**, 709-710.

15. Hedrick, P.W., *Genetics of population*. Jones & Bartlett Publ, Boston, 1985, 629 pp.

16. Ranasinghe, L.B.E., Role of synergists in the selection of specific organophosphate resistance mechanisms in *Culex pipiens quinquefasciatus* Say (Diptera: Culicidae). Ph.D., University of California, Riverside, 1976, 122 pp.

17. Comins, H.N., The development of insecticide resistance in the presence of migration. *J. Theor. Biol.*, 1977, **64**, 177-197.

18. Taylor, C.E. and Georghiou, G.P., Suppression of insecticide resistance by alteration of gene dominance and migration. *J. Econom. Entomol.*, 1979, **72**, 105-109.

19. Tabashnik, B.E., Rosenheim, J.A. and Caprio, M.A., What do we really know about management of insecticide resistance? *This volume*.

20. Raymond *et al.*, in preparation.

21. Magnin, M., Résistance aux insecticides organophosphorés, détection,caractérisation, génétique et dynamique dans les populations naturelles. Thèse de Doctorat, Université Paris VI, Paris, 1986.

22. Beyssat-Arnaouty, V., Etude biochimique et moléculaire de la résistance aux insecticides organophosphorés, chez les moustiques du complexe *Culex pipiens* L. Thèse de Doctorat, Université de Montpellier II, Montpellier, 1989, 226 pp.

23. Callaghan, A., Genetic and biochemical studies of elevated esterase electromorphs in *Culex pipiens*. Ph.D thesis, University of London, 1989, 246 pp.

24. Raymond, M., Callaghan, A., Fort, P. and Pasteur, N., Worldwide migration of amplified insecticide resistance genes in mosquitoes. *Nature*, 1991, **350**, 151-153.

25. Curtis, C.F. and Pasteur, N., Organophosphate resistance in vector populations of the complex *Culex pipiens* L. Diptera: Culicidae . *Bull. Ent. Res.*, 1981, **71**, 153-161.

26. Curtis, C.F. and White, G.B., Plasmodium falciparum transmission in England, entomological and epidemiological data relative to cases in 1983. *J. Trop. Med. Hyg.*, 1984, **87**, 101-114.

27. Mouchès, C., Pasteur, N., Bergé, J.B., Hyrien, O., Raymond, M., Robert de Saint Vincent, B., de Silvestri, M. and Georghiou, G.P., Amplification of an esterase gene is responsible for insecticide resistance in a California *Culex* mosquito. *Science*, 1986, **233**, 778-780.

28. Raymond, M., Beyssat-Arnaouty, V., Sivasubramanian, N., Mouchès, C., Georghiou, G. P. and Pasteur, N., Diversity of the amplification of various esterases B responsible for organophosphate resistance in *Culex* mosquitoes. *Biochem. Genet.*, 1989, **27**, 417-423.

29. Pasteur, N., Raymond, M., Pauplin, Y., Nancé, E. and Heyse, D., Role of gene amplification in insecticide resistance. In *Pesticides and Alternatives*, ed J.E. Casida, Elsevier Science Publishers, London, 1990, pp. 439-47.

30. Anxolabéhère, D., Kai, H., Nouaud, D., Périquet, G. and Ronsseray, S., The geographical distribution of *P-M* hybrid dysgenesis in *Drosophila melanogaster*. *Génét. Sél. Evol.*, 1984, **16**, 15-26.

31. Simmons, G.M., Kreitman, M.E., Quattlebaum, W.F. and Miyashita, N., Molecular analysis of the alleles of alcohol dehydrogenase along a cline in *Drosophila melanogaster*. I. Maine, north Carolina, and florida. *Evolution*, 1989, **43**, 393-409.

32. Villani, F., Urbanelli, S., Gad, A., and Bullini, L., Electrophoretic variations of *Culex pipiens* from Egypt and Israel. *Biol. J. Lin. Soc.*, 1986, **29**, 49-62.

33. Georghiou, pers. com., Raymond *et al.* and Chevillon *et al.* unpublished data.

HERBICIDE RESISTANCE IN THE WEED *ALOPECURUS MYOSUROIDES* (BLACK-GRASS): THE CURRENT SITUATION.

STEPHEN MOSS

AFRC Institute of Arable Crops Research, Rothamsted Experimental Station, Harpenden, Herts AL5 2JQ, UK

ABSTRACT

Chlorotoluron-resistant *Alopecurus myosuroides* was first detected in England in 1982 and has now been found on 46 farms in 19 counties. Resistance is mainly associated with winter cereal crops established by non-ploughing cultivation techniques, where herbicides have been applied regularly, on average 1.6 applications per year. The degree of resistance varies considerably between populations but substantial reductions in herbicide performance at field recommended doses can occur. Most populations show cross-resistance to many other herbicides but resistance is not related directly to chemical grouping or mode of action. The patterns of cross-resistance are not consistent between populations, either in terms of the specific herbicides affected or the degree of resistance. Resistance to substituted-urea herbicides such as chlorotoluron tends to increase quite slowly from one year to the next. Resistance development can be more rapid to diclofop-methyl, and possibly other aryloxyphenoxypropionate herbicides, at least in some fields.

INTRODUCTION

Alopecurus myosuroides Huds. (black-grass) is an annual grass weed mainly associated with autumn sown crops, especially cereals. It is a competitive weed and populations can build up rapidly. Infestations are influenced by the cultural system used but herbicides have been viewed as the main method of controlling this weed in winter cereals for about the past 25 years. A high level of control is needed to prevent the weed increasing [1] and consequently many fields have received successive annual applications of herbicides for many years. Although 11 different active ingredients are currently available for the control of *A. myosuroides* in cereals [2], the most widely used herbicides have been chlorotoluron and isoproturon, which were introduced into the UK in the early 1970's. A survey in 1988 of pesticide usage in England and Wales indicated that chlorotoluron was applied to 10%, isoproturon to 45% and formulated isoproturon mixtures to 17% of the winter cereal area treated with herbicides which control *A. myosuroides* [3]. The repeated use of the same herbicide type is one of the key factors implicated in the development of resistant weed populations.

DISTRIBUTION OF RESISTANCE

A population showing partial resistance to chlorotoluron was first detected near Faringdon, Oxfordshire in 1982 [4]. Since 1982, seed samples from 491 farms in England have been tested for resistance, and the presence of chlorotoluron-resistant *A. myosuroides* confirmed on 46 farms [5]. These farms were widely distributed in 19 counties of England - Bedfordshire (1 farm), Buckinghamshire (3), Cambridgeshire (3), Dorset (1), Essex (13), Gloucestershire (1), Hertfordshire (1), Kent (1), Leicestershire (1), Lincolnshire (2), Norfolk (1), Northamptonshire (1), Nottinghamshire (1), Oxfordshire (6), Suffolk (5), Surrey (1), Sussex (2), Warwickshire (1), Worcestershire (1).

Resistance is not absolute, but is quantitative at the population level and it has been necessary to devise a resistance rating system based on the response of plants to chlorotoluron in glasshouse screening tests [6]. Ratings range from S, (susceptible, no evidence of resistance), through 1*, (marginal, but usually not statistically significant reductions in herbicidal activity in pot bioassays), to 5*, where the ED_{50} value for chlorotoluron is over 10 times that of a standard susceptible population. A key element in this ranking system is the inclusion of three standard reference populations: Rothamsted (susceptible), Faringdon (partially resistant 2*) and Peldon (highly resistant 5*). At present only populations classified as 2* or more are deemed to be resistant to chlorotoluron. The use of chlorophyll fluorescence techniques on detached leaves shows considerable promise as an alternative technique to spraying plants in pots for detecting resistance to photosynthetic inhibitor herbicides [7].

The most severe cases of resistance (4*-5*) occur in the Peldon area of Essex. Most of the other resistant populations are at the 2* level at present, although 3*-5* levels of resistance have been detected in

Buckinghamshire (1 farm), Essex (3), Leicestershire (1), Lincolnshire (1), Oxfordshire (2), and Suffolk (1). There is some concern regarding the future development of resistance by the relatively high proportion of populations (16%) at the 1* level detected in ADAS random field surveys [5].

There is evidence of resistance to some aryloxyphenoxypropionate herbicides which, at least in some cases, does not appear to be associated with resistance to chlorotoluron. This is a recent development and the extent of this type of resistance is being studied.

CULTURAL AND HERBICIDE HISTORY OF FIELDS WITH RESISTANT *A. MYOSUROIDES*

To determine whether there was an association between resistance in *A. myosuroides* and the cultural and herbicide history, information was obtained for 20 fields with resistant populations. Information for 264 crop years was obtained, which represented the period of 6-20 years before resistance was detected.

All the fields were in a predominantly winter cereal rotation - winter cereals being grown in 89% of the crop years (range = 64-100% for individual fields). Eight fields were in continuous winter cereals. There was a low frequency of mouldboard ploughing - fields were ploughed in only 27 (10%) of the crop years (range 0-56%). Herbicides to control *A. myosuroides* were applied frequently on all fields, on average 1.7 applications per year (range 1.0 - 3.3). The main herbicides used were chlorotoluron or isoproturon (56% of applications) and diclofop-methyl (15%). A range of other herbicides was also used, often in sequences or mixtures and sometimes at lower than recommended doses. Every field had a different herbicide history.

The use of herbicides on fields where resistant populations were found was not atypically high, and many fields where *A. myosuroides* was still susceptible had received a similar intensity of herbicide treatment. It is therefore not possible to relate the development of resistance to chlorotoluron to a particular pattern of herbicide use. Most of the herbicides used to control *A. myosuroides* are soil acting and their performance is influenced greatly by climatic and other environmental factors. Thus the selection pressure imposed by these herbicides may be poorly correlated with the amount applied.

Thus resistance to chlorotoluron was mainly associated with continuous, or near continuous, winter cereal cropping, non-inversion tillage systems and regular, but not atypically high use of herbicides to control *A. myosuroides*.

TABLE 1
Herbicides assessed for activity on a population of *A. myosuroides* from Peldon, Essex [8,9,10,11; Moss, unpublished]

GROUP A. Resistance can cause reductions in the performance of these herbicides:

CHLOROTOLURON, DICLOFOP-METHYL, PENDIMETHALIN, TERBUTRYN, barban, chlorsulfuron, cyanazine, diuron, flamprop-M-isopropyl, imazamethabenz-methyl, linuron, methabenzthiazuron, metoxuron, metribuzin, simazine.

GROUP B. Resistance can reduce the performance of these herbicides, although moderate to good levels of control can be achieved in the field when other conditions are favourable for herbicide activity.

FENOXAPROP-ETHYL, <u>FLUAZIFOP-P-BUTYL</u>, ISOPROTURON, <u>QUIZALOFOP-ETHYL</u>, TRALKOXYDIM, TRI-ALLATE, metazachlor, SMY 1500.

GROUP C. Resistance does not appear to affect the performance of these herbicides.

<u>ETHOFUMESATE</u>, <u>PROPYZAMIDE</u>, <u>SETHOXYDIM</u>, TRIFLURALIN, <u>carbetamide</u>, glyphosate, paraquat.

Note: 1.   Herbicides in UPPER CASE have been used in at least two experiments, including trials under simulated or actual field conditions. The other herbicides have been used in fewer experiments, so we can be less confident about their ranking.

2.   Herbicides <u>underlined</u> have given over 90% control of *A. myosuroides* from Peldon when used at the field recommended rate under simulated field conditions.

CROSS RESISTANCE

A major concern is the extent of cross-resistance to other herbicides with different modes of action. The population studied in greatest detail, Peldon, shows varying degrees of resistance to 23 herbicides (Table 1). Resistance is not related in any simple way to chemical grouping or mode of action.

Resistance to chlorotoluron and isoproturon in the Peldon population appears to be due mainly to rapid herbicide degradation via the oxidative processes of N-dealkylation and ring-alkyl oxidation associated with cytochrome P450 [8]. Cross-resistance to other herbicides may be the result of detoxification via similar oxidative processes. There is some supporting evidence for this. There is a high degree of resistance to the dinitroaniline herbicide pendimethalin, but cross-resistance does not extend to trifluralin, which is structurally related to pendimethalin but lacks available ring-methyl substituents [8,9]. Thus resistance in the Peldon populations appears to be due mainly to the enhanced ability of plants to metabolise herbicides. The critical factor appears to be the degree to which the herbicide can be metabolised, and hence detoxified. This depends on molecular structure but is not related to the conventional grouping of herbicides.

Patterns of cross-resistance were not consistent between populations, either in terms of the specific herbicides affected or to the degree of resistance. The results of a dose response experiment in which plants of five *A. myosuroides* populations were treated with a range of doses of two herbicides with contrasting modes of action are summarised in Figure 1.

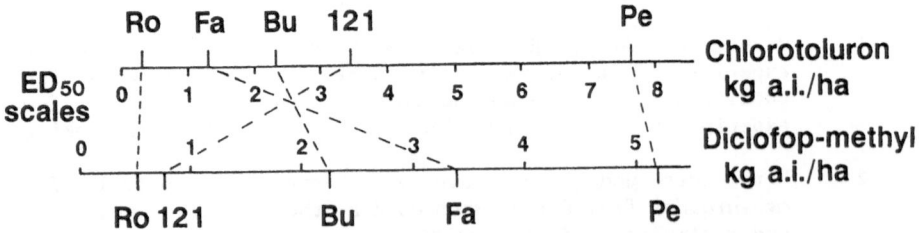

Figure 1. Response of five populations of *A. myosuroides* to chlorotoluron and diclofop-methyl. (Ro - Rothamsted; Fa - Faringdon; Bu - Bucks C; 121 - H/121 (Suffolk); Pe - Peldon)

The herbicides used were chlorotoluron, a photosynthetic inhibitor, and diclofop-methyl, an inhibitor of acetyl-coenzyme A carboxylase (ACCase) activity (an enzyme associated with fatty acid biosynthesis). The dose response data (foliage fresh weight) was analysed and ED_{50} values calculated. The Peldon and Bucks C populations showed similar degrees of resistance to both herbicides. However, the H/121 population showed resistance to chlorotoluron, but no resistance to diclofop-methyl. In contrast the Faringdon population showed a much higher degree of resistance to diclofop-methyl than to chlorotoluron. The results demonstrate clearly that the degree of resistance to the two herbicides differed between populations, but it has not been possible so far to relate these differences to past herbicide use.

There is increasing evidence that these results are not unrepresentative, and that many other populations differ in their relative level of response to different herbicides. It now seems probable that other resistance mechanisms, apart from enhanced metabolism, exist in some populations. The evidence suggests that resistance in *A. myosuroides* is as complex as in *Lolium rigidum* in Australia, where many permutations of cross-resistance pattern exist [12,13].

EFFECT OF RESISTANCE ON HERBICIDE ACTIVITY IN THE FIELD

It is important that the effects of resistance are measured at doses recommended for application in the field so that the likely impact of resistance can be assessed. No herbicide treatments or sequences gave consistently good control of *A. myosuroides* in field trials conducted at Peldon, where the most resistant populations occur [6,14]. While such field experiments are useful, they have the limitation that it is difficult to determine how much poor herbicide performance is due to resistance and how much to other unrelated factors [10]. This is a particular problem with the soil acting herbicides used for *A. myosuroides* control, as the performance of these is greatly influenced by soil and climatic conditions [15].

An alternative method of assessing comparative performance of herbicides is to grow plants of different populations in a standard soil in containers at one site. Thus soil and climatic variables are eliminated. We have used such a technique to simulate field conditions, which involves sowing seeds in soil in perforated plastic containers (27 x 18 x 10 cm deep) buried to the rim in the field, and treated with herbicides at the normal time and rates of application [10]. Fresh weight of foliage is recorded as a measure of herbicide activity.

Results are presented in Table 2 for an experiment in which seeds were sown on 5 October 1987, pre-emergence herbicides were applied on 6 October, post-emergence herbicides on 4 January 1988 when *A. myosuroides* plants had 3 leaves and foliage weight was assessed on 13 March.

TABLE 2

The effect of herbicides on two *A. myosuroides* populations grown in containers in the field in 1987/1988

Herbicide	Rate of application (kg a.i./ha)	% reduction in foliage fresh weight	
		Rothamsted	Peldon
Pre-emergence			
Cyanazine	2.25	47	31
Ethofumesate	1.40	99	99
Simazine	1.15	84	73
Metazachlor	1.25	99	82
Metribuzin	1.05	42	31
Trifluralin	1.20	43	47
Post-emergence			
Barban	0.31	81	43
Carbetamide	2.10	94	90
Chlorotoluron	3.50	100	20
Fluazifop-P-butyl[1]	0.094	97	93
"	0.188	98	95
Isoproturon	2.50	100	63
Propyzamide	0.35	96	92
"	0.70	90	91
Quizalofop-ethyl[2]	0.125	98	97
Sethoxydim[3]	0.34	98	97
Tralkoxydim[1]	0.20	95	58
SMY 1500	1.75	100	63

S.E. ± 3.9

1 = a non-ionic wetter ('Agral') was used with these treatments
2 & 3 = adjuvant oil (2 ='Fyzol', 3 ='Adder') was used with these treatments.

The Rothamsted population (susceptible standard) was well controlled by many herbicides, especially those applied post-emergence. All herbicides gave some control of the Peldon populations but significantly ($P \leq 0.05$) poorer control, compared with Rothamsted, was achieved with barban, chlorotoluron, cyanazine, isoproturon, metazachlor, metribuzin, simazine, tralkoxydim and SMY 1500. In contrast, carbetamide, ethofumesate, fluazifop-P-butyl, propyzamide, quizalofop-ethyl, sethoxydim and trifluralin gave similar levels of control of both populations.

With the exception of trifluralin, all the herbicides that can be used in cereals showed reduced activity on the Peldon population. In contrast, several of the herbicides that can only be used in non-cereal crops, e.g. ethofumesate, fluazifop-P-butyl, propyzamide, quizalofop-ethyl, and sethoxydim gave good control. Thus A. myosuroides at Peldon should be easier to control by herbicides in non-cereal than in cereals crops.

A limitation of experiments involving single dose rates is that excellent levels of control achieved by some herbicides can mask small differences in sensitivity. This has been confirmed in subsequent studies with fluazifop-P-butyl, which have demonstrated that this herbicide is marginally less effective on the Peldon population [5]. It is important to emphasise that the results obtained with a single resistant population may not be generally applicable, as other populations may have different cross-resistance patterns.

Results of a series of experiments in simulated field conditions, in which populations with differing chlorotoluron resistance ratings were included, are summarised in Table 3. The * ratings for chlorotoluron resistance, obtained from pot screening tests were well correlated with the activity of this herbicide in simulated field conditions. Reductions in the activity of isoproturon, diclofop-methyl and fenoxaprop-ethyl were also clearly demonstrated. The H/121 population from Suffolk had shown a high degree of resistance to chlorotoluron, but no resistance to diclofop-methyl in previous pot experiments (see previous section). In contrast, the Norfolk A population was shown to be susceptible to chlorotoluron in two previous screening tests [6,16], but had shown a low level of resistance to diclofop-methyl. In a dose response experiment conducted in the glasshouse, the diclofop-methyl ED_{50} value for the Norfolk A population was 3.8 fold higher than that of the Rothamsted standard (1.18 v 0.31 kg a.i./ha).

Thus, these contrasting patterns of cross-resistance were confirmed in simulated field conditions. In addition, the results for trifluralin supported other studies that showed that the Peldon population was not resistant to this herbicide [11].

TABLE 3
The efficacy of four herbicides applied at recommended field rates to a range of *A. myosuroides* populations grown in outdoor containers [5,9,10,11, Moss unpublished] or in field plots at one site [5], 1985-1991.

Population[1]	% reduction in foliage fresh weight							
	Roth.	Norf.	Box.	Far.	Tip.	H/121	Bucks	Peld.
Resistance rating[2]	susc.	susc.	1*	2*	3*	4*	4*	5*
Chlorotoluron 3.5 kg ai/ha	82-100 (5)	- -	61 (1)	42-50 (2)	55 (1)	38 (1)	4 (1)	0-30 (5)
Isoproturon 2.5 kg ai/ha	66-100 (4)	- -	95 (1)	84 (1)	- -	45 (1)	23 (1)	15-63 (5)
Diclofop-methyl 1.14 kg ai/ha	53-100 (4)	83 (1)	46 (1)	30-51 (2)	- -	85 (1)	- -	8-55 (4)
Fenoxaprop-ethyl 0.12 kg ai/ha	98 (1)	- -	- -	- -	- -	- -	46 (1)	61 (1)
Trifluralin 1.2 kg a.i./ha	43-87 (3)	- -	- -	- -	- -	- -	- -	47-86 (3)

1 = Roth.= Rothamsted; Norf.= Norfolk A; Box.= Boxworth; Far.= Faringdon;
    Tip.= Tiptree A; H/121 = Suffolk; Bucks = Bucks C; Peld.= Peldon
2 = Rating for resistance to chlorotoluron based on glasshouse tests [6].
( ) = Number of experiments

The experiments showed that, while resistance does not cause complete inactivity of herbicides, substantial reductions in activity of a wide range of herbicides can occur. The * ratings obtained from glasshouse screening experiments were well correlated with chlorotoluron and isoproturon activity in simulated field conditions. However, the relationship between * rating for chlorotoluron and the performance of the other herbicides was poorer, for reasons that are only partially understood.

RATE OF DEVELOPMENT OF RESISTANCE

Studies have demonstrated that herbicide resistance in *A. myosuroides* has developed during the last 12 years in several fields, and that current screening tests are not simply detecting long-standing differences between populations [17]. However, the rate of development of resistance to chlorotoluron has been much slower than the analogous evolution of resistance to a range of other herbicides in *Lolium rigidum* [13,18], despite continued, and in some cases increasing, use of herbicides to control *A. myosuroides*. The use of several herbicides with contrasting modes of action, as commonly occurs in winter cereal fields in England, may result in a differential selection pressure for resistant individuals, especially if more than a single resistance mechanism is present in a population. Thus resistance to different herbicides may not develop at the same rate.

Preliminary studies have supported this suggestion. In two fields with contrasting initial levels of resistance (Faringdon and Peldon), resistance to chlorotouron increased very slowly during a period of 4-7 years, despite both fields receiving 8-10 applications of chlorotoluron or isoproturon during this period. In contrast, resistance to diclofop-methyl increased very dramatically at Faringdon, but changed very little at Peldon. This difference cannot be linked simply to intensity of herbicide use as diclofop-methyl was applied 3 - 5 x during this period in both fields.

It is probable that different resistance mechanisms are involved in the Faringdon and Peldon populations. At Peldon a high degree of resistance to diclofop-methyl was associated with a high degree of resistance to chlorotoluron. At Faringdon, in the 1990 samples, an even higher degree of resistance to diclofop-methyl was associated with a much lower degree of resistance to chlorotoluron. At Peldon, resistance to both herbicides appears to be approaching a maximum level, despite the continued intensive use of herbicides. If the major resistance mechanism in the Peldon population is enhanced metabolism, then this may impose an upper limit to the degree of resistance which can occur. The mechanisms involved with resistance at Faringdon have not been determined yet.

The results do demonstrate that, at least in one field, resistance to diclofop-methyl, and possibly other aryloxyphenoypropionate herbicides, has developed much more rapidly than resistance to substituted-urea herbicides such as chlorotoluron. It is possible that there will be a rapid development of resistance elsewhere to the aryloxyphenoxypropionate and cyclohexanedione herbicides, which have a similar mode of action and are being used increasingly for grass weed control in a variety of crops. In Australia rapid development of resistance to these herbicides has occurred in *Lolium rigidum* [13,19]. The foliar acting nature and high intrinsic activity of such herbicides may impose a higher selection pressure than the soil acting herbicides chlorotoluron and isoproturon, whose activity is more affected by environmental conditions.

Different herbicides appear to be selecting for resistance at differing rates. We need determine the significance of these differences and to assess their implications on the recommendations for control and prevention of resistance.

THE PREVENTION AND CONTROL OF RESISTANT POPULATIONS

It is important that a major emphasis is placed on cultural control measures in order to reduce the reliance on herbicides and minimize *A. myosuroides* populations by non-chemical means. The cultural control measures being advocated include ploughing, crop rotation, spring cropping, delayed autumn drilling, improved stubble hygiene and growing more competitive crops. However, it is unlikely that cultural control measures alone will be sufficient to contain *A. myosuroides* populations in rotations of predominantly autumn sown crops. In cereal crops, the use of herbicides which are least affected by resistance and which have not shown a propensity to cause a rapid escalation of resistance is advised. Recommended herbicides include trifluralin, tri-allate and isoproturon. In non-cereal crops a wider choice of effective herbicides is available.

Provisional guidelines for an integrated control strategy have been formulated, but further information on the factors determining the rate of development of resistance is needed. More definitive recommendations are being prepared, not only for the control of existing resistant populations, but more importantly, to minimize the risk of resistance developing more widely.

ACKNOWLEDGEMENTS

I wish to thank my sandwich students, Julia Balmford, Glenys Williams and Manenkeu Ndoping for technical assistance, and the Home-Grown Cereals Authority for funding their placements out of the Research and Development Levy.

REFERENCES

1. Moss, S.R., The seed cycle of *Alopecurus myosuroides* in winter cereals: a quantitative analysis. Proceedings European Weed Research Society Symposium: Integrated Weed Management in Cereals, 1990, 27-36.

2. Ivens, G.W., The UK Pesticide Guide, CAB International/British Crop Protection Council, 1991.

3. Davis, R.P., Garthwaite, D.G. and Thomas, M.R., Arable farm crops 1988, Survey Report Pesticide Usage England and Wales No. 35., 1990, Ministry of Agriculture, Fisheries and Food, London.

4. Moss, S.R. and Cussans, G.W., Variability in the susceptibility of *Alopecurus myosuroides* (black-grass) to chlorotoluron and isoproturon. Aspects of Applied Biology 9, The Biology and Control of Weeds in Cereals, 1985, 91-98.

5. Clarke, J.H. and Moss, S.R., The occurrence of herbicide resistant *Alopecurus myosuroides* (black-grass) in the United Kingdom and strategies for its control. Proceedings 1991 Brighton Crop Protection Conference - Weeds, 1991, (in press).

6. Clarke, J.H. and Moss, S.R., The distribution and control of herbicide resistant *Alopecurus myosuroides* (black-grass) in central and eastern England. Proceedings 1989 Brighton Crop Protection Conference - Weeds, 1989, 301-308.

7. Rubin, B., Qualitative and quantitative methods for identifying and monitoring herbicide-resistant weeds. In Achievements and Developments in Combating Resistance, eds I. Denholm, A.L. Devonshire and D.W. Holloman, Elsevier Applied Science Publishers, London, 1992, (in press).

8. Kemp, M.S., Moss, S.R. and Thomas, T.H., Herbicide resistance in *Alopecurus myosuroides*. In Managing Resistance to Agrochemicals: from Fundamental Research to Practical Strategies, eds M.B. Green, H.M. LeBaron and W.K. Moberg, American Chemical Society, Washington, 1990, pp. 376-393.

9.   Moss, S.R. and Cussans, G.W., Detection and practical significance of herbicide resistance with particular reference to the weed *Alopecurus myosuroides* (black-grass). In Combatting Resistance to Xenobiotics: Biological and Chemical Approaches, eds M.G. Ford, D.W. Holloman, B.P.S. Khambay and R.M. Sawicki, Ellis Horwood, Chichester (England), 1987, pp. 200-213.

10. Moss, S.R., Herbicide resistance in black-grass (*Alopecurus myosuroides*). Proceedings 1987 British Crop Protection Conference - Weeds, 1987, 879-886.

11. Moss, S.R., Herbicide cross-resistance in slender foxtail (*Alopecurus myosuroides*). Weed Science, 1990, **38**, 492-496.

12. Heap, I.M. and Knight, R., Variations in herbicide cross-resistance among populations of annual ryegrass (*Lolium rigidum*) resistant to diclofop-methyl. Australian Journal of Agricultural Research, 1990, 41, 121-128.

13. Powles, S.B. and Matthews, J.M., Multiple herbicide resistance in annual rye-grass (*Lolium rigidum*). A driving force for the adoption of integrated weed management. In Achievements and Developments in Combating Resistance, eds I. Denholm, A.L. Devonshire and D.W. Holloman, Elsevier Applied Science Publishers, London, 1992, (in press).

14. Orson, J.H. and Livingston, D.B.F., Field trials on the efficacy of herbicides on resistant black-grass (*Alopecurus myosuroides*) in different cultivation regimes. Proceedings 1987 British Crop Protection Conference - Weeds, 1987, 887-894.

15. Orson, J.H., Variability in the control of black-grass with herbicides in winter cereals - harvest years 1987 - 1989, HGCA Project Report No. 27., 1991, Home Grown Cereals Authority, London.

16. Moss, S.R. and Orson, J.H., The distribution of herbicide-resistant *Alopecurus myosuroides* (black-grass) in England. Aspects of Applied Biology 18, Weed Control in Cereals and the Impact of Legislation on Pesticide Application, 1988, 177-185.

17. Moss, S.R. and Cussans, G.W., The development of herbicide-resistant populations of *Alopecurus myosuroides* (black-grass) in England. In Herbicide Resistance in Weeds and Crops, eds J.C. Caseley, G.W. Cussans and R.K. Atkin, Butterworth-Heinemann, Oxford, 1991, pp. 45-55.

18. Heap, I.M., Resistance to herbicides in annual ryegrass (*Lolium rigidum*) in Australia. In Herbicide Resistance in Weeds and Crops, eds J.C. Caseley, G.W. Cussans and R.K. Atkin, Butterworth-Heinemann, Oxford, 1991, pp. 57-66.

19. Heap, J. and Knight, R., A population of ryegrass tolerant to the herbicide diclofop-methyl. The Journal of the Australian Institute of Agricultural Science, 1982, **48**, 156-157.

IRAC FRUIT CROPS WORKING GROUP SPIDER MITE RESISTANCE MANAGEMENT STRATEGY

P. K. LEONARD

DowElanco, Letcombe Laboratory, Letcombe Regis, Wantage,
Oxfordshire, OX12 9JT, UK

ABSTRACT

The Insecticide Resistance Action Committee (IRAC) was created
to combat the threat of pesticide resistance. IRAC's main
objective is to develop and implement technically sound
strategies to delay or prevent the onset of resistance.
IRAC's Fruit Crops Working Group (FCWG) proposed a spider mite
resistance management strategy in 1988. This was based on a
rotation of products which were not thought to be subject to
the same cross resistance mechanisms. The strategy was to a
large extent pragmatic as there was limited evidence of
product cross resistance. To support the implementation of
this strategy, proposed resistance monitoring methods were
published.

Feedback on these proposals was actively sought for a
period of two and a half years from within the Industry,
experts, advisors, and from Agrochemical distributors. An
IRAC-funded research project was conducted at Cornell
University to obtain preliminary data on cross resistance
patterns for 17 acaricides.

The original proposed strategy and product groupings were
reviewed in light of the results of the Cornell study and with
the combined feedback from other sources. A revised strategy
and product grouping is presented along with plans for further
research and implementation.

INTRODUCTION

The Insecticide Resistance Action Committee (IRAC) was
created, under the umbrella of the International Group of
National Associations of Agrochemical Manufacturers (GIFAP),
to combat the threat of pesticide resistance. The formation
and structure of IRAC was described by Voss [1]. IRACs' main

objective is to develop and implement technically sound strategies to delay or prevent the onset of resistance.

IRACs' Fruit Crops Working Group (FCWG) held its first meeting in 1985 and has been active throughout the last six years. One of the Groups' first achievements was to propose a strategy for management of spider mite resistance, the development of which was described by Lemon [2]. The purpose of this paper is to present a revised version of the Groups' strategy for spider mite resistance management.

IRAC Fruit Crops Working Group - members 1991:

C. Erdelen	Bayer AG
D. Giles	Schering AG
A. C. Grosscurt	Duphar B.V.
D. Highwood	Shell Research Ltd.
P. Leonard	DowElanco (Chairman)
T. Merriam	American Cyanamid Co
H. P. Streibert	Ciba-Geigy AG
A. Waltersdorfer	Hoechst AG
P. Wege	I.C.I. Agrochemicals (secretary)

1988 SPIDER MITE RESISTANCE MANAGEMENT STRATEGY

The original strategy was based on a rotation of acaricides which are not subject to the same resistance mechanisms. Acaricides were grouped according to known or expected cross resistance patterns. In order to achieve this the collective wisdom of the member companies represented was pooled together with a literature search and advice from a number of leading experts. It was, however, understood that our knowledge of the many and complex cross resistance patterns was incomplete. It was clear that a considerable research effort would be required to clarify the situation. The original strategy was therefore regarded, to a large extent, as being pragmatic.
Available acaricides were grouped as follows:

Group A Organotins   [3,4]
Group B Clofentezine, hexythiazox   [5]
Group C Bridged diphenyl compounds (Bromopropylate and
    dicofol)
Group D pyrethroids
Group E Flubenzimine
Group F Tetradifon
Group G Amitraz
Group H Propargite
Group I Quinomethionate
Group J Benzoximate
Group K Dinobuton

The following guidelines were developed for use with these product groupings:

1. Not more than one compound from any one group should be

applied to the same crop in the same season.

2. Any one compound should be used only once per season on any one crop. *

3. Compounds from the same group should not be mixed.

4. Compounds should be used in such a way that detrimental effects on predatory insects and mites are minimized.

5. Use compounds only at manufacturer's recommended rates and timings.

6. Monitoring should be conducted to detect early signs of resistance. **

* Because of specific activity against certain life stages, some compounds may be recommended for two successive applications.

** Resistance monitoring methods are proposed for two spotted spider mite (_Tetranychus_ _urticae_) and the fruit tree red spider mite (_Panonychus_ _ulmi_) adults and for _P. ulmi_ and _Tetranychus_ spp. eggs [6].

The strategy was presented at the 1988 Brighton Crop Protection Conference [2], and attracted feedback from a variety of sources around the world. One of the most important effects was that it stimulated discussion within the Industry. The strategy provided us with a set of guidelines where they were most needed; where marketing decisions were being made for new compounds. These decisions are often difficult as there are many potential conflicts of interest. The need to target greater short term sales to justify escalating research and development costs has to be balanced by a need to promote long term market potentials. The latter may only be achieved by managing the threat of resistance. These goals have traditionally been hard to reconcile. The spider mite resistance management strategy provided us with an "independent" and technically sound framework with which to balance these apparently opposed market requirements.
Outside the industry a number of specialists, advisors and distributors were asked for their views on the proposed strategy. This process was actively pursued for two years. Each comment was considered by the Group and a number of additions and alterations to the strategy resulted.

IRAC FUNDED RESEARCH

In addition to extensive consultation, a research project was initiated at Cornell University to investigate the validity of the proposed product groupings. The plan was to compare the responses of an unselected and six field-selected strains of

T. urticae to 17 commercially available acaricides. The protocol included 101 chemical/strain combinations, each bioassay consisting of four replicates and five concentrations. Every effort was made to standardize bioassay procedures to maximize the value of between-product comparisons.

From the outset it was clear that the resolution of such a broad study would be limited. In particular, when populations exhibited resistance to more than one chemical, it would not be possible to differentiate between cross resistance conferred by a single mechanism and multiple resistance with two or more independent but coexisting mechanisms. The primary goal, therefore, was to identify putative cases of cross resistance that could subsequently be studied in more depth.

The project commenced in January 1990 and results were presented to the FCWG in April 1991 by Dr. T. J. Dennehy, the principal researcher at Cornell University. A detailed account of methods adopted and results obtained will be published separately.

FCWGS' REVISED STRATEGY FOR SPIDER MITE RESISTANCE MANAGEMENT

Results of the Cornell study were, to a large extent, consistent with the original product groupings, and as a result the original groupings remained largely unchanged. However, two revisions were made to accomodate products not included in the 1988 recommendations:

1. Formetanate is placed in a group on its own. Although there are similarities between the structures of formetanate and amitraz, there was experimental evidence to support leaving these two acaricides in separate groups.

2. Abamectin is placed in a group on its own as there was no indication of cross resistance.

Other revisions to the product groupings were made as a result of discussions within the FCWG.

3. The acaricidal acylureas flufenoxuron and flucycloxuron are grouped with clofentezine and hexythiazox. This decision was supported by experimental results generated by a number of the companies represented on the FCWG.

4. Flubenzimine was withdrawn as it is not commercially available.

5. Acaricidal organophosphates will be included as an additional group.

The revised 1991 product groupings and use guidelines are shown in full below:

Group A Organotins
Group B Clofentezine, hexythiazox,
        flufenoxuron and flucycloxuron
Group C Bridged diphenyl compounds (Bromopropylate and
        dicofol)
Group D pyrethroids
Group E Tetradifon
Group F Amitraz
Group G Propargite
Group H Quinomethionate
Group I Benzoximate
Group J Dinobuton
Group K Abamectin
Group L Organophosphates
Group M Formetanate

Product groupings will continue to be upgraded and revised as new products enter the market and as our understanding of acaricide cross resistance improves.

The following guidelines are recommended for use with the above product groupings:

1. Not more than one compound from any one group should be applied to the same crop in the same season.

2. Any one compound should be used only once per season on any one crop.

3. Mixtures of acaricides from different groups may be used but use of mixtures of products from the same group is not recommended.

4. Compounds should be used in such a way that detrimental effects on predatory insects and mites are minimized.

5. Use compounds only at manufacturer's recommended rates and timings.

6. Monitoring should be conducted to detect early signs of resistance. *

* Resistance monitoring methods are proposed for T. urticae and P. ulmi adults and eggs [6].

The footnote, "Because of specific activity against certain life stages, some compounds may be recommended for two successive applications" was deleted as it was no longer regarded as being necessary.

These guidelines, like the product groupings, will continue to be reviewed as further data on factors affecting selection of acaricide resistance become available. At present they represent a concensus of opinion amoung companies represented on the FCWG.

FURTHER RESEARCH

Interestingly, one of the most important results of the Cornell study was to focus attention on potential interactions of acaricidal pyrethroids and organophosphates on other key acaricides. Future work by the FCWG will aim to investigate these potential interactions as a high priority.

Traditionally the vast majority of laboratory research on spider mite resistance has been conducted using <u>Tetranychus</u> spp. as they are relatively easy to culture. By comparison little experimental work has been conducted using <u>P. ulmi</u> as it is more difficult to culture. The FCWG is keen to investigate the predictive value of using experimental results obtained using <u>T. urticae</u> when developing strategies for management of both <u>Tetranychus</u> and <u>Panonychus</u> species. Proposals for work on both of these areas of research are currently being considered by the FCWG.

A literature survey to provide additional information on cross resistance between acaricidal organophosphates will be conducted. Experimental work on this question will be considered when results of the survey are known.

IMPLEMENTATION

The ultimate objective of IRAC, "To <u>implement</u> strategies to delay or prevent the onset of resistance," has been difficult to achieve. The FCWG has proposed resistance monitoring methods for adult mites and winter eggs and has published a strategy for spider mite resistance management. This, however, only represents the first few steps toward achieving our ultimate objective of implementation.

We plan to work toward achieving this objective by pursuing three lines of "attack".

1. By publication and presentation of the strategy to ensure that the widest possible audience is accessed. This has probably been the most successful mechanism so far.
2. By promotion within the Industry. This is regarded as being of a high priority as a number of new chemical groups advance to the market.
3. By working alongside other organizations with compatible objectives such as the recently established International Organization for Resistance Pest Management (IORPM). The development of IORPM offers a unique opportunity for an independent organization to promote technically sound strategies through to practice. It is for this reason that the FCWG is keen to maintain its representation on IORPMs' Orchards and Tetranychid Working Group.

ACKNOWLEDGEMENTS

Sincere thanks to Mr. R. Lemon of Schering for his role as chairman of the Fruit Crops Working Group during the initial development of the strategy, to Dr. T. Dennehy, Dr. G. Sterk and the late Dr. R. Sawicki of Cornell, Gorsem and Rothamsted, respectively for their expert advice and to all past members of the Fruit Crops Working Group for their contribution to the development of the strategy.

REFERENCES

1. Voss, G., Insecticide/Acaricide Resistance: Industry's Efforts and Plans to Cope. Pesticide Sci., 1988, 23, 149-156.

2. Lemon, R. W., Resistance monitoring methods and strategies for resistance management in insect and mite pests of fruit crops. Proceedings of the Brighton Crop Protection Conference - Pests and Diseases., 1988, 9B-2, 1089.

3. Edge, V. E., James, D. G., Organotin resistance in two-spotted mite, Tetranychus urticae Koch (Acaria Tetranychidae) in Australia. Proceedings of the 10th International Congress of Plant Protection, 1983, 2, 639.

4. Bolevski, A., Resistance of the European red spider mite Panonychus ulmi Koch; (Tetranychidae) to the chemical Plictran (Cyhexatin) and to other organotin acaricides. Gradinarska Lozarska Nauka, 1983, 20 (5), 29-37.

5. Gough, N., Chemical control of two-spotted mite on field roses in S. Queensland. Proceedings of the symposium on mite control in horticultural crops, Orange, Department of Agriculture, New South Wales, 1987, July, 82-86.

6. IRAC., Proposed insecticide/acaricide susceptibility tests developed by Insecticide Resistance Action Committee, 1990, Bulletin OEPP/EPPO, Bulletin 20, 389-404.

FUNGICIDE RESISTANCE STRATEGIES IN WINTER WHEAT IN THE NETHERLANDS

MAARTEN A. DE WAARD
Wageningen Agricultural University, Department of Phytopathology,
P. O. Box 8025, Binnenhaven 9, 6700 EE Wageningen, the Netherlands

ABSTRACT

Disease control in winter wheat in the Netherlands may be based on a seed treatment, a spray against eyespot and two foliar sprays against leaf and ear diseases. Fungicides which inhibit sterol biosynthesis (SBIs) are rarely used in seed treatments but are used to a limited extent against eyespot (prochloraz) and are essential for control of leaf and ear diseases (triazoles and morpholines). Exclusive use of triazoles for mildew control up to 1985 led to reduced sensitivity to these fungicides in *Erysiphe graminis* f.sp. *tritici*. However, their use continued since field performance of most triazoles remained adequate. Use of fenpropimorph began in 1984. From 1986 onwards, the main strategy to prevent or retard resistance development to both groups of fungicides was based on their alternating use. Nevertheless, in 1989 and 1990 a limited number of fields in the province of Limburg revealed isolates with reduced sensitivity to fenpropimorph. Resistance levels up to 13 were noted. The resistance level of these isolates to triadimenol varied from 18-132. In 1989 and 1990, field performance of the fungicides was difficult to judge because of a low mildew incidence. An evaluation of the resistance strategy used is discussed.

INTRODUCTION

Winter wheat is an important arable crop in the Netherlands. In most areas of the country cropping is in rotation with potatoes, sugar beet and sometimes with pulses (mainly peas), grass seed, oil seed rape and flax. In the province of Groningen continuous wheat cropping is common practice. Winter wheat is the dominant small grain with an annual area of about 120,000 ha. It is mainly cropped on marine clay soils in the northern, central and south-western parts of the country. One of the minor areas is located on loess soils in the south in the province of Limburg. The adoption of short-strawed wheat cultivars, application of growth regulators, increased use of fertilizers and pesticides and the introduction of new cultivars gradually increased yield per hectare from 5 (metric) ton in 1974 to 8 ton in 1986 [1]. Occasionally, yields up to 11 ton are achieved.

The contribution of fungicides to the increasing yield can be attributed to a seed treatment and 3 sprays. The first spray is to control eyespot. During 1974 - 1983 and 1983 - 1987, this spray was only applied in about 11.6 and 7.5% of the fields respectively, and especially in Groningen were continuous wheat cropping is practiced [2]. The second and third sprays are to control leaf diseases before ear emergence (42% of the fields) and leaf and ear diseases after ear emergence (94% of the fields), respectively [1, 3]. This spray frequency compares favourably with the relatively higher number of sprays in wheat and barley in surrounding countries such as Germany and Great Britain. Reasons for this difference are the general use of crop rotation systems by which eyespot severity often remains below the threshold level, the heavy clay soil by which mildew severity remains relatively low and the use of the EPIPRE advisory system.

MAJOR DISEASES

Advisory models for major diseases in winter wheat in the Netherlands are described by Drenth and Stol [2].

Brown rust (*Puccinia recondita*) is a common disease of wheat. An epidemic may reduce yield by about 20%. Control is based on the use of resistant cultivars and the application of fungicides. Under Dutch conditions epidemics occurring after heading (GS 51) are important. One fungicide application should be sufficient for control and is, if necessary, applied before the end of flowering (GS 69).

Yellow rust (*Puccinia striiformis*) causes significant crop losses in the Netherlands in one out of 5 years and may reduce yield in such years by 40%. Development is optimal under wet conditions at 10-15 °C. Control is based on the use of resistant cultivars and on application of fungicides. Again, when required one application is generally sufficient if applied between pseudo-stem erection (GS 30) and the end of flowering (GS 69).

Mildew (*Erysiphe graminis* f.sp. *tritici*) is a common cereal disease and can cause an annual crop loss of 30%. Its development is optimal between 15 and 20 °C. Extremely dry or wet weather conditions may inhibit its development. The disease occurs most frequently on sandy and loess soils (*e.g.* Limburg) as compared to clay soils. Control measures include the use of resistant cultivars and fungicides applications. Chemical control is often necessary from early in the season (GS 31) up to flowering (GS 69). If the first spray is applied before emergence of the flag leaf (GS 39), a second spray may be necessary. In such cases the second spray is also aimed to control all other leaf and ear diseases present.

Mycosphaerella graminicola (*Septoria tritici*) is the causal agent of leaf spot. Development of the disease is optimal between 15-18 °C but may also spread at much lower temperatures. *Leptosphaeria nodorum* (*Septoria nodorum*) is the causal agent of leaf spot and glume blotch. Its development is optimal under wet conditions at temperatures between 20-25 °C. Between 1983 and 1987, EPIPRE advised a control of *Septoria* spp in 48% of all fields. Usually, the optimal timing is the middle of the flowering period (DC 65). A single spray is sufficient and should also eradicate simultaneously other leaf and ear diseases.

Favourable conditions for spread of eyespot (*Pseudocercosporella herpotrichoides*) occur in early spring at temperatures between 8-10 °C and a relative humidity above 80%. Losses occur at the end of the season (lodging) but for effective control the fungicide should reach the stem base and stop infection at an early stage. Therefore, chemical control is possible up to growth stage GS 32 but remains according to EPIPRE standards rather restricted.

Fusarium culmorum, F. graminearum and *Gerlachia nivalis* may incite seedling, foot and ear diseases. Only losses caused by ear diseases may be high. Fungicides for control of *Fusarium* diseases are not available.

FUNGICIDE HISTORY FOR SEED TREATMENT

Various pathogens may cause seedling diseases. The following fungicides are licensed in the Netherlands for seed disinfection. Up to 1980 the mixture sodium dimethyldithiocarbamate/fuberidazole dominated the market. Emergence of benzimidazole resistant isolates of *G. nivalis* [4] reduced its use to about 40% of the seed being disinfected. Other products currently used are, guazatine (52%), bitertanol/fuberidazole (5%) and guazatine/imazalil (3%). Although registered, the combination product triadimenol/fuberidazole is seldom used. Carboxin is registered for use against loose smut. However, its present use is rare because of reduced efficacy. This may be a non-confirmed case of resistance development. Because of the low profitability of wheat production, some farmers harvest their own seed and do not apply any seed desinfection. This is estimated to be about 15-20% of total seed production.

FUNGICIDE HISTORY FOR CONTROL OF FOOT, LEAF AND EAR DISEASES

Fungicides used for control of foot, leaf and ear diseases are listed by Daamen [1]. The spectrum of antifungal activity of most of these fungicides is given in Table 1.

Benzimidazoles (benomyl, carbendazim) and thiophanate-methyl were used from the 1970's onwards for control of leaf and ear diseases and eyspot. Their use against leaf and ear diseases have been replaced by triazoles and morpholines. Benzimidazoles are still used for control of eyespot, despite of the fact that in one confined area (Groningen) resistance to carbendazim in *P. herpotrichoides* was detected in 1984 [5]. The imidazole prochloraz is currently also used for control of this disease. In addition, this fungicide has a good activity agains *Septoria* diseases. Prepacked mixtures of benzimidazoles and prochloraz were never registered.

Triazole fungicides are used for control of various leaf and ear diseases. Triadimefon and propiconazole were applied from 1978 and 1982 respectively, but their use increased sharply in 1983. In 1985 triadimefon was replaced by triadimenol because of its better field performance. Because these triazoles only have a weak or moderate effect against *Septoria* diseases, propiconazole is often applied as a mixture with prochloraz and triadimenol with anilazine

The morpholine tridemorph has been used to some extent for mildew control. However, introduction of the triazoles almost abolished its use. Fenpropimorph was registered in 1983 and intensively used for disease control from 1984 onwards. It is now regarded as the fungicide with the highest curative activity against mildew. In order to extend its spectrum of antifungal activity to *Septoria* spp, mixtures with chlorothalonil or prochloraz are being applied.

On a large proportion of the winter wheat area, triazole and morpholine sprays are also combined with maneb in order to increase control of leaf and ear diseases. Experiments repeatedly demonstrated that combinations with this compound do not result in a reliable increase in disease control and yield. Proposals are under discussion to forbid the use of maneb [6].

Two other registered fungicides with a selective activity against mildew are pyrazophos and sulphur. Pyrazophos is marketed as a combination product with carbendazim for control of leaf and ear diseases. Its use remained rather restricted. Use of sulphur is also neglectable. Fungicides which had a registration in the past are captafol (*Septoria*) and ethirimol (mildew).

TABLE 1

Activity of fungicides registered for used in the Netherlands against foliar pathogens of wheat.

Active ingredient	Activity[1,2]					
	brown rust	yellow rust	mildew	*Septoria*	*Fusarium*	eyespot
Anilazin	-	-	-	++	-	-
Benomyl	-	-	+	+	-	++
Carbendazim	-	-	+	+	-	++
Chlorothalonil	-	-	-	++	-	-
Fenpropimorph	+++	++	+++	+	-	-
Maneb	-	-	-	+	-	-
Propiconazole	++	+++	++±	++	-	-
Prochloraz	-	-	+	++	-	++
Pyrazophos	-	-	++	-	-	-
Sulphur	-	-	+	+	-	-
Triadimefon	++	+++	+++	-	-	-
Triadimenol	+++	+++	+++	+	-	-
Tridemorph	-	+	++	-	-	-
Triforine	-	-	+±	+	-	-
Thiophanate-methyl	-	-	+	+	-	++

[1] -: no activity; +, +±, ++, ++±, +++: low, moderate, medium, high and very high activity, respectively.
[2] Symbols indicate activity against wild-type population.

CHEMICAL ANTI-RESISTANCE STRATEGIES

Before 1984, chemical anti-resistance strategies to prevent or retard resistance development to triazoles in wheat pathogens were seldom applied since this was the only effective group of fungicides available. Further, resistance risks to fungicides which inhibit sterol 14α-demethylation (DMIs) were considered low. After 1984, several reports from some European countries described decreased sensitivity of barley and wheat powdery mildew to triazoles [7, 8]. Decreased sensitivity in *E. graminis* f. sp. *tritici* was also noted in the Netherlands [9]. Therefore, the introduction of fenpropimorph in 1983 came just too late to devise optimal anti-resistance strategies. However, these strategies were implemented from 1986 onwards.

The first measure with respect to seed treatments was to discourage use of DMIs with systemic action in emerging seedlings in order to restrict the selection period for foliar pathogens. This measure became quite effective. A second measure was to confine the control of eyespot to prochloraz only, since this DMI (an imidazole) exerts little selection pressure on most foliar pathogens, except for diseases caused by Septoria spp. With respect to control of foliar diseases six optional strategies were available (Table 2). Options 1 and 2 were not recommended in practice since by 1984 triazoles lacked sufficient curative action against mildew to be used in the EPIPRE advisory system. Morpholine/triazole mixtures (option 6) also did not become common practice since the triazole component in such a mixture, exclusively aimed to control mildew (first foliar spray), was not regarded as relevant. Another major reason for this point of view is that governmental policy in the Netherlands discourages the use of pre-packed mixtures in general since this would lead to an increased use of pesticides. Pre-packed mixtures of morpholines and triazoles were never registered in the Netherlands. Other considerations not to use successive mixtures of morpholine/triazole were that these may not prevent erosion of resistance to triazoles, but favour further stepwise development of resistance to triazoles, and lead to an increased selection pressure of morpholines, especially in mildew [10]. Options 4 and 5 imply two morpholine applications a year and were therefore not recommended. In consequence, the officially recommended strategy for disease control in winter wheat in Limburg became option 3. Mixtures used in option 3 for the second spray were predominantly triazole combinations with anilazin, chlorothalonil or prochloraz. In this option, prochloraz was accepted as a companion, because of its selective action against Septoria spp. The strategy became readily adopted by most farmers in Limburg, although a limited number did not adhere to the recommendations of advisory officers and used instead mixtures of fenpropimorph and chlorothalonil, or prochloraz, as the second spray (option 5).

TABLE 2

Optional anti-resistance strategies in control of foliar pathogens of wheat in the Netherlands, based on alternating use of fungicides and/or the use of fungicide combinations.

Option	First spray	Second spray
1	triazole	morpholine/non-triazole[1]
2	triazole	morpholine/triazole
3	morpholine	triazole/non-morpholine[2]
4	morpholine	triazole/morpholine
5	morpholine	morpholine/non-triazole
6	morpholine/triazole	morpholine/triazole

[1] Non-triazole: chlorothalonil or prochloraz
[2] Non-morpholine: anilazin, clorothalonil or prochloraz

Farmers in other provinces of the country may have followed completely different strategies. This is due to the disease situation in particular areas. For instance, mildew incidence on mildew-resistant varieties of wheat on the Groningen clay soil is minor and occurs relatively late in the season. Instead, in June Septoria or yellow rust may be predominant and as a consequence, the first spray is carried out with prochloraz or triazoles, respectively. Fenpropimorph/prochloraz or triazole mixtures are used in the second spray for control of foliar and ear diseases. It may be that in future in these areas one single foliar spray with a "strong" morpholine/DMI mixture timely applied just after flag leaf emergence results in adequate control of all leaf and ear diseases. In such situations the use of morpholine/DMI mixtures should be acceptable.

SURVEYS TO MONITOR FUNGICIDE SENSITIVITY

Surveys to monitor the sensitivity of *E. graminis* f. sp. *tritici* to fungicides which inhibit sterol biosynthesis (SBIs) were carried out from 1982 to 1985 for triadimefon and from 1984 onwards for fenpropimorph. In these studies, wheat leaves with powdery mildew symptoms were collected from fields in Limburg and occasionally from other parts of the country. EPIPRE, a computer based pest and disease advisory system, was used to collect relevant data on spray history and mildew development. Usually, 3-6 weeks after SBI applications, six mildew samples per field were collected and propagated on wheat seedlings cv Okapi. Conidia of the second generation of these isolates were used to inoculate wheat seedlings sprayed in foliar spray tests with a range of fungicide solutions, made from the formulated products. Details of the procedure have been described [9]. Since 1988, the following modifications in the procedure were made in order to increase reproducibility: (1) seedlings were sprayed under standardized conditions in an automated spray cabinet; (2) mildew assessments of seedlings was based on the disease simulation progamme "Distrain"; (3) the software package "Statgraphics" was used for linear regression and statistical analyses. The degree of resistance was expressed as the ratio between the EC_{50} of the field isolates and the EC_{50} of the reference isolate LH (Q-value). Q-values were used to compare the dynamics of fungicide resistance from 1982 onwards.

DYNAMICS OF SENSITIVITY TO TRIAZOLES

At the start of the survey in 1982 the sensitivity of mildew isolates from Limburg was compared with isolates from a confined wheat growing area around Lienden (Gelderland). Up to this time, triazoles were never used in the province of Gelderland and isolates from Lienden were, therefore, assumed to have baseline sensitivity. The average EC_{50} of these isolates indeed did not differ from an existing wild type. One of the Lienden isolates was taken as a standard reference (LH). The mean Q-value of isolates from Limburg was 3.8 (Table 3). Obviously, four years of use of triazoles did lead to a small but significant decrease in sensitivity. In subsequent years (1983-1985), this value did not change significantly, irrespective of the fungicide regime applied. Apparently, the initial source of inoculum for all fields is the same and the fungicide regimes do not readily select for populations with differential triazole-sensitivity within one growing season. Frequency distributions of EC_{50} values of field isolates collected before and after triazole treatment have been published [9]. The reduced sensitivity of the populations is probably the decisive factor for the reduction in curative action of triazoles in mildew control over these years. Isolate 67 was one of the isolates with the highest Q-values in 1983 (4.1). This isolate was maintained over years on seedlings sprayed with sub-lethal rates of triadimefon and used in later years as a second reference isolate. The Q-value of this isolate appeared to be 6.0

and 10.5 in 1984 and 1985, respectively, suggesting a gradual decrease in sensitivity to triadimefon.

TABLE 3

Sensitivity to triadimefon of isolates of *Erysiphe graminis* f.sp. *tritici* collected in 1982-1985 in Lienden (Gelderland) and Limburg using foliar spray tests.

Year	Isolate (reference or provence)	Fungicide history	Number of isolates (or tests)[1]	Triadimefon	
				EC_{50} ($\mu g\ ml^{-1}$)	Q^2
1982	Lienden	none	10	1.1 ± 0.9	-
	Limburg	triazoles	28	4.2 ± 2.0[3]	3.8
1983	Isolate LH	-	(15)	0.7 ± 0.7	-
	Limburg	none	63	1.4 ± 1.6[3]	2.0
	Limburg	triazoles	102	1.8 ± 1.5[3]	2.6
1984	Isolate LH	-	(19)	0.7 ± 0.4	-
	Isolate 67	-	(11)	4.2 ± 1.6[3]	6.0
	Limburg	none	46	2.8 ± 3.9[3]	4.0
	Limburg	triazoles	12	3.0 ± 2.7[3]	4.2
	Limburg	morpholines and triazoles	36	2.9 ± 2.3[3]	4.1
1985	Isolate LH	-	(15)	2.0 ± 1.6	-
	Isolate 67	-	(15)	21.0 ± 19.9[3]	10.5
	Limburg	morpholines	39	6.2 ± 2.1[3]	3.1
	Limburg	morpholines and triazoles	29	6.2 ± 1.5[3]	3.1

[1] Between brackets: number of tests with reference isolates LH and 67.
[2] EC_{50} field isolate : EC_{50} reference isolate (Lienden or LH).
[3] Significantly different from isolate LH (P = 0.05), tested in the same year.

DYNAMICS OF SENSITIVITY OF FENPROPIMORPH

The first survey to monitor the sensitivity to fenpropimorph (Corbel) began in 1984, immediately after its introduction in 1983. Isolates from non-treated and Corbel-treated fields in Limburg had approximately similar mean EC_{50} values to those of the reference isolates LH and 67 and were therefore regarded to have baseline sensitivity (Table 4). Up to 1986 this situation did not change. The variation in EC_{50} values over these years could mainly be ascribed to variations in test conditions since the Q-value of isolate 67 always was about 1. In 1987, the mean EC_{50} value of all the isolates tested showed a marginal difference with the reference isolates. This was due to the fact that 5 out of 16 fields tested yielded isolates with Q-values between 2.0 and 2.8. Differences in field performance of Corbel could not be detected [11]. Results of 1988 were similar to those of 1987, except for one field which yielded isolates with a mean Q-value of 2.9. This field had by far the highest mildew incidence of all fields surveyed. Therefore, in surveys during 1989 and 1990, special attention was paid to fields in the neighbourhood of this suspicious field. In 1989 and 1990, the mean Q-values of all isolates tested increased to 6.7 and 4.2, respectively (Table 4). Mean Q-values of isolates from some individual fields were 10 or even higher. The frequency distribution of Q-values in 1988, 1989 and 1990 also demonstrated these changes in the sensitivity of the population (Fig. 1). It can also be noted that the median class of the Q-values increased from 0-2 in 1989 to 2-4 in 1990. It is assumed that these changes in sensitivity mark the beginning of resistance development.

TABLE 4

Sensitivity to morpholines of *Erysiphe graminis* f.sp. *tritici* isolates collected in 1984 - 1990 from Limburg using foliar spray tests.

Year	Isolate (reference or province)	Fungicide history	Number of isolates (or tests)[1]	Fenpropimorph	
				EC_{50} (μg ml^{-1})	Q
1984	Isolate LH	-	(4)	0.9 ± 0.5	-
	Isolate 67	-	(4)	0.8 ± 0.2	0.9
	Limburg	none	15	1.0 ± 0.3	1.1
	Limburg	triazole	20	1.0 ± 0.3	1.0
	Limburg	morpholines	18	0.8 ± 0.3	0.9
1985	Isolate LH	-	(12)	1.4 ± 0.6	-
	Isolate 67	-	(9)	1.2 ± 0.5	0.9
	Limburg	morpholines	84	1.9 ± 0.6	1.4
1986	Isolate LH	-	(14)	2.3 ± 1.0	-
	Isolate 67	-	(14)	2.4 ± 1.3	1.0
	Limburg	morpholines	102	2.8 ± 1.0	1.2
1987	Isolate LH	-	(16)	0.4 ± 0.1	-
	Isolate 67	-	(16)	0.3 ± 0.1	0.8
	Limburg	morpholines	96	0.6 ± 0.2^2	1.5
1988	Isolate LH	-	(16)	1.0 ± 0.4	-
	Isolate 67	-	(16)	1.2 ± 0.3	1.2
	Limburg	morpholines	96	1.4 ± 0.5^2	1.4
1989	Isolate LH	-	(19)	0.9 ± 0.4	-
	Isolate 67	-	(19)	1.0 ± 0.3	1.1
	Limburg	morpholines	114	5.8 ± 6.0^2	6.7
1990	Isolate LH	-	(17)	1.4 ± 0.3	-
	Isolate 67	-	(17)	1.2 ± 0.2	0.9
	Limburg	morpholines	126	6.1 ± 4.0^2	4.2

[1] Between brackets: number of tests with reference isolates LH and 67.
[2] Significantly different from isolate LH (P - 0.05), tested in the same year.

The crucial question whether the reduced sensitivity observed was correlated with reduced field performance of Corbel can not be answered. Climatic conditions in 1989 and 1990 did not allow severe mildew epidemics and as mildew isolates were collected from farmers fields proper control treatments were not available. Use of wheat varieties with low susceptibility to mildew increased in these years, and may have masked also reduced activity of Corbel. Other factors which make a comparison of Corbel performance difficult are the wide variation in field and crop conditions (fertilization), fungicide history, initial infection levels and the long-distance wind dissemination of mildew conidia. Another obvious reason may be that field rates of Corbel are still sufficient to control populations with reduced sensitivity. All these factors may explain why farmers, so far, have not complained about poor mildew performance of Corbel.

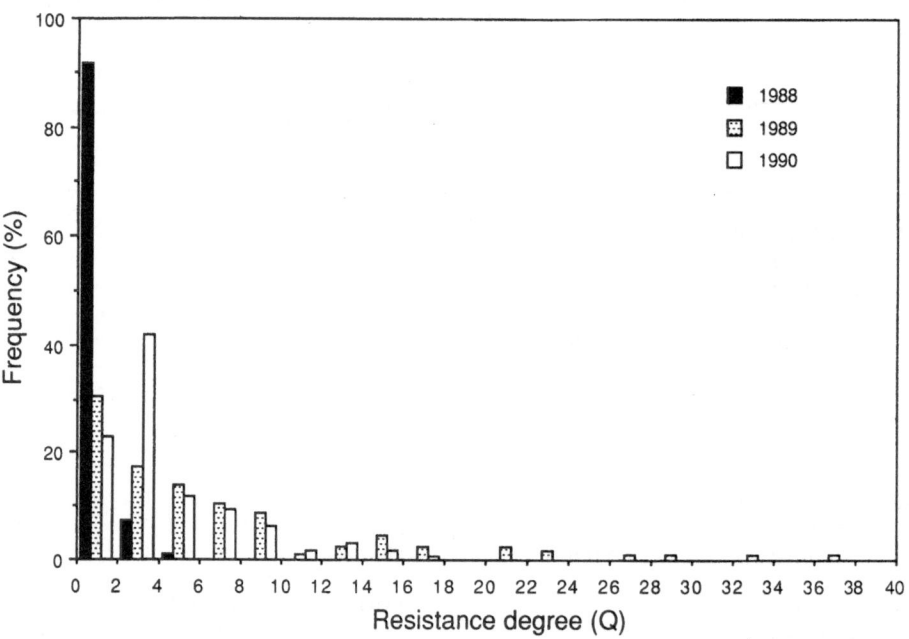

Figure 1. Frequency distribution of resistance degrees to fenpropimorph of *Erysiphe graminis* f.sp. *tritici* isolates collected in Limburg (The Netherlands) in 1988 (n=96), 1989 (n=114), and 1990 (n=126). The resistance degree (Q) is the ratio between the EC_{50} values of field isolates and the mean EC_{50} values of the wild-type isolate LH, determined in the respective years.

CONCLUDING EXPERIMENTS

In view of the 1989 results there was a need to confirm the reduced sensitivity to fenpropimorph of some of the field isolates, and to study their sensitivity to triadimenol. Tests were carried out after about 10 generations on non-treated seedlings (sensitive isolates) or seedlings treated with 10 μg fenpropimorph ml^{-1} (isolates with decreased sensitivity to fenpropimorph). Most field isolates tested appeared to have Q-values for fenpropimorph ranging from about 8 to 13 (Table 5). Q-values of the same isolates to triadimenol ranged from about 20 to 130; the one of isolate 67, maintained from 1983 onwards on seedlings sprayed with sublethal rates of triadimefon, was as high as 185. No obvious correlation between Q-values for both fungicides was observed. These data clearly confirm the results found in the 1989 survey for fenpropimorph and also indicate that the 1989 Q-values for triadimenol are now much higher than the mean 1985 Q-values for triadimefon.

TABLE 5

Sensitivity to fenpropimorph and triadimenol of some field isolates of *Erysiphe graminis* f.sp. *tritici* collected in 1989 from Limburg using foliar spray tests.

Isolate	Fenpropimorph		Triadimenol	
	EC$_{50}$ (μg ml^{-1})	Q	EC$_{50}$ (μg ml^{-1})	Q
LH	3.3 ± 0.5[1]	-	0.15± 0.1[1]	-
67	3.5 ± 1.0	1	27.8 ± 6.6[3]	185
1Ac	25.5 ± 10.8[2]	8	2.9 ± 0.1[3]	22
3a	44.6 ± 14.7[2]	13	10.8 ± 2.1[3]	79
8Ad	27.1 ± 9.8[2]	9	11.4 ± 1.9[3]	83
8Bb	32.0 ± 11.9[2]	9	18.9 ± 4.6[3]	132
11a	4.9 ± 1.4	2	14.5 ± 7.8[3]	123
13e	27.2 ± 7.9[3]	8	8.8 ± 1.0[3]	66
14c	33.8 ± 10.4[2]	10	10.1 ± 3.2[3]	69
16c	39.1 ± 13.7[2]	12	2.7 ± 1.4[3]	18

[1] Mean and standard deviation of four replicates.
[2] Significantly different from LH at P=0.01; [3] idem P=0.001.

DISCUSSION AND EVALUATION

The results of the surveys indicate that *E. graminis* f. sp. *tritici* may become less sensitive to triazole and morpholine fungicides. These data do not give support for the presence of positively or negatively correlated cross resistance to these fungicides. The simultaneous decreased sensitivity to both types of fungicides, as observed in the 1989 isolates, is probably a case of multiple resistance. The relatively high level of resistance to triadimenol in most 1989 isolates as compared to triadimefon in the 1985 isolates may be explained in two ways: either Q-values of triadimenol are higher than for triadimefon or Q-levels to triazoles increased from 1985 onwards.

The consequence of the reduced sensitivity of mildew to both classes of fungicides in relation to their field performance remains obscure. Field experiments should confirm whether field rates are still sufficient to recommend these chemicals as fungicides for mildew control. This is possibly the case since farmers have not reported serious reductions in activity against mildew. In the past, field performance studies were common practice in the Netherlands but currently governmental organizations do not give priority for their continuation. This is difficult to understand since it is also the policy of the Netherlands and other governments to reduce the use of pesticides, especially in debatable cases [6].

An evaluation of the advised anti-resistance strategy, which is also in line with FRAC recommendations [12], is difficult since no results for other strategies applied under comparable conditions are available. However, it seems justifiable to say that the alternating use of morpholines and triazoles, both restricted to one application per year (including seed treatment) may have contributed to a relatively slow development of resistance to triazoles. This statement is based on reports describing relatively high levels of resistance in the mid 1980's in barley powdery mildew in Germany and Great Britain in association with decreased disease control [7, 8]. The statement is also supported by results of Limpert [13] who found a relatively low frequency of triadimenol-resistant isolates of *E. graminis* f. sp. *hordei* in the Netherlands as compared to surrounding countries.

Evaluation of the Dutch strategy to limit the use of Corbel to one application per season is even more difficult. In surrounding countries application frequencies of the fungicide in wheat and barley were significantly higher, yet no isolates with the same degree of decreased sensitivity have been detected. This may be a consequence of the use of other strategies such as the use of morpholine/triazole mixtures. However, this explanation is difficult to comprehend since reduced sensitivity of mildew to triazoles would make these fungicides less effective in mixtures as an anti-resistance strategy for morpholines. An alternative explanation may be that long distance dispersal of conidia of *E. graminis* f. sp. *tritici* is, in part, responsible for the start of the mildew epidemics in Limburg. If so, the sensitivity of isolates from Limburg is not a reflection of the local fungicide history. The relevance of this hypothesis is supported by Limpert *et al.* [14] since they found isolates of *E. graminis* f. sp. *hordei* with a high frequency of virulence genes (*Va6* and *Va9*) in this province where these virulences were of no importance for the cultivars grown. The best explanation for this phenomenon was wind drift from adjacent areas in Germany and the British Isles. These findings may also explain why the sensitivity of the mildew population in Limburg to triazoles in 1985 was independent of the fungicide history for specific fields, and why 1989 and 1990 isolates with reduced sensitivity to fenpropimorph did not originate from the same locations (results not shown). If this hypothesis is relevant, it remains to be explained why isolates with such a high degree of decreased sensitivity to fenpropimorph were not reported in the surrounding countries. The reason might be a difference in the method of testing. It may be that the foliar spray test is a more sensitive method for the detection of low levels of resistance as compared with vapour phase tests in Petri-dishes, a method often used by other investigators.

Present results suggest that the resistance risk for morpholine fungicides may be underestimated. This can also be illustrated from the fact that FRAC has working groups for all important classes of fungicides at risk except for morpholines. The assertion that morpholines have a relatively low resistance risk may be true to some extent since morpholines inhibit sterol biosynthesis at two sites and fenpropimorph possibly at three [15]. On the other hand, *Nectria haematococca* f. sp. *cucurbitae* readily developed polygenic resistance to fenpropimorph *in vitro* [16]. Morpholines should, therefore, not too readily be chosen as a companion in mixtures with triazoles. Instead protectant partners should be considered [7]. If such products lack sufficient mildew activity, one should consider also combinations with specific mildew fungicides such as pyrazophos. The feasibility of such combinations is highly dependent of the presence of certain yield-damaging pathogens.

If, in view of the disease situation, triazole/morpholine mixtures are indispensable, one has to consider the concentration of the companion fungicides in the product. In pre-packed mixtures the concentration of morpholines is often reduced for obvious reasons. This may not be a good anti-resistance strategy since such mixtures may not give adequate control of mildew populations with a low level of resistance. Since resistance to fenpropimorph is probably polygenic [16], it is supposed that reduced rates may also favour stepwise development of resistance [10]. Unfortunately, computer models either confirm this hypothesis [17], or suggest that evolution of polygenically controlled resistance will only be weakly dependent on the concentration applied [18].

Fungicide resistance management in wheat in the Netherlands is highly focused on mildew. Although the strategy adopted for mildew is also relevant for pathogens like brown rust, yellow rust and *Septoria* spp, hardly any attention is paid to resistance development in these pathogens. There is no reason to believe that these pathogens would react differently from mildew with respect to resistance development to triazoles and morpholines. Funds are lacking to support surveys, or even to establish baseline sensitivities, and any future resistance problems will be noticed just because of lack of performance. It is hoped that surveys in surrounding countries may give a timely warning for the situation in the Netherlands.

ACKNOWLEDGEMENT

The author thanks Dr.R. Maag AG, Switzerland and BASF AG, Germany for financial support of part of the study and Dr. R.A. Daamen, Dr. S.W. Ellis and Ing. K. Smant for a critical reading of the manuscript.

REFERENCES

1. Daamen, R.A., Surveys of cereal diseases and pests in the Netherlands. 1. Weather and winter wheat cropping during 1974-1986, Neth. J. Pl. Path., 1990, **96**, 227-236.

2. Drenth H. and Stol, W., Het EPIPRE-adviesmodel, PAGV reports, 1990, **97**, 1-130.

3. Daamen, R.A. and Stol, W., Surveys of cereal diseases and pests in the Netherlands. 2. Stem-base diseases of winter wheat, Neth. J. Pl. Path., 1990, **96**, 251-260.

4. Hartke, S. and Buchenauer, H., Effect of mercury-free seed dressings and the included active substances against carbendazim-sensitive and carbendazim-resistant *Gerlachia nivalis* strains on winter wheat, J. Plant Diseases and Protection, 1985, **92**, 247-257.

5. Sanders, P.L., De Waard, M.A. and Loerakker W.M., Resistance to carbendazim in *Pseudocercosporella herpotrichoides* from Dutch wheat fields, Neth. J. Pl. Path., 1986, **92**, 15-20.

6. Anonymous, Meerjarenplan Gewasbescherming, Ministerie van Landbouw, Natuurbeheer en Visserij, The Netherlands, 1990, pp. 133.

7. Heany, S.P., Population dynamics of DMI fungicide sensitivity. In Fungicide Resistance in North America, ed. C.J. Delp, APS Press, St Paul, Minnesota, USA, 1988, pp. 89-92.

8. Scheinpflug, H., History of DMI fungicides and monitoring for resistance. In Fungicide Resistance in North America, ed. C.J. Delp, APS Press, St Paul, Minnesota, USA, 1988, pp. 77-78.

9. De Waard, M.A., Kipp, E.M.C., Horn, N.M. and Van Nistelrooy, J.G.M., Variation in sensitivity to fungicides which inhibit ergosterol biosynthesis in wheat powdery mildew, Neth. J. Pl. Path., 1986, 92, 21-32

10. De Waard, M.A., Interactions of fungicide combinations. In Fungicide Resistance in North America, ed. C.J. Delp, APS Press, St Paul, Minnesota, USA, 1988, pp. 98-100.

11. De Waard, M.A. and Kipp, E.M.C., Sensitivity of Erysiphe graminis f.sp. tritici to fenpropimorph in the Netherlands, ISPP Chemical Control Newsletter, 1988, 11, 32-33.

12. Anonymous, Fungicide Resistance Action Committee (FRAC) Report, GIFAP Bulletin, 1990, 16: 1-12.

13. Limpert, E., Frequencies of virulence and fungicide resistance in the European barley mildew population in 1985, J. Phytopath., 1987, 119, 298-311.

14. Limpert, E., Andrivon D. and Fischbeck G., Virulence patterns in populations of Erysiphe graminis f. sp. hordei in Europe in 1986, Plant Pathol., 1990, 39, 402-415.

15. Ziogas B.N., Oesterhelt G., Masner, P., Steel, C.C. and Furter, R., Fenpropimorph: a three site inhibitor of ergosterol biosynthesis in Nectria haematococca var. cucurbitae, Pestic. Biochem. Physiol., 1991, 39, 74-83.

16. Demakopoulou, M.G., Ziogas, B.N., and Georgopoulos, S.G., Evidence for polygenic control of fenpropimorph resistance in laboratory mutants of Nectria haematococca f.sp. cucurbitae, ISPP Chemical Control Newsletter, 1989, 12, 34-35.

17. Josepovits, G., A model for evaluating factors affecting the development of insensitivity to fungicides. Crop Protection, 1989, 8, 106- 113

18. Shaw, M.W., A model of the evaluation of polygenically controlled fungicide resistance, Plant Pathol., 1989, 38, 44-55.

EVOLUTION AND MANAGEMENT OF RESISTANCE IN THE COLORADO POTATO BEETLE, *LEPTINOTARSA DECEMLINEATA*

RICHARD T. ROUSH and WARD M. TINGEY
Department of Entomology, Comstock Hall, Cornell University, Ithaca, NY 14853 USA

ABSTRACT

The Colorado potato beetle is notorious for its rapid development of insecticide resistance, even compared to other major pests, such as those on cotton. The causes of this rapid evolution appear to include: (1) capacity for rapid population growth; (2) high fraction of the population being treated each generation (due to the relative scarcity of populations on untreated alternative host plants) ; (3) selection across all active life stages; (4) the existence of a single mechanism capable of causing broad cross-resistance; and (5) use of soil insecticides. We have developed an integrated resistance management program for the Colorado potato beetle that includes: (1) crop rotation, scouting and border sprays; (2) the replacement of soil insecticides with alternative foliar insecticides for control of aphids and leafhoppers; and (3) intergenerational rotation of pesticides, including those from *Bacillus thuringiensis*, that do not show cross-resistance and preserve predators. Other refinements, including trap crops, propane flame technologies, and genetically resistant host plants are under investigation. With adequate grower education, resistance in the Colorado potato beetle seems manageable.

INTRODUCTION

The Long Island, New York, population of the Colorado potato beetle (CPB), *Leptinotarsa decemlineata* (Say), is often cited as an example of the severity that resistance problems can reach [1]. Currently on Long Island, the principal methods for control of CPB are the use of propane flamers against adults [2] and an inorganic insecticide (cryolite, sodium fluoaluminate) against the larvae; rotenone and BT are also available, although more expensive and less effective. The failure of new insecticides seemed to occur at a rate of about once every 3 years [1,3,4]. This seems to be an extraordinarily rapid rate compared to other major agricultural insect pests, such as many of those on cotton, especially since the Colorado potato beetle has fewer than three complete generations per year. The purpose of this paper is to discuss the factors that have led to the severity of resistance problems in the CPB, particularly compared to resistance development in the cotton bollworms (*Helicoverpa armigera,* discussed at this conference by Forrester and Fitt, and *Heliothis virescens*), another pest complex for which some similar data are available. This analysis suggests that, in retrospect, the evolution of resistance in CPB has not been extraordinarily rapid, but is no less severe. Further, at least some of the causes of resistance are avoidable, which suggests avenues to delay resistance in those areas and for those compounds where resistance is not already too severe.

Causes of Resistance in the Colorado Potato Beetle

Several factors are clearly involved in the rapid development of resistance in the CPB, but it is not possible to rank their significance with any precision. Thus, the order of presentation is arbitrary.

Capacity for Rapid Population Growth
If not checked, the CPB appears to be easily capable of population growth rates in excess of 40 fold per generation, with overwintering survival exceeding 60% [5]. In contrast, population growth rates for bollworms appear to be about 5-10 fold per generation, with overwintering mortalities often less than 5% [6]. As a consequence of rapid population growth, complete defoliation of the crop is possible. Thus, there has been a tendency for intensive insecticide use, not only for economically damaging densities, but also as insurance against the fear that CPB populations might get out of control. The agronomic investment in a potato crop may exceed US$3500 and still gross $6000 per hectare, for which growers can easily justify insecticide costs of $250-750 per hectare.

Because resistance in the CPB from Long Island is so well known, it is tempting to speculate that there is something unique about the area (e.g., surrounding ocean and urbanization limit immigration of susceptible beetles) that promotes resistance development in CPB. However, more recent surveys have shown that problems at least as severe can be found in other intensive but even smaller potato producing areas of New York, especially the Savannah muck [7,8], about 40 km west of Syracuse, and nearly as severe in many other locations throughout northeastern North America [4]. The more extensive recognition of the problems on Long Island may be attributed to careful and extensive documentary efforts [e.g, 3,4]. On the other hand, there are many populations of CPB that are still fairly susceptible to at least some compounds, often within 10-20 km of highly resistant populations, throughout the eastern USA [8,9,10]. There are many ecological and climatic differences among potato producing areas in New York, but the features that appear to be most common to the areas where resistance is severe are those that promote rapid population growth in the early season, including: (1) relatively warm early season temperatures (on Long Island, this is due to the moderating influence of the ocean; at Savannah due to the black soils that make up the muck) and (2) intensive potato production in a limited area, often either without rotation to other crops (e.g., Savannah) or where rotation is limited to relatively short distances (e.g., Long Island). Rotation of potatoes with other crops, usually oats, clover, wheat and field corn (maize), delays colonization by CPB adults, which allows the crop to grow to a less sensitive stage, and reduces the total density of overwintered adults that find the crop [11, 12].

Because the adults are older before they start oviposition, rotation probably also reduces the total number of eggs laid. Not only does crop rotation reduce the number of insecticide applications needed [8,11], the frequency of insecticide applications and resistance are strongly correlated [8].

Proportion of the Population Treated Each Generation

Although there are alternative host plants reported in the northeastern USA [13,14], it is unusual to find CPB on plants other than potato, tomato, or egg plant, at least in New York. Thus, in contrast to bollworms, which have a wide range of hosts where a large portion of the population may escape exposure [e.g., 15, 16], CPB is concentrated on crops where all generations may be subject to selection. Strictly speaking, there are host plants where CPB are not exposed to pesticides, particularly where potatoes are rotated with field corn. As potatoes are not effectively controlled by the herbicides used in corn, there are often volunteer plants from tubers that escaped harvest the previous season. These are nearly always heavily infested with CPB, but account for only a very small number of plants compared to those in potato fields. Censusing these plants and the larvae on them suggests that they may account for only about 2-3% of the CPB populations in areas using a corn-potato rotation (unpublished data), which is probably not enough to delay resistance significantly. Further, most of the volunteers are completely defoliated by the first generation of larvae; these individuals may then face selection as adults when they move into adjacent potato fields.

Finally, even if there were more alternate host plants, the limited dispersal tendencies of CPB, critical to the effectiveness of crop rotation, would probably minimize the benefits. In several mark-release-recapture studies, more than 80% of the recaptured adult CPB have been found within 10 meters of their release, or at least within the same field, even after periods of a week or more [17,18,19, Fig.1]. Secondly, mate location in CPB seems to occur only at short distances (less than a meter), probably by visual detection [20] without the use of volatile pheromones [21]. CPB mate repeatedly, perhaps 20 times or more in a lifetime [20,21]. Even after three to eight matings to males of one genotype, roughly a third of the offspring, potentially 10 to 20 per day [22], will bear the genotype of the last mate [21]. Thus, especially when sprays are directed against the adults, the biology of the insects and selection favor the intermating of individuals that have survived sprays within the same field. In contrast, bollworms may move an average of about 3-6 km in a lifetime [23], have an efficient sex pheromone system that encourages panmixia across a large area, and, although there may be some selection, treatment is less intensively directed against the adults.

Selection Across All Active Life Stages

Not only are all active life stages exposed to pesticides, but resistant adults and all larval instars survive insecticide applications better than susceptible CPB. For example, when placed on foliage sprayed at field application rates of four representative insecticides (fenvalerate, carbofuran, azinphosmethyl, and endosulfan), mortality of susceptible second through fourth instar CPB larvae from Iowa ranged from 85-100%, but was only 0-70% for F_1 (heterozygous) larvae of crosses between resistant Long Island and susceptible CPB (unpublished data). Thus, in contrast to other species such as bollworms (e.g., [24]), selection (as indicated by discrimination between resistant and susceptible genotypes, [25])

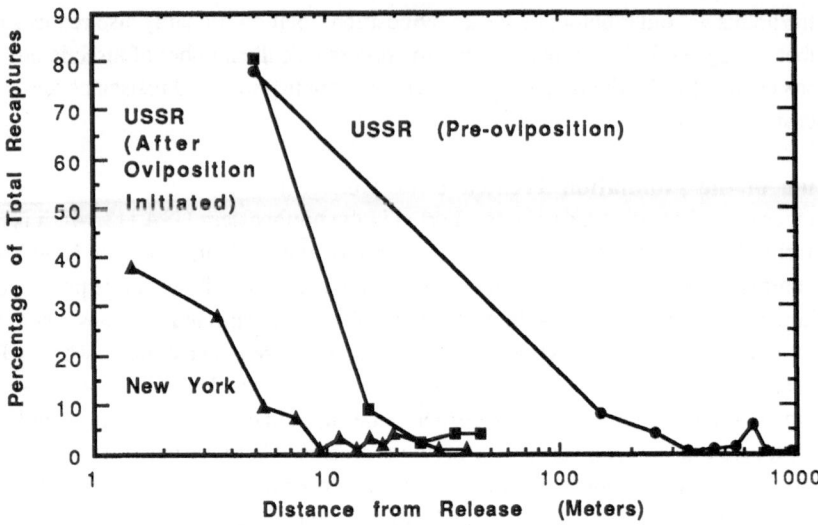

Fig. 1. Dispersal of adult CPB in mark-release-recapture studies in Moldavia (southern USSR, from [17]), and three independent studies at two locations in New York (967 beetles released, 95 recovered; they were marked either where collected in the field with honey bee tags or in the laboratory with fluorescent dyes mixed with balsam; unpublished data).

in the CPB appears to occur even in young larvae. The window of intense selection in bollworms is relatively brief (perhaps one week), resulting in perhaps only one episode of selection for any given individual when a weekly spray program is followed, and even 20% of those may not receive a lethal dose [26]. In contrast, when weekly spray programs are used against CPB, an individual may be exposed to selection twice as a larva, and then 3-4 times as an adult (we have collected marked beetles more than three weeks after release) and mortality of susceptible populations usually exceeds 90% under normal application conditions. Even though there may be only two or three generations of CPB each year, each individual can be selected several times for resistance. Thus, as noted more generally by Rosenheim and Tabashnik [27], it should not be surprising that CPB can develop resistance more rapidly than other species such as bollworms in spite of fewer generations per year.

Existence of a Single Mechanism Causing Broad Cross-Resistance

Although supplemental genetic and biochemical data are always helpful, the key information needed for managing resistance is the effective dominance of resistance and cross-resistance among resistance genes [25]. In reciprocal F_1 crosses, we found that resistance to several insecticides seemed to be nearly completely dominant, but that resistance to fenvalerate, a pyrethroid insecticide, seemed to be sex-linked and recessive in females. Similar results for resistance to another pyrethroid, permethrin, were obtained by Argentine et al. [28] who showed that the gene resembled *kdr* (knock-down resistant nerve insensitivity; [29,30]) in houseflies.

Most populations of CPB are resistant to several insecticides, but this may be due to multiple (several co-existing resistance genes due to multiple pesticide use) rather cross-resistance (one resistance gene conferring resistance to several compounds). To clarify cross-resistance patterns, we isolated individual resistance genes in an isogenic background by backcrossing (see [31] for a discussion of this approach) a resistant strain from Long Island to a susceptible strain from North Carolina for five generations (F_1 through BC_4). In separate lines, we selected with carbofuran, azinphosmethyl, or endosulfan every second generation with a dose that was just sufficient to kill about 95% of susceptible larvae. When assayed with topical applications of technical material or with formulated material on treated plants, all of the resulting strains showed a similar pattern of cross-resistance, ranging up to about 30 fold at the LD_{50}, including (in approximate order of declining resistance ratios) carbofuran, parathion, DDT, azinphosmethyl, phosmet, phorate, endosulfan, dimethoate, methamidophos, and about 3X resistance to pyrethroids, but not oxamyl. Synergism with piperonyl butoxide (PBO) was greater against the resistant strain than the susceptible strain for several of the compounds, suggesting the involvement of increased metabolism through a microsomal mixed function oxidase [30]. In addition, strains selected with endosulfan also showed cross-resistance to dieldrin (perhaps due to a gene similar to one described in other species [29,30] for target site insensitivity to cyclodienes), consistent with results from the first backcross generation suggesting that resistance to endosulfan was not due to a single gene. Collectively, these results suggest that there are at least three major genes for resistance in the CPB from Long Island: a single major microsomal oxidase gene that confers broad cross-resistance to a wide range of compounds, a sex-linked gene (similar to kdr) for resistance to pyrethroids, and a gene that confers resistance to dieldrin and contributes to resistance to endosulfan.

For resistance management purposes, these results (and data from other populations that suggest the involvement of altered acetylcholinesterases [32] and OP-metabolizing esterases [28]) suggest that there are only two groups of currently available classical insecticides that will generally show little cross-resistance: (1) the pyrethroids and (2) organophosphates, carbamates, and endosulfan. This conclusion is consistent with the patterns of resistance shown in field populations [32].

Further, the cross-resistance patterns suggest a possible chronology for the evolution of resistance genes in the CPB. Selection for the presumed major microsomal oxidase gene might have started with DDT (using the dates given by [1,3], introduced in 1945, with first reports of field failures in 1952), which would have led to the more rapid failures of carbaryl (1959-1963), azinphosmethyl (1959-1964), monocrotophos (1973-1973), phosmet (1973-1973), phorate (1973-1974), and carbofuran (1974-1976). Early use of dieldrin (1954-1957) would have predisposed the populations for resistance to endrin (1957-1958) and endosulfan (the latter possibly already affected by the use of DDT). Exposure to DDT might also have predisposed the populations to the rapid evolution of resistance to permethrin and fenvalerate through a kdr-like gene. (Although disulfoton was reported as showing resistance (in 1974) after just one year of use [3], we have never found it to be effective, even against susceptible CPB populations, and have omitted it here. Further, we have not determined the mechanism of resistance to oxamyl, reported in its first year of use (1978), but note that oxamyl is very similar to aldicarb, which was used for many years prior to the introduction of oxamyl; although control failures for aldicarb were never reported, resistance levels did increase in field populations [3], perhaps through an altered acetylcholinesterase.) Thus, even though CPB has become resistant to at least 12 insecticides between 1945 and 1983, about one every three years, it has become resistant to only three (or four) classes of insecticides, as measured by the genes used by the insect. This average of about one class every ten years may be even slower than found for cotton bollworms [33,34].

Use of Soil Insecticides

To compare the effects of foliar and soil formulations of insecticide in selecting for resistance, CPB larvae were exposed to soil or foliar treatments in the field in accordance with normal practice. Except for pyrethroids (which provided control about ten days), all foliar applications of insecticides lost effectiveness (stopped killing even susceptible larvae, and presumably therefore no longer selecting for resistance) within a week of application. In contrast to foliar insecticides, the soil insecticides phorate and carbofuran killed susceptible larvae for up to a month and caused little or no mortality of larvae heterozygous for the resistance genes. As there is cross-resistance between phorate and carbofuran and most of the registered foliar compounds, the use of these soil insecticides has probably significantly increased resistance even to foliar insecticides.

RESISTANCE MANAGEMENT

An understanding of the causes of resistance in the CPB suggests various methods to slow the further degradation of pesticide efficacy, which we have tried to integrate into a program that growers can find acceptable. Although we cannot change the fact that there is a single mechanism capable of causing broad cross-resistance, we can minimize all of the other factors that contribute to resistance.

Crop Rotation, Predators, and Other Non-insecticidal Controls

For the CPB as for many other pests, one of the most effective resistance management strategies is to reduce the number of insecticide applications. Perhaps the most important single feature of any resistance management program for CPB is to take advantage of its one weakness: its limited dispersal tendencies. As discussed above, crop rotation is one of the most effective ways to slow down and delay the growth of CPB populations. Without crop rotation, frequent spraying (or flaming, as discussed below) is virtually inevitable.

However, crop rotation does not in itself reduce pesticide use if a calendar spray program is still followed. CPB is relatively easy to detect; the infestations are often patchy within a field, but somewhat predictable because they will be immigrating from the previous year's potato fields and surrounding hedge rows. Thus, a major component of our resistance management program is to encourage scouting before spraying. Further, we encourage growers not only to focus their scouting attention on the edges of the fields, but to only spray these edges during the early part of the season. Not only does this practice (which has led to widespread adoption) reduce pesticide costs, but the untreated portion of the field provides a refuge for susceptibility for any CPB that disperse there (thus decreasing the fraction of the population that is treated).

Replacement of Soil Insecticides with Foliar Insecticides

In addition to the CPB, the potato leafhopper, *Empoasca fabae*, and the green peach aphid, *Myzus persicae*, are serious insect pests of potatoes in the northeastern United States. Even where soil insecticides are no longer efficacious for CPB control (as seems to be true throughout most of New York), soil insecticides are often used to control leafhoppers and aphids, even though they are not always effective. After the registration of the soil formulation of carbofuran was recently cancelled for environmental reasons (birds often ate the granules), only phorate and disulfoton remain available for this use. Although phorate appears to select strongly for resistance in CPB (as discussed above) and may therefore maintain the high frequencies of the presumed mixed function oxidase gene, disulfoton appears to cause little or no mortality of CPB and is therefore expected to select less strongly for resistance, but is unfortunately also less effective than phorate against leafhoppers.

Given continuing environmental concerns about ground water contamination from soil insecticides, we have also tested three foliar insecticides (dimethoate, parathion, and methamidophos), which proved to be as effective (with proper scouting) and cheaper than soil insecticides for the control of leafhoppers. In addition, methamidophos provided excellent control of aphids. As noted above, the presumed mixed function oxidase gene caused significant levels of resistance to parathion. However, the levels of resistance to dimethoate and methamidophos are much lower, at least in part because dimethoate and methamidophos are not very effective even against susceptible CPB. In trials on sprayed foliage, dimethoate and methamidophos killed only early instar larvae of the susceptible strains, but because leafhoppers generally disperse into fields near the end of the first larval generation, small larvae are generally not present when leafhopper sprays are necessary. Thus, parathion does not appear to be a good choice for leafhopper control because of its effects on resistance development in CPB. Further, although resistance has not been reported in the potato leafhopper, dimethoate and parathion confer high levels of resistance in *Myzus persicae* [35, A. L. Devonshire, pers. com.], whereas acephate (metabolized to methamidophos as the active form) does not [36]. A similar pattern of cross-resistance appears to occur in New York, because methamidophos is one of the few insecticides that still controls these aphids. Thus, when resistance to aphids and CPB is considered, methamidophos is the preferred alternative to soil insecticides for managing potato insects and their resistance.

Rotation of Pesticides

On the basis of simple computer simulation models that show the benefits of pesticide rotation across generations [25], cross-resistance patterns, and the effects of the currently registered insecticides on pests and beneficials, the resistance management program attempts to fit each pesticide into one time interval each year where it is most effective and least damaging to beneficial species. In so doing, the chance that any one generation gets treated twice with the same pesticide group is reduced, such that selection for any given resistance gene does not occur across all active life stages.

After tuber initiation, potato plants can withstand considerable defoliation [37]. However, not only can overwintering CPB adults cause severe damage to the emerging potatoes (if in sufficient densities), their offspring can cause damage either as late instar larvae or "summer" adults (Fig. 2 illustrates the approximate timing of these generations).

Thus, it is important to develop a management program that allows insecticidal control of both adults and larvae, when needed.

No. Adult Colorado Potato Beetles per 50 Plants

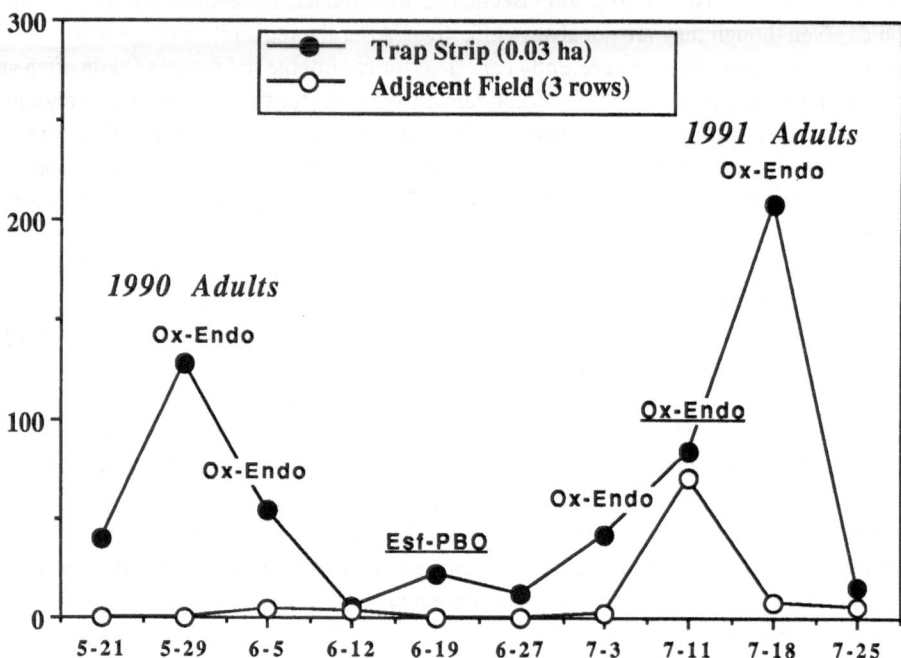

Fig. 2. Phenology of CPB adults and efficacy of two rows (a "trap strip") of early planted potatoes in arresting their dispersal. Two rows were planted in the spring of 1991 along the edge of the field planted to potatoes 1990 and 10 m from the edge of the 1991 field. Density data are given for the trap strip and the first three rows of the 4 ha 1991 potato field. The "1990 Adults" were those that overwintered; "1991 Adults" matured in the spring of 1991. Pesticide applications (a mixture of oxamyl and endosulfan or esfenvalerate and piperonyl butoxide) were applied to both trap and the crop (by the grower, underlined) or only the trap on the dates indicated.

As noted above, there are really only two groups of classical insecticides that consistently show little cross-resistance in the CPB, (1) the pyrethroids and (2) organophosphates, carbamates, and endosulfan. Of these, only the pyrethroids and a mixture of oxamyl and endosulfan are generally effective in those areas of New York where classical insecticides still work at all [8]. In addition, novel insecticidal endotoxins produced by *Bacillus thuringiensis tenebrionis* (*Btt*) are effective against small (first and second instar) CPB larvae [38,39] and the inorganic stomach poison cryolite (sodium fluoaluminate) is effective against all larval instars. Although we have not detected any resistance against cryolite, resistance to *Btt* has been reported from a laboratory population of the CPB [40]. Thus, to maintain the efficacy of the classical insecticides, which are the only compounds effective against the adults, growers are encouraged to use them only against economically damaging populations of adults and to use *Btt* and cryolite against larvae.

The next problem is to assign the optimal generations for the use of pyrethroids and oxamyl-endosulfan against adults. The pyrethroids are generally most effective when used in cooler (< 25°C) temperatures [41]. Another consideration is the effect on beneficial species. Perhaps the most important beneficial species is the lady beetle *Coleomegilla maculata*, a predator of CPB egg masses and aphids [42,43]. When bioassayed on treated foliage, pyrethroids proved to be highly toxic to *C. maculata*, but oxamyl and endosulfan were much less so. Although these predators are present on potatoes in the early part of the season, they appear to be much more abundant late in the season. Finally, pyrethroids confer high levels of resistance (and therefore select strongly) in *Myzus persicae* [36], which are much more abundant in the latter part of the season than earlier. Thus, the optimal time for the use of pyrethroids, when CPB densities warrant, is against the overwintered adults, thereby reserving oxamyl-endosulfan for summer adults.

Under good pest management practices, not all of these insecticides will be used in any given year (with good crop rotation, it may not be necessary to treat for CPB at all). Where only generation per year must be treated, the two groups of classical insecticides might best be used in alternate years. In principle, the grower could determine which pesticide was likely to be most effective and use only that pesticide during the year. Toward that end, we developed a simple bioassay technique, the dip test (much like that developed by [44]), that uses readily available materials and formulated insecticides to estimate pesticide resistance in the CPB within 24 hours (Roush, Beckley, Ferro, Hoy, and Tingey, unpublished). Unfortunately, the growers cannot seem to find time to use it, so we now recommend a general pattern of rotation that will work in most places.

Further Refinements

To further reduce the use of classical insecticides, and thereby manage resistance even more effectively, we have investigated other non-chemical methods for control. In untreated fields, aphids are often controlled to a high degree by parasitic Hymenoptera and lady beetles, including *C. maculata* and *Coccinella semptempunctata*. Thus, the only major pest for which non-chemical controls are not currently available is the potato leafhopper, but plants genetically resistant to all three pests are under development [45]. In addition, especially in those areas where resistance is already so severe that classical insecticides are ineffective, growers have mounted propane burners on tractors to kill the overwintered adults with heat without causing crop loss [2]. This tactic proves to be cost effective compared to insecticides, but can only be used during good weather (when the insects are up on the plants) and when the plants are still small (i.e., less than about 12 cm). A third approach is to attract and concentrate the beetles with a trap crop for control by flaming methods or insecticides. Overwintered beetles are readily attracted to potatoes in the early season; if potatoes along a border emerge before the main field, CPB adults will aggregate there, and may even do so throughout the season. As illustrated in Fig. 2, even where insecticides are applied to the trap, they are applied to a tiny fraction of the total crop acreage, thereby increasing the refuge for susceptible beetles even more effectively than border sprays (which generally cover 5-10 times as much area as the border strip).

CONCLUSIONS

By exploiting the major weakness of the CPB, its limited dispersal tendencies, and making the optimal choices of insecticides while reducing their use, resistance in the CPB is manageable. Further, most of the steps that are favored for resistance management, including eliminating the use of soil insecticides, using biological alternatives such as *Btt*, and reducing the overall number of pesticide applications, are consistent with good pest management and societal goals to reduce pesticide contamination of the environment. The most important limitation to the implementation of these goals seems to be education of the growers and their crop scouts that the program is effective. Nonetheless, this research shows that resistance management and good environmental management can be compatible.

Acknowledgments. We thank Nada Carruthers and Patti Beckley for technical assistance with much of the work described here and Kate Aronstein for translating [17].

REFERENCES

1. Georghiou, G. P., The magnitude of the resistance problem, In Pesticide Resistance: Strategies and Tactics for Management, ed. National Academy of Sciences, National Academy Press, Washington, D.C., 1986, pp. 14-43.

2. Moyer, D., Kujawski, R., Derksen, R., Moeller, R., Sieczka, J. B. and Tingey, W. M., Development of a Propane Flamer for Colorado Potato Beetle Control. Cornell Cooperative Extension, Suffolk County, Riverhead, New York (Mimeographed, also on videotape), 1991, 7 pp.

3. Forgash, A. J., History, evolution, and consequences of insecticide resistance. Pestic. Biochem. Physiol., 1984, **22**, 178-186.

4. Forgash, A. J., Insecticide resistance in the Colorado potato beetle. In Proc. Symp. Colorado Potato Beetle, XVIIth Int. Cong. Entomol., eds. D. N. Ferro and R. H. Voss, University of Massachusetts Res. Bull. No. 704, Amherst, 1985, pp. 33-51.

5. Harcourt, D. G., Population dynamics of *Leptinotarsa decemlineata* (Say) in eastern Ontario. II. Population and mortality estimation during six age intervals. Can. Entomol. 1964, **96**, 1190-98.

6. Fitt, G. P., The ecology of *Heliothis* species in relation to agroecosystems. Annu. Rev. Entomol., 1989, **34**, 17-52.

7. Goh, K. S., Tingey, W. M., Gibson, R. L., Semel, M., Clarke,T., Moyer, D., MacNeil, C., Young, D. , Kubecka, T. and Lemaire, R., Insecticide resistance in New York populations of the Colorado potato beetle, 1986 and 1987. Insect. Acar. Tests, 1988, **13**, 142-3.

8. Roush, R. T., Hoy, C.W., Ferro, D. N. and Tingey, W.M., Insecticide

resistance in Colorado potato beetles (Coleoptera:Chrysomelidae): Influence of crop rotation and insecticide use. J. Econ. Entomol., 1990, **83**, 315-9.

9. Heim, D. C., Kennedy, G. G. and van Duyn, J. W., Survey of insecticide resistance among North Carolina Colorado potato beetle (Coleoptera: Chrysomelidae) populations. J. Econ. Entomol., 1990, **83**, 698-705.

10. Tisler, A. M. and Zehnder, G. W. Insecticide resistance in the Colorado potato beetle (Coleoptera: Chrysomelidae) on the eastern shore of Virginia. J. Econ. Entomol. 1990, **83**, 666-71.

11. Wright, R. J., Evaluation of crop rotation for control of Colorado potato beetles (Coleoptera: Chrysomelidae) in commercial potato fields on Long Island. J. Econ. Entomol., 1984, **77**, 1254-9.

12. Lashomb, J. H., and Ng, Y. S., Colonization by Colorado potato beetle, *Leptinotarsa decemlineata* (Say) (Coleoptera: Chrysomelidae), in rotated and non-rotated potato fields. Environ. Entomol. 1984, **13**, 1352-56.

13. Hare, J. D. and Kennedy, G.G., Genetic variation in plant-insect associations: Survival of *Leptinotarsa decemlineata* populations on *Solanum carolinense*. Evolution, 1986, **40**, 1031-43.

14. Hare, J. D., Ecology and management of the Colorado potato beetle. Annu. Rev. Entomol., 1990, **35**, 81-100.

15. Schneider, J.C., and Roush, R.T., Genetic differences in oviposition preference between two populations of Heliothis virescens. In Evolutionary Genetics of Invertebrate Behavior: Progress and Prospects, ed. M.D. Huettel, Plenum, New York, 1986, pp. 163-71.

16. Gunning, R.V. and Easton, C. S., Pyrethroid resistance in *Heliothis armigera* (Hubner) collected from unsprayed maize crops in New South Wales in 1983-1987. J. Aust. Entomol. Soc., 1989, **28**, 57-61.

17. Chigarev, G. A. and Molchanova, V. A., Application of radio-active labeling method in studying the efficiency of spring potato survey and spreading of over-wintered Colorado beetles. Trudy Vsesoyuznogo Instit. Zashchity Rastenil, 1967, **27**, 75-81.

18. Bach, C., The influence of plant dispersion on movement patterns of the Colorado potato beetle, *Leptinotarsa decemlineata* (Coleoptera: Chrysomelidae). Great Lakes Entomol. 1982, **15**, 247-52.

19. Voss, R. H. and Ferro, D. N., Phenology of flight and walking by Colorado potato beetle (Coleoptera: Chrysomelidae) adults in western Massachusetts. Environ.

Entomol., 1990, **19**, 117-22.

20. Szentesi, A., Behavioral aspects of female guarding and inter-male conflict in the Colorado potato beetle. In Proc. Symp. Colorado Potato Beetle, XVIIth Int. Cong. Entomol., eds. D. N. Ferro and R. H. Voss, University of Massachusetts Res. Bull. No. 704, Amherst, 1985, pp. 127-37.

21. Boiteau, G., Sperm utilization and post-copulatory female-guarding behavior in the Colorado potato beetle, *Leptinotarsa decemlineata*. Entomol. Exp. Appl., 1988, **47**, 183-7.

22. Peferoen, M., Huybrechts, R. and de Loof, A. Longevity and fecundity in the Colorado potato beetle, *Leptinotarsa decemlineata*. Entomol. Exp. Appl. 1981, **29**, 321-9.

23. Schneider, J.C., Roush, R.T., Kitten, W.F. and Laster, M.L., Movement of *Heliothis virescens* (F.) (Lepidoptera: Noctuidae) in Mississippi in the spring: Implications for area-wide management. Environ. Entomol., 1989, **18**, 438-46.

24. Daly, J., Fisk, J.H. and Forrester, N.W., Selective mortality in field trials between strains of *Heliothis armigera* (Lepidoptera: Noctuidae) resistant and susceptible to pyrethroids: functional dominance of resistance and age class. J. Econ. Entomol., 1988, **81**, 1000-7.

25. Roush, R. T., Designing resistance management programs: How can you choose ? Pest. Sci., 1989, **26**, 423-41.

26. Wolfenbarger, D. A., Harding, J. A. and Robinson, S. Tobacco budworm (Lepidoptera: Noctuidae): Variations in response to methyl parathion and permethrin in the subtropics. J. Econ. Entomol., 1984, **77**, 701-5.

27. Rosenheim, J. A. and Tabashnik, B. E., Evolution of pesticide resistance: Interactions between generation time and genetic, ecological, and operational factors. J. Econ. Entomol. 1990, **83**, 1184-93.

28. Argentine, J. A., Clark, J. M. and Ferro, D. N., Genetics and synergism of resistance to azinphosmethyl and permethrin in the Colorado potato beetle (Coleoptera: Chrysomelidae). J. Econ. Entomol. , 1989, **82**, 698-705.

29. Oppenoorth, F. J., Biochemistry and genetics of insecticide resistance. In Comprehensive Insect Physiology, Biochemistry, and Pharmacology, Vol. 12. eds. G.A. Kerkut and L.I. Gilbert, Pergamon, New York, 1985, pp. 731-773.

30. Soderlund, D. M., and Bloomquist, J. R., Molecular mechanisms of insecticide resistance. In Pesticide Resistance in Arthropods, eds. R. T.Roush and

B. E.Tabashnik, Chapman and Hall, New York, 1990, pp. 58-96.

31. Roush, R. T., and Daly, J. C., The role of population genetics in resistance research and management. In Pesticide Resistance in Arthropods, eds. R. T.Roush and B. E.Tabashnik, Chapman and Hall, New York, 1990, pp. 97-152.

32. Ioannidis, P. M., Grafius, E. and Whalon, M. E., Patterns of insecticide resistance to azinphosmethyl, carbofuran, and permethrin in the Colorado potato beetle (Coleoptera: Chrysomelidae). J. Econ. Entomol., 1991, **84**, 1417-23.

33. Sparks, T. C., Development of insecticide resistance in *Heliothis zea* and *Heliothis virescens* in North America. Bull. Entomol. Soc. Am., 1981, **27**, 186-92.

34. Plapp, F. W., Jr., Campanhola, C., Bagwell, R. D. and McCutcheon, B. F., Management of pyrethroid-resistant tobacco budworms on cotton in the United States. In Pesticide Resistance in Arthropods, eds. R. T.Roush and B. E.Tabashnik, Chapman and Hall, New York, 1990, pp. 237-60.

35. Sawicki, R. M., Devonshire, A. L., Rice, A. D., Moores, G. D., Petzing, S. M. and Cameron, A., The detection and distribution of organophosphorous and carbamate insecticide-resistant *Myzus persicae* (Sulz.) in Britain in 1976. Pestic. Sci. 1978, **9**, 189-201.

36. Sawicki, R. M., and Rice, A. D., Response of susceptible and resistant peach-potato aphids *Myzus persicae* (Sulz.) to insecticides in leaf dip bioassays. Pestic. Sci., 1978, **9**, 513-6.

37. Ferro, D. N., Morzuch, B. S. and Margolies, D., Crop loss assessment of the Colorado potato beetle (Coleoptera: Chrysomelidae) on potatoes in western Massachusetts. J. Econ. Entomol., 1983, **76**, 349-56.

38. Ferro, D. N. and Gelernter, W. D., Toxicity of a new strain of *Bacillus thuringiensis* to Colorado potato beetle (Coleoptera: Chrysomelidae). J. Econ. Entomol., 1989, **82**, 750-5.

39. Zehnder, G. W. and Gelernter, W. D., Activity of the M-ONE formulation of a new strain of *Bacillus thuringiensis* against Colorado potato beetle (Coleoptera: Chrysomelidae): Relationship between susceptibility and insect life stage. J. Econ. Entomol., 1989, **82**, 756-61.

40. Whalon, M. E, Miller, D. L., Hollingworth, R. M., Grafius, E. J. and Miller, J. R., Laboratory selection of a resistant Colorado potato beetle (Col.: Chrysomelidae) strain to the CryIIIA coleopteran specific delta endotoxin of *Bacillus thuringiensis*. J. Econ. Entomol., 1992 (in press).

41. Grafius, E. Effects of temperature on pyrethroid toxicity to Colorado potato beetle

(Coleoptera: Chrysomelidae). J. Econ. Entomol. 1986, **79**, 588-91.

42. Groden, E., Drummond, F. A., Casagrande, R. A. and Haynes, D. L., *Coleomegilla maculata* (Coleoptera: Coccinellidae): Its predation upon the Colorado potato beetle (Coleoptera: Chrysomelidae) and its incidence in potatoes and surrounding crops. J. Econ. Entomol., 1990, **83**, 1306-15.

43. Hazard, R. V., Ferro, D. N., van Driesche, R. G. and Tuttle, A. F., Mortality of eggs of the Colorado potato beetle (Coleoptera: Chrysomelidae) from predation by *Coleomegilla maculata* (Coleoptera: Coccinellidae). Environ. Entomol., 1991, **20**, 841-8.

44. Watkinson, I. A., Wiseman, J. and Robinson,J., A simple test kit for field evaluation of the susceptibility of insect pests to insecticides. In Proceedings 1984 British Crop Protection Conference, Brighton. British Crop Protection Council, Croydon, England, 1984, pp. 559-64.

45. Tingey, W. M., Potato glandular trichomes: Defensive activity against insect attack. In Naturally Occurring Pest Bioregulators, ed. P. A. Hedin, American Chemical Society Symposium Series, No. 449, Washington, D. C., 1991, pp. 126-35.

MULTIPLE HERBICIDE RESISTANCE IN ANNUAL RYEGRASS (*Lolium rigidum*): A DRIVING FORCE FOR THE ADOPTION OF INTEGRATED WEED MANAGEMENT

STEPHEN B. POWLES & JOHN M. MATTHEWS
Department of Crop Protection
Waite Agricultural Research Institute
University of Adelaide
Glen Osmond.   South Australia 5064

ABSTRACT

Ryegrass (*Lolium rigidum*) is an abundant, polymorphic, wind-pollinated, self-incompatible annual grass that is ubiquitous in cropping regions of southern Australia. Over the past decade hundreds of herbicide resistant ryegrass populations have appeared in response to selection pressure from herbicides. A striking feature of resistance in ryegrass is the development of multiple herbicide resistance. Currently there are biotypes of ryegrass that are resistant to herbicides within ten different chemical classes. There are varying patterns of resistance between different biotypes. Ryegrass populations that have developed resistance to one herbicide can rapidly develop resistance to new herbicides that are effective initially. Genetic diversity in ryegrass and the frequent exposure of huge numbers of plants to herbicide results in the selection of one to many different resistance genes.

Multiple herbicide resistance in ryegrass occurs as a result of multiple resistance mechanisms. The spectrum of multiple resistance is determined by the number of resistance mechanisms possessed by individual plants in the population.

It is difficult to control resistant ryegrass biotypes with selective herbicides. Integrated weed management practices are required in which herbicides are only one of the control methods used. Multiple herbicide resistance will likely develop in weed species other than ryegrass and therefore, as a general principle, herbicides should be used in accordance with biological and evolutionary realities. Management of herbicides is required so as to minimise the likelihood for major herbicide resistance developments. Integrated weed management practices and herbicide resistance management programmes will need to be established if agriculture is to retain the benefits of herbicides in the long term.

1) Development of Multiple Herbicide Resistance in Ryegrass

The most abundant *Lolium* species in Australia is *Lolium rigidum* (ryegrass) which is a wind-pollinated, self-incompatible diploid grass of Mediterranean origin with vigorous vegetative growth.   Ryegrass is a useful component of pastures in much of the Mediterranean climatic zone of Australia and is a prolific seed producer which ensures its spontaneous re-establishment as an annual pasture species. These same characteristics make ryegrass a crop weed which can severely limit yields (1, 2).

Ryegrass is ubiquitous in cereal cropping regions and is therefore a major target for control by herbicides. Prior to the introduction of selective herbicides, ryegrass was managed by a combination of techniques involving biological (pasture rotation with grazing animals) and physical control (cultivation, burning). The era of herbicide control of ryegrass commenced in the early 1970's following release of the pre-emergent herbicide trifluralin. In 1978 the cereal-selective aryloxyphenoxypropionate herbicide, diclofop-methyl, became available for the post-emergent control of ryegrass and wild oats in cereal and dicotyledonous crops. Diclofop-methyl is widely used throughout southern Australia and in many situations has substituted for the physical and biological methods traditionally used to control ryegrass. In excess of 5 million hectares of cropping land are currently treated annually with diclofop-methyl. Other aryloxyphenoxypropionate and cyclohexanedione herbicides that also control ryegrass have been progressively introduced throughout the 1980's.   Since 1982 the cereal-selective sulfonylurea herbicides chlorsulfuron and triasulfuron have become widely used for ryegrass control.

The appearance of herbicide resistance in ryegrass has followed the widespread usage of selective herbicides. Table 1 shows the history of the development of resistance in one ryegrass biotype (SLR 31). Within three years of diclofop-methyl use resistance was evident (3) with chlorsulfuron and then sethoxydim failing later (Table 1). This biotype now has resistance to herbicides from at least four different chemical classes and therefore exhibits multiple resistance (multiple resistance is defined as resistance to two or more distinct classes of herbicides after selection by members of those two or more classes of herbicides). Over the past few years, hundreds of multiple herbicide resistant ryegrass populations have been detected in cropping areas of southern Australia and some biotypes are resistant to many herbicides within ten different chemical classes (4, 5, 6, 7, 8, 9, Table 2).   A feature of this multiple resistance is that there are varying patterns of resistance between different ryegrass biotypes (6).

In Israel, ryegrass has developed resistance to triazine herbicides and diclofop-methyl (10) and in the United States populations of *Lolium multiflorum*, have developed resistance to diclofop-methyl (11).

TABLE 1
Herbicide history and development of resistance in one population of *Lolium rigidum* (SLR 31)

Herbicide selection pressure	First used	Resistance evident	Resistant to
Trifluralin	1970	1977	Trifluralin
Diclofop-methyl	1977	1980	Aryloxyphenoxypropionates
Chlorsulfuron	1983	1984	Sulfonylureas/Imidazolinones
Sethoxydim	1986	1988	Cyclohexanediones

TABLE 2
Herbicide groups to which biotypes of *Lolium rigidum* exhibit resistance+

Aryloxyphenoxypropionates
Chloroacetanilides
Cyclohexanediones
Dinitroanilines
Imidazolinones
Substituted ureas
Sulfonylureas
Triazines
Triazinones
Triazoles

+ different biotypes can be variably resistant within and between herbicide groups, with some biotypes resistant to all of the above herbicides.

---

2) An Explanation for Multiple Herbicide Resistance in Ryegrass

Herbicide resistance results from genetic variability within plant populations. It is well established that both self-pollinated and cross-pollinated plant species show substantial genetic variability for many characters (12, 13) including response to herbicides (14, 15). Somody et al (16) showed that 908 populations of *Avena fatua* displayed significant inter-population variability when treated with diclofop-methyl.

There are biological and agro-ecological factors which help to explain the development of multiple herbicide resistance in ryegrass in Australia. The adaptable and polymorphic nature of the species has resulted in its widespread presence in different soil types and microclimates of the temperate cropping regions of Australia. Ryegrass is endemic in large numbers over vast areas frequently exposed to herbicides and is often present at very high numbers early in the growing season (20,000 seedlings m^{-2}). When a large ryegrass population (susceptible) is treated with an effective herbicide a small but significant percentage of the population survives. Many of these survivors possess resistance genes that enable them to survive the dose of herbicide used. As ryegrass is an obligate cross-pollinated species there is gene exchange between different surviving plants (within the range of pollen access) and accumulation of resistance genes is likely in the progeny.

A herbicide exerts strong selection pressure favouring individual plants possessing genes endowing any mechanisms enabling survival in the presence of that herbicide. When a herbicide is applied to a large, polymorphic population there are plants which survive because they possess one or more (different) mechanisms enabling survival at the dose of herbicide used. This means that there can be survivors because of different resistance genes. There is no reason to believe, a priori, that only one resistance mechanism will be exclusively selected from a panmictic population. Similarly, there is no a priori reason to believe that mechanisms selected from geographically diverse populations will be the same, or present at the same frequency before, during or after the selective process.

The resistance mechanisms possessed by ryegrass biotypes can explain multiple herbicide resistance. Our studies reveal that ryegrass biotypes can have a number of different resistance mechanisms that contribute to multiple herbicide resistance (Table 3). Biotype SLR31 is resistant to aryloxyphenoxypropionate and cyclohexanedione herbicides despite possessing acetyl co-enzyme A carboxylase (ACCase) susceptible to these herbicides (17,18) whereas another biotype has an insensitive acetyl CoA carboxylase (Holtum & Powles, unpublished data). Biotype SLR31 has enhanced capacity to metabolise the sulfonylurea herbicide chlorsulfuron (5), slightly enhanced capacity to metabolise diclofop-methyl (18) and altered membrane polarisation properties allowing membranes to recover from herbicide treatment (19). All ryegrass biotypes resistant to the aryloxyphenoxypropionate and cyclohexanedione herbicides show the ability to recover membrane potential following treatment with herbicide whereas susceptible biotypes cannot (19). Resistance to sulfonylurea herbicides can be due to an insensitive acetolactate synthase (9) or endowed by changes other than a modification at the enzyme target site (5, 17). Resistance to triazine and urea herbicides is due to an increased rate of metabolic detoxification (21). These studies of mechanisms endowing multiple herbicide resistance in ryegrass, while still under active investigation, reveal that there are a number of different resistance mechanisms (Table 3). It is possible for individual ryegrass plants to possess one or more mechanisms and this determines the spectrum of multiple herbicide resistance. Ryegrass biotypes resistant to a wide range of herbicides (Table 2) probably possess all of the resistance mechanisms summarised in Table 3 and probably other mechanisms yet to be identified.

The reason that a resistant ryegrass population can so rapidly extend the spectrum of herbicide resistance when treated with a new herbicide is the result of the presence of a number of different resistance mechanisms in the population. The possession of a number of different resistance mechanisms provides the springboard enabling rapid development of multiple resistance. When treated with a chemically dissimilar herbicide a proportion (dependent on gene frequencies) of the resistant population can possess one or more resistance mechanisms effective on this herbicide. Such plants will survive and produce seed and if the herbicide is used repeatedly this genotype will rapidly increase to the point that weed control fails. This is what has occured with biotype SLR 31 (Table 1). Thus, multiple herbicide resistance in ryegrass is due to a number of resistance mechanisms being selected in response to classical evolutionary selection pressure imposed by a herbicide(s). Whether, or not, ryegrass in the presence of herbicides has genetic features which accelerate genetic change, such as an increased rate of transposon related mutations (20), remains to be investigated.

TABLE 3
Resistance mechanisms identified in multiple herbicide resistant ryegrass

Herbicide Group	Resistance Mechanism	Biotype	Ref.
Aryloxyphenoxypropionate and Cyclohexanediones			
	ACCase mutants	WLR 96	Holtum &Powles unpublished
	Non-ACCase	SLR 31	18,19
Sulfonylureas and Imidazolinones			
	ALS mutants	WLR 1	9
	Metabolism mutants	SLR3 1	5
Triazines and Substituted ureas			
	Metabolism mutants	WLR 2	20

There are other examples where herbicide resistant grass weed species exhibit multiple herbicide resistance. In cereal growing regions of England and Germany, biotypes of *Alopecurus myosuroides* treated mainly with chlortoluron display multiple herbicide resistance (22). The resistant *Alopecurus myosuroides* can metabolise chlortoluron and isoproturon at a faster rate than susceptible plants and as this metabolism can be inhibited by cytochrome P450 inhibitors, it seems that increased mono-oxygenase activity is central to multiple resistance in this species (23). Whether this is the sole mechanism endowing multiple resistance in *Alopecurus myosuroides* has yet to be established. Triazine resistant biotypes of *Phalaris paradoxa* and *Alopecurus myosuroides* were shown to be cross-resistant to diclofop-methyl (11, 24). Similarly, biotypes of *Avena fatua* which developed resistance following selection with triallate displayed cross-resistance to diclofop-methyl (25).

Multiple insecticide resistance is well known in many insect pest species which possess mechanisms conferring multiple resistance to a range of chemically dissimilar insecticides (26, 27, 28). Brattsten (29) has demonstrated that insects with multiple resistance possess a number of different resistance mechanisms, including the capacity for increased rates of metabolism, decreased target site sensitivity and decreased rates of insecticide uptake.

3) The Selection of Resistance Mechanisms

The dynamics of the development of herbicide resistance will reflect the number and frequency of genes that are contributing to resistance, the biochemical efficiency of specific resistance mechanisms as well as the dose and duration of herbicide usage (Table 4). For example, one trait may confer a modest degree of resistance (e.g. increased metabolism) but be present at a high initial gene frequency whereas a second trait may confer high level resistance (e.g. target site insensitivity) but be present at a low initial gene frequency.

TABLE 4
Factors influencing the dynamics of herbicide resistance development

1. The number of individuals treated
2. The initial frequency and inheritance of resistance genes
3. The biochemical efficiency of resistance mechanisms
4. The duration of selection pressure from a herbicide
5. The dosage of herbicide used
6. The particular agricultural practices
7. The fitness of resistant individuals

As some resistance mechanisms are more efficient than others in conferring resistance, the frequency of the genes conferring these resistance mechanisms will vary in a dynamic way throughout the period of resistance evolution (dependent on initial gene frequencies and the length and characteristics of the herbicide selection pressure). Whilst physiological/biochemical studies may identify a major resistance mechanism there may be other factors that are contributing to the resistance, or have contributed at an earlier stage in the evolution of the resistance, and are maintained at a relatively high gene frequency. There are examples of this with triazine resistant weed species; while triazine resistance is mostly due to an alteration at the 32 kdalton herbicide binding protein such that triazine herbicides cannot bind effectively at photosystem II (30) resistant plants can also display an increased rate of metabolic detoxification of triazine herbicide (31). This increased rate of metabolism may be inconsequential in comparison to the modified 32 kdalton protein in conferring triazine resistance but was selected, and may have been of importance, earlier in the evolution of triazine resistance. With multiple resistant ryegrass biotypes it is clear that a population can carry different resistance mechanisms for a herbicide group and that there will be changes in the frequencies and importance of various resistance mechanisms throughout a period of herbicide selection.

Little evidence is available as to the influence of the herbicide dose used throughout a selection period on the dynamics of development of herbicide resistance and the resistance mechanisms selected. Changes in herbicide dose may select for different resistance mechanisms. Frequently the first response to the appearance of herbicide resistance is to increase the herbicide dose applied. It is probable that herbicide dose has a major effect as low dosages may select for different mechanisms than high dosages. Such changes in resistance mechanisms in response to changes in dose have been shown in a study with multiple resistant *Culex pipiens* in which changes in the frequency of particular resistance mechanisms occured as the use rates of insecticide increased (32). Research on herbicide resistant weeds is needed in this area.

4) The Control of Multiple Herbicide Resistant Weed Species

Although herbicide resistance is a worldwide phenomenon (see 33, 34, 35) there has been, until now, little need for concerted control strategies because control of resistant biotypes has usually been possible with alternative herbicides. Management factors, especially the choice and combination of herbicides, have changed in response to the appearance of resistance. For example, triazine resistant weeds in the northern hemisphere, although geographically and numerically widespread (34), have been mainly controlled by alternative herbicides (albeit at greater cost).

Management of a multiple resistant weed ryegrass population cannot be achieved simply by changing herbicides. There are virtually no selective herbicides available in Australia which can control biotypes with the degree of resistance evident in Table 2.

Ryegrass and *A. myosuroides* (22) should not be regarded as unique in displaying multiple herbicide resistance, they are probably just the first cases in prominent weed species. Multiple resistance is becoming evident in populations of wild oats in Australia (36) and Canada (37) which exhibit resistance to a range of aryloxyphenoxypropionate and cyclohexanedione herbicides. Control of multiple herbicide resistant ryegrass, and other such weeds, requires an integrated weed management programme.

INTEGRATED WEED MANAGEMENT (IWM)

IWM is the planned and integrated use of biological, chemical and physical methods to control weeds. Such IWM strategies have long been used to control recalcitrant weed species throughout the world (although control has rarely been defined as IWM). The spectacular successes of herbicides over the past two decades has meant that IWM practices, in many countries, have ceased to be directed at species easily controlled by herbicides. The appearance of multiple herbicide resistant weed biotypes will force adoption of IWM because multiple resistance patterns such as shown in Table 2 prevent the use of selective herbicides. We are undertaking substantial research to identify IWM practices for the control of multiple herbicide resistant ryegrass populations.

A) Control by biological means

Crucial to success of an IWM programme is an understanding of ryegrass biology so as to identify periods in the life cycle in which intervention is most likely to be effective. It is also imperative to consider the agricultural systems in the area under study. For ryegrass there are two critical features which can be exploited:

1. Short-term seed life. Ryegrass is highly fecund and under favourable conditions produces large numbers of seed. The great majority of seed germinates in the year after maturity (38). A crucial fact is that ryegrass seed does not have a long-term residual seedbank life with little seed remaining viable in the soil after 3 years (I. Heap & R. Knight, pers. comm. 1989, J. Matthews & S. Powles, unpublished). It is therefore possible to consider management practices which reduce the seedbank population of multiple resistant ryegrass over a short time period.

2. Value as a pasture plant. Throughout southern Australia, cropping phases are interspersed with pasture phases in which ryegrass is highly valued. Following the appearance of multiple herbicide resistance in ryegrass the simplest way to achieve control is to return the land area to a pasture phase. In a pasture, various management options to prevent or minimise ryegrass seed-set can be practiced so as to reduce the ryegrass soil seedbank reserves to very low levels. These include use of a fallow, heavy grazing and/or haymaking (39) together with the use of non-selective herbicides (glyphosate/paraquat) to minimise viable seed-set in the pasture phase. It must be emphasised that minimising

seed-set will not change the resistance status of the residual seedbank but represents a means of greatly reducing overall numbers.

In addition to reducing the numbers of multiple resistant ryegrass by controlling seed-set we are investigating a novel approach to change the resistance status of the residual seedbank population of ryegrass. This involves the introduction of tetraploid, herbicide-susceptible ryegrass into fields in which seedbank reserves of the diploid multiple resistant ryegrass has been reduced to low levels. This approach relies on the fact that ryegrass is an obligate cross-pollinated species and that a tetraploid hybridising with a diploid (resistant) will produce a sterile triploid (40,41). The tetraploid susceptible ryegrass will be sown into fields containing the resistant diploid ryegrass such that there will be a considerable excess of tetraploid ryegrass and therefore a high probability that resistant diploid ryegrass will hybridise with the tetraploid ryegrass, producing no viable seed. If successful this practice offers a means of dramatically changing the resistance status of a ryegrass population and is, in essence, a form of biological control. This technique may be employed with ryegrass because it is a valuable pasture plant but would not be of use for weeds with few positive features. Of course the tetraploid ryegrass population will have to be managed so as to avoid intense herbicide selection for resistance genes.

B) Physical control

Over the past decade there has been a major trend in Australian agriculture towards minimum soil disturbance by a reduction in soil cultivation. In many situations herbicides have substituted for cultivation as a means of controlling ryegrass and other weed species. The appearance of multiple herbicide resistant ryegrass populations may mean that judicious cultivation will again be used as a means of managing herbicide resistant populations.

A large portion of the ryegrass seed remaining on the plant at the end of the growing season, or lodged on the soil surface can be destroyed by burning the plant residue (39) and this represents a means of minimising the amount of viable seed entering the seedbank.

C) Herbicide management strategies to minimise resistance

It is clear that multiple herbicide resistant ryegrass populations (Table 2) result from multiple resistance mechanisms (Table 3) and that ryegrass has biological features that enable it to readily develop herbicide resistance when challenged with a new herbicide. Consequently; any new herbicide, effective initially on resistant ryegrass, has a high probability of failing if persistently used because of the further extension of resistance.

Therefore, any new herbicide should be managed so as to minimise the rate at which resistance can develop.

Thus far, herbicides have been promoted with little recognition that their persistent use may lead to resistance. For multiple herbicide resistant weed populations and as a general principle in the future, herbicides must be used in accordance with biological and evolutionary realities. Ryegrass (Table 2) and *A. myosuroides* (22) are just the first two of what will probably be many cases of multiple herbicide resistance in prominent weed species. What is required is management of herbicides to minimise resistance developments. This will inevitably involve a reduction in the amount of herbicide applied to a given land area and will involve restraint by marketing arms of herbicide manufacturers and a great deal of co-operation between industry and public and private sector researchers and advisors. Extension programmes are being launched in southern Australia which advise the adoption of IWM techniques. The first cases of herbicide management strategies (re-active) have been recently implemented for the sulfonylurea/imidazolinone herbicides in North America and Australia. The agrochemicals industry already has experience of insecticide resistance management: Since 1983 an insecticide resistance management strategy has been in operation in summer crop regions of Australia (42). Herbicides and weeds seem no different and the lessons evident with multiple resistance in ryegrass in Australia will apply equally to weeds in other parts of the world which are likely to develop multiple herbicide resistance

CONCLUSIONS

Ryegrass is an abundant, genetically variable, cross-pollinated diploid with the ability to rapidly develop resistance to a wide range of herbicides. In Australia some biotypes of ryegrass are resistant to almost all selective herbicides available for ryegrass control. Multiple resistance is endowed by different resistance genes and it is possible for individual ryegrass plants to possess multiple co-occuring resistance mechanisms. It is therefore apparent that strategies relying solely on selective herbicides are unlikely to give long-term ryegrass control.

The repercussions of the appearance of multiple herbicide resistance is that herbicides must be used in accordance with biological and evolutionary realities. Integrated weed management practices and herbicide resistance management programmes will need to be established if agriculture is to retain the benefits of herbicides in the long term.

REFERENCES

1.   Reeves, T.G., Effect of annual ryegrass (*Lolium rigidum* Gaud.) on yield of
      wheat. Weed Res., 1976, **16**, 57-63.

2.   Medd, R.W., Auld, B.A., Kemp, D.R. and Murison, R.D., The influence of
      wheat density and spatial arrangement on annual ryegrass, *Lolium rigidum*,
      competition., Aust. J. Agric. Res., 1985, **36**, 361-71.

3   Heap, J. and Knight, R., A population of ryegrass tolerant to the herbicide
      diclofop-methyl., J. Aust. Inst. Ag. Sci., 1982, **48**, 156-7.

4.   Heap, I. and Knight, R., The occurence of herbicide cross-resistance in a
      population of annual ryegrass, *Lolium rigidum*, resistant to diclofop-methyl.,
      Aust. J. Agric.Res., 1986, **37**, 149-56.

5.   Christopher, J.T., Powles, S.B., Liljegren, D.R. and Holtum, J.A.M., Cross-
      resistance to herbicides in annual ryegrass (*Lolium rigidum*). II. Chlorsulfuron
      resistance involves a wheat-like detoxification system., Plant Physiol., 1991, **95**,
      1036-43.

6.   Heap, I.M. and Knight, R., Variations in herbicide cross-resistance among
      populations of annual ryegrass (*Lolium rigidum*) resistant to diclofop-methyl.,
      Aust. J. Agric. Res., 1990, **41**, 121-28.

7.   Powles, S.B. and Howat, P.D., Herbicide resistant weeds in Australia., Weed
      Tech., 1990, **4,** 178-85.

8.   Burnet, M.W., Hildebrand, O.B., Holtum, J.A.M. and Powles, S.B., Amitrole,
      triazine, substituted urea, and metribuzin resistance in a biotype of rigid ryegrass.,
      Weed Sci., 1991, **39**. In press.

9.   Christopher, J.T., Powles, S.B., Holtum, J.A.M., Cornwall, G. and Davis, R.,
      Sulfonylurea resistance in rigid ryegrass induced by chlorsulfuron differs from
      sulfonylurea cross-resistance induced by diclofop-methyl., Weed Tech., 1991. In
      press.

10.   Yaacoby, T., Schonfeld, M. and Rubin, B., Characteristics of atrazine resistant
      biotypes of three grass weeds, Weed Sci., 1986, **34**, 181-84.

11.   Stanger, C.E. and Appleby, A.P., Italian ryegrass (*Lolium multiflorum*) accessions
      tolerant to diclofop., Weed Sci., 1989, **37**, 350-52.

12.   Allard, R.W., Jain, S.K. and Workman, P.L., The genetics of inbreeding
      populations, Adv. Genet., 1968, **14**, 50-5.

13.   Brown, A.D.H., Enzyme polymorphisms in plant populations., Theor. Pop.
      Biol., 1979, **15,** 1-42.

14.   Jana, S. and Naylor, M., Adaptation for herbicide tolerance in populations of
      Avena fatua., Can. J. Bot., 1982, **60**, 1611-17.

15.   Price, S.C., Hill, J.E. and Allard, R.W., Genetic variability for herbicide reaction
      in plant populations., Weed Sci., 1983, **31**, 652-57.

16.   Somody, C.N., Nalewaja, J.D. and Miller, S.D. Wild oat (*Avena fatua*) and
      Avena sterilis morphological characteristics and response to herbicides., Weed
      Sci., 1984,   **32**, 353-59.

17.   Matthews, J.M., Holtum, J.A.M., Liljegren, D.R., Furness, B. and Powles,
      S.B., Cross-resistance to herbicides in annual ryegrass (*Lolium rigidum*). 1.
      Properties of the herbicide target enzymes acetyl-coenzyme A carboxylase and
      acetolactate synthase., Plant Physiol., 1990, **94**, 1180-86.

18. Holtum, J.A.M., Matthews, J.M., Liljegren, D.R. and Powles, S.B. Cross-resistance to herbicides in annual ryegrass (*Lolium rigidum*). III. On the mechanism of resistance to diclofop-methyl., Plant Physiol., 1991. In press.

19. Hausler, R., Holtum, J.A.M. and Powles, S.B., Cross-resistance to herbicides in annual ryegrass (*Lolium rigidum*). IV. Correlation between membrane effects and resistance to graminicides. Plant Physiol., 1991. In press.

20. Kingman, A.J., Chater, K.F. and Kingsman, F.M., Transposons, Cambridge University Press, 1988.

21. Burnet, M.W.M., Loveys, B.R., Holtum, J.A.M. and Powles, S.B., A mechanism of simazine resistance in *Lolium rigidum*., Plant Physiol., 1991. Submitted.

22. Moss, S.R. and Cussans, G.W., Detection and practical significance of herbicide resistance with particular reference to the weed *Alopecurus myosuroides* (Black-grass). In M. Ford, D. Hollomon, B. Khambay and R. Sawicki, eds. Biological and Chemical Approaches to Combating Resistance to Xenobiotics. Society of Chemical Industry, London, 1987, pp. 201-13.

23. Kemp, M.S., Moss, S.R. and Thomas, T.H., Herbicide resistance in *Alopecurus myosuroides* .In Managing Resistance to Agrochemicals: From Fundamental Research to Practical Strategies, Eds. M.B.Green, H.M. LeBaron and W.K. Moberg, American Chemical Society, Washington DC, 1990, pp. 376-93.

24. Rubin, B., Yaacoby, T. and Schonfeld, M., Triazine resistant grass weeds: Cross resistance with wheat herbicide, a possible threat to cereal crops, Brit. Crop Prot. Conf. 1985, **9B-4**, 1163-70.

25. Thai, K.M., Jana, S. and Naylor, J.M., Variability for response to herbicides in wild oats (*Avena fatua*) populations, Weed Sci., 1985, **33**, 829-35.

26. Siegfried, B.D., Scott, J.G., Roush, R.T. and Zeichner, B.C., Biochemistry and genetics of chlorpyrifos resistance in the german cochroach, *Blattella germanica* (L)., Pest. Biochem. Physiol., 1990, **38**, 110-21.

27. Hemingway, J., Miyamoto, J. and Herath, P.R.J., A possible novel link between organophosphorus and DDT insecticide resistance genes in *Anopheles*: Supporting evidence from fenitrothion metabolism studies, Pest. Biochem. Physiol., 1991, **39**, 49-56.

28. Yu, S.J., Insecticide resistance in the fall armyworm, *Spodoptera frugiperda*., Pest. Biochem. Physiol., 1991, **39**, 84-91.

29. Brattsten, L.B. Resistance mechanisms to carbamate and organophosphate insecticides. In Managing Resistance to Agrochemicals: From Fundamental Research to Practical Strategies, ed. M.B.Green, H.M. LeBaron and W.K. Moberg, American Chemical Society, Washington DC, 1990, pp. 42-60.

30. Arntzen, C.J., Pfister, K. and Steinback, K.E., The mechanism of chloroplast triazine resistance: Alterations in the site of herbicide action. In. Herbicide Resistance in Plants., Eds. H.M. LeBaron and J. Gressel, J Wiley & Sons, New York, 1982, pp. 185-214.

31.   Gressel, J., Regev, Y., Malkin, S., and Kleifeld, Y., Characterization of an
       atrazine resistant biotype of *Brachypodium distachyon.*, Weed Sci., 1983, **31**,
       450-56.

32.   Villani, F. and Hemingway, J., The detection and interaction of multiple
       organophosphorus and carbamate insecticide resistance genes in field populations
       Culex pipiens from Italy. Pest. Biochem. Physiol., 1987, **27**, 218-228.

33.   H.M. LeBaron and J. Gressel., Herbicide resistance in plants, J Wiley & Sons,
       1982.

34.   LeBaron, H.M. and McFarland, J.E., Herbicide resistance in weeds and crops:
       An overview and prognosis. In Managing Resistance to Agrochemicals: From
       Fundamental Research to Practical Strategies, Eds. M.B.Green, H.M. LeBaron
       and W.K. Moberg, American Chemical Society, Washington DC, 1990, pp. 336-
       52.

35.   J.C. Caseley; G.W. Cussans and R.K. Atkin., Herbicide resistance in weeds and
       crops. Butterworth Heinemann, Oxford, 1991.

36.   Heap, I.M., Morrison, I.N. and Friesen, L.F., Cross-resistance to
       aryloxyphenoxypropionate and cyclohexanedione herbicides in *Avena fatua* (Wild
       oat) populations in Western Canada. Weed Science Society of America Abstracts,
       1991, **31**, 54.

37.   Mansooji, A., Holtum, J.A.M., Matthews, J.M. and Powles, S.B., Multiple
       herbicide resistance in wild oats (*Avena sterilis*). Weed Science Society of
       America Abstracts, 1991, **31**, 78.

38.   McGowan, A.A., Comparative germination patterns of annual grasses in north-
       eastern Victoria., Aust. J. Exp. Agric. An. Husb., 1970, **10**, 401-4.

39.   Reeves, T.G. and Smith, I.S., Pasture management and cultural methods for the
       control of annual ryegrass (*Lolium rigidum*) in wheat. Aust. J. Exp. Agric. Anim.
       Husb., 1975, **15,**   527-30.

40.   Breese, E.L., Lewis, B.J. and Evans, G.M., Interspecies hybrids and
       polyploidy. Phil. Trans. R. Soc. London., 1981, **B292,**   487-97.

41.   Stebbins, G.L. Chromosomal Evolution in higher plants. Edward Arnold Ltd.
       London, 1971.

42.   Forrester, N.W. Designing, implementing and servicing an insecticide resistance
       management strategy. Pest. Sci., 1990, **28,**   167-79.

MANAGEMENT OF INSECTICIDE RESISTANCE IN *HELIOTHIS ARMIGERA* IN AUSTRALIA - ECOLOGICAL AND CHEMICAL COUNTERMEASURES

NEIL W. FORRESTER[1] & GARY P. FITT [2]

[1] New South Wales Agriculture & Fisheries   [2] CSIRO Division of Entomology
Agricultural Research Station, Narrabri, NSW 2390, Australia.

ABSTRACT

The background and structure of the Australian Insecticide Resistance Management (IRM) Strategy is described. This Strategy aims to integrate both chemical and non-chemical control measures to broaden the range of mortality factors and avoid focussing selection pressure on any one control measure. Chemical countermeasures include:- rotation of unrelated chemical groups on a per generation basis; ovicide/larvicide mixtures; mixtures of the bacterial pathogen *Bacillus thuringiensis* with endosulfan (and in the future pyrethroids); targetting pyrethroids to egg hatch; frequent and thorough scouting and adherence to thresholds to minimise the need for sprays and to ensure their maximum effectiveness through optimal timing and the use of synergists or (possibly in the future) metabolically refractory pyrethroids. Ecological countermeasures include:- the use of 'soft' chemicals early (and possibly later) to gain maximum benefit from natural enemies; the pursuit of early crops (through better irrigation and nitrogen management, judicious use of plant growth regulators and early sowing) to avoid late season *Heliothis armigera* populations ; avoidance of unfavourable cropping or rotation programmes; the destruction of overwintering pupae and the adoption of host plant resistance wherever possible eg. okra leaf cotton.

The Australian Strategy is presented as a successful working example of an IRM Strategy complementing good IPM (Integrated Pest Management) practice. The Strategy has not overcome the resistance problem but has proven to be a successful delaying tactic extending the useful life of the pyrethroids and buying time to allow the discovery, development and implementation of alternative control measures, both chemical and non-chemical.

INTRODUCTION

In January 1983, pyrethroids failed to give satisfactory field control of *Heliothis* (= *Helicoverpa*) *armigera* (Hübner) at Emerald in central Queensland. Resistance was quickly shown to be the problem (1) and an Insecticide Resistance Management (IRM) Strategy was introduced within six months of the first field failures. The Strategy aims to manage not only pyrethroid resistance but also potential resistance to other chemical groups such as carbamates and organophosphates. It also aims to prevent reselection of historical endosulfan resistance

(2). This voluntary Strategy has been in place now for eight seasons and has been well accepted by cotton and field crop growers, with all but total compliance. Monitoring of the impact of the Strategy has indicated its success as a delaying tactic in extending the useful life of the pyrethroids, with greater success against endosulfan resistance (Table 1, Figs. 1-3).

STRUCTURE OF THE AUSTRALIAN STRATEGY

The Australian IRM Strategy incorporates both chemical and ecological countermeasures. This integrated approach was designed to broaden the range of mortality factors to avoid focussing selection pressure on any one control measure.

Chemical Countermeasures

The basis of the Australian Strategy is the rotation of unrelated chemical groups between successive generations of *Heliothis*. The cropping season is divided into three stages with the use of pyrethroids (a maximum of three sprays) restricted to a 35 day period (42 days up to 1988/89) during the middle of the season, Stage 2 (3). This 35 day period corresponds to the minimum time required for the development of one generation of *H. armigera* in the field (4). Thus, pyrethroid selection pressure is restricted to one of the four to five generations per season. Endosulfan is recommended to be used either early or in the middle of the season (Stages 1& 2) but not late (Stage3). Thus, endosulfan selection pressure is restricted to three of the four to five generations per season. The restrictions on endosulfan were less severe than those for the pyrethroids as the resistance problem was not considered to be as acute. With endosulfan being used principally early in the season, selection pressure on *H. armigera* was expected to be much reduced since this species is less abundant on cotton at this time. These restrictions were applied to all the crop hosts of *H. armigera* and also to management of coincident pest species such as midge on sorghum and Rutherglen bugs on sunflowers (3). The other major recommendation of the Strategy is to add an ovicide (now principally methomyl, formerly chlordimeform) to pyrethroids and/or endosulfan when egg densities are high. The use of insecticidal mixtures has been widely recommended as a resistance management tool (5-7), though in Australia this has been restricted to larvicide/ovicide mixtures as no economically acceptable larvicide combinations could be found. However, recently the biological insecticide *Bacillus thuringiensis* has been increasingly used in mixtures with endosulfan and in the future, with pyrethroids. The aims of this approach are twofold; firstly to allow a reduction in the use of the environmentally sensitive endosulfan and secondly to act as a "safety net" to kill resistant *H. armigera* larvae surviving pyrethroid or endosulfan sprays. Wider adoption of these mixtures is currently hindered by the high cost of an effective rate of the biological component.

% SURVIVING DISCRIMINATING DOSE

STUDY AREA	SEASON	FENVALERATE			ENDOSULFAN		
		I	II	III	I	II	III
Namoi/Gwydir	1983/84	9.3	9.5	14.6	-	-	-
	84/85	7.5	12.9	27.9	-	-	-
	85/86	7.8	13.0	44.5	-	-	-
	86/87	32.2	36.7	42.9	7.1	16.7	20.1
	87/88	19.8	30.1	38.4	7.3	17.6	23.0
	88/89	19.6	42.4	60.7	8.8	13.2	10.6
	89/90	24.7	45.3	62.5	9.2	14.8	15.9
	90/91	55.7	61.1	61.5	12.2	22.7	31.3
Emerald	1985/86	6.8	17.1	14.4	-	-	-
	86/87	8.8	26.5	29.8	7.7	20.6	17.3
	87/88	15.9	27.1	27.0	9.5	14.3	13.7
	88/89	19.8	38.7	44.3	8.1	13.6	7.1
	89/90	27.9	44.6	54.6	3.1	21.0	20.9
	90/91	24.7	52.2	34.5	10.1	37.1	16.0
Inverell	1987/88	10.2	20.4	19.0	11.3	10.5	5.8
	88/89	21.9	28.9	41.7	9.4	4.8	5.4
	89/90	22.1	32.7	38.2	4.0	5.2	7.1
	90/91	47.8	34.6	45.1	3.4	8.5	10.8

Table 1. Average pyrethroid and endosulfan resistance levels in *Heliothis armigera* for each Stage(I, II & III) of the Resistance Management Strategy, for three study areas (the Namoi and Gwydir river valleys of northern New South Wales, the Emerald Irrigation Area of central Queensland and a sample of the unsprayed refugia area centred on Inverell in northern NSW). Results expressed as the percentage of larvae (reared from field collected eggs) surviving the fenvalerate or endosulfan discriminating dose (0.2 & 10 µg per 30-40 mg larva, respectively). Stage II pyrethroid window 42 days duration to 1988/89 season, thereafter 35 days. Piperonyl butoxide introduced into commercial use in 1990/91 season.

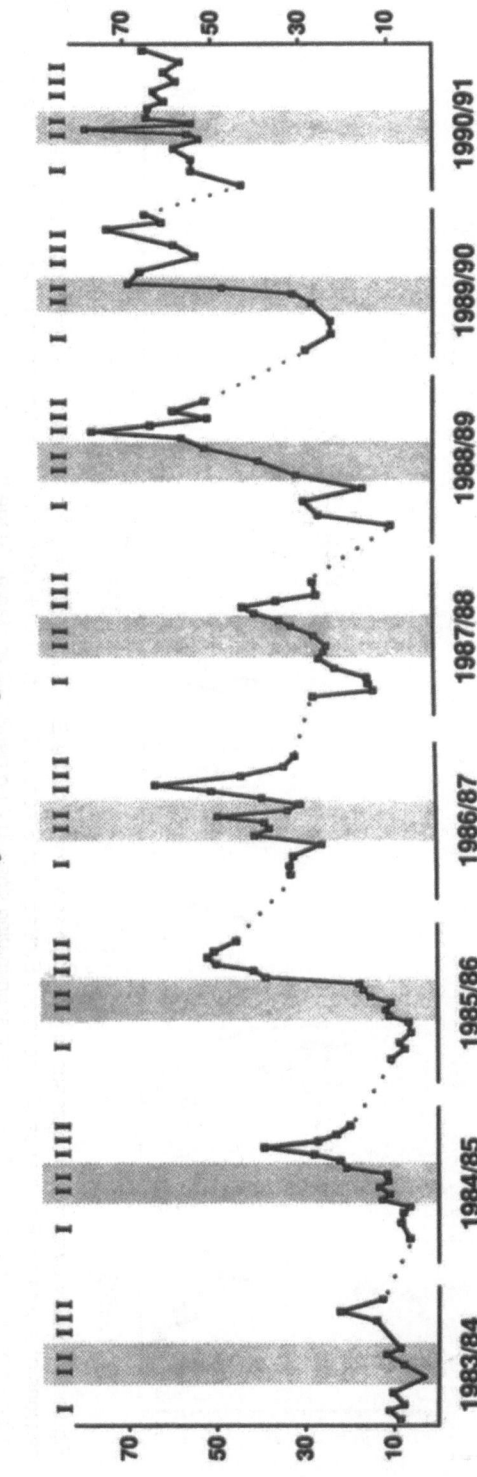

Figure 1. Weekly pyrethroid resistance in *Heliothis armigera* from the Namoi and Gwydir river valleys of northern New South Wales, Australia, for the 8 seasons since the introduction of a curative Resistance Management Strategy (for Stages I, II & III). Results expressed as the percentage of larvae (reared from field collected eggs) surviving the fenvalerate discriminating dose (0.2 μg per 30-40 mg larva). Stage II pyrethroid window 42 days duration to 1988/89 season, thereafter 35 days. Piperonyl butoxide introduced into commercial use in 1990/91 season.

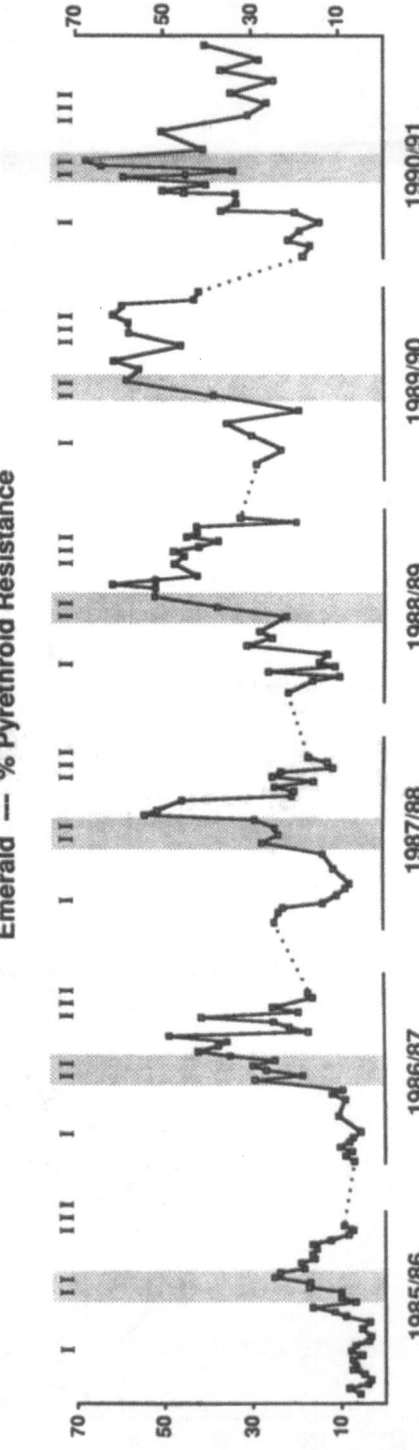

Figure 2. Weekly pyrethroid resistance in *Heliothis armigera* from the Emerald Irrigation Area of central Queensland, Australia, for the past 6 seasons of a curative Resistance Management Strategy (for Stages I, II & III). Results expressed as the percentage of larvae (reared from field collected eggs) surviving the fenvalerate discriminating dose (0.2 μg per 30-40 mg larva). Stage II pyrethroid window 42 days duration to 1988/89 season, thereafter 35 days. Piperonyl butoxide introduced into commercial use in 1990/91 season.

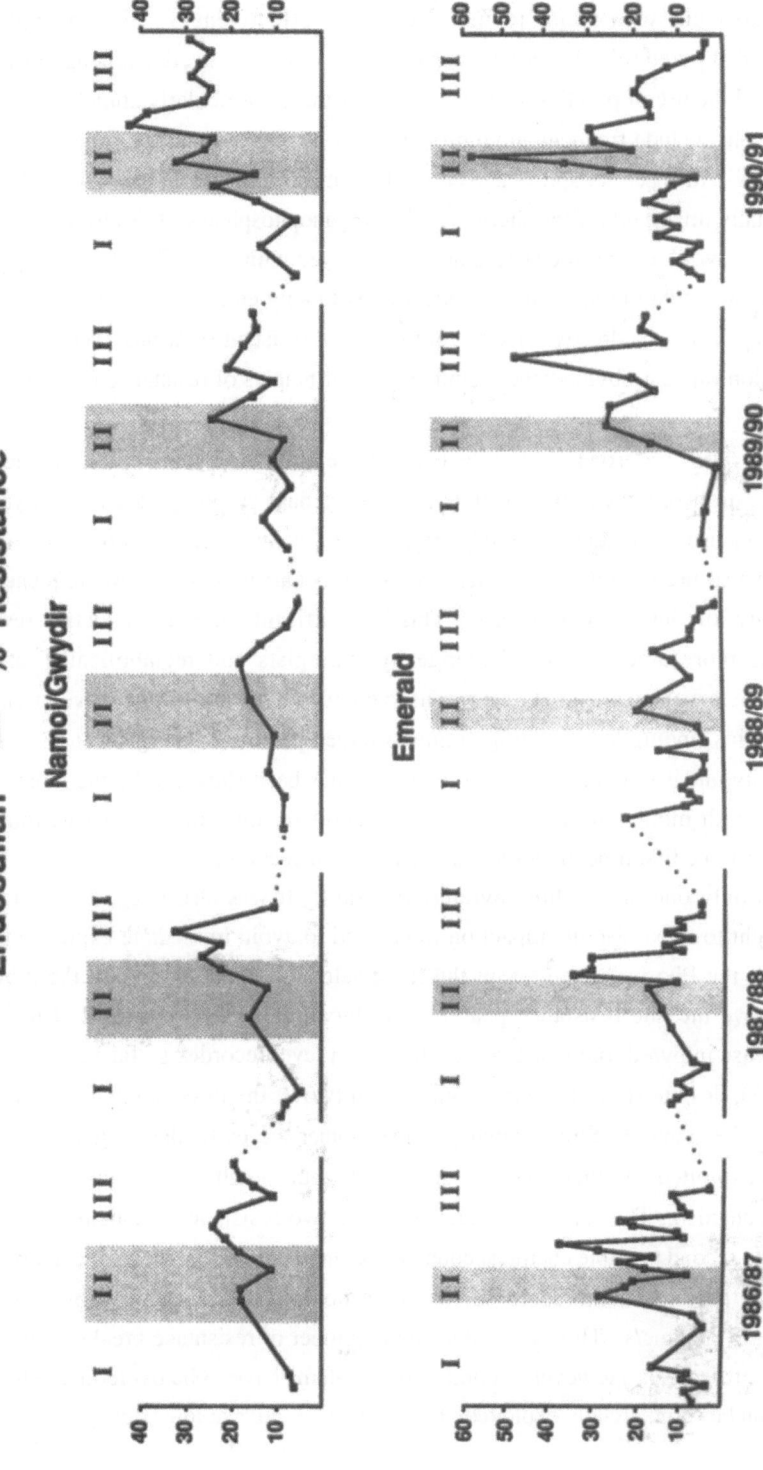

Figure 3. Weekly endosulfan resistance in *Heliothis armigera* from the Namoi and Gwydir river valleys of northern New South Wales and the Emerald Irrigation Area of central Queensland, Australia, for the past 5 seasons of a curative Resistance Management Strategy (for Stages I, II & III). Results expressed as the percentage of larvae (reared from field collected eggs) surviving the endosulfan discriminating dose (10 μg per 30-40 mg larva). Stage II pyrethroid window 42 days duration to 1988/89 season, thereafter 35 days.

There are other recommendations within the Strategy which are not strictly chemical countermeasures but which aim to maximise their effectiveness. For example, it is recommended that pyrethroids be targetted on egg hatch to avoid selection among older larvae (8) and to avoid the use of pyrethroids when *H. armigera* is particularly abundant (3). Other recommendations include frequent and thorough scouting of crops and spraying on threshold. This can minimise the need for sprays and ensure their maximum effectiveness through optimal timing (especially important for the shorter residual organophosphates). It is also recommended that a suspected pyrethroid failure should not be resprayed with a pyrethroid or if a pyrethroid is used to control sorghum midge, not to follow up with a pyrethroid for *Heliothis* control, as the midge spray will have already selected for pyrethroid resistant *Heliothis*. While the above recommendations appear obvious, these simple basic principles of resistance management are often ignored.

Studies with piperonyl butoxide (Pbo) have indicated that oxidative metabolic detoxification, probably via a polysubstrate monooxygenase system, is the major pyrethroid resistance mechanism in field populations of *H. armigera* in Australia (9). A nerve insensitivity mechanism is also present but is much less important occurring at only low frequencies and conferring only low levels of resistance. This is significant since the oxidative resistance mechanism is more amenable to challenge by synergists and metabolically refractory pyrethroids, whereas there are no known means to overcome the intractable nerve insensitivity mechanism. This finding led to a programme to screen potential synergists and assess their residual activity under field conditions.To date, Pbo has been shown to be the most suitable synergist (on both moths and larvae) despite its short residual life. Pbo was introduced commercially for the first time in 1990/91 season and was recommended to be used at 250-350 g a.i./ha with only one of the three pyrethroid sprays. It was also suggested that Pbo be sprayed at night to maximise its impact on moths and to avoid immediate exposure to light. The impact of the Pbo was to interrupt the rapid selection of moths within the pyrethroid window, remove the late resistance peak (due to larval selection) and to limit the overall seasonal increase in pyrethroid resistance to the lowest level recorded so far in all three study areas (Table 1), despite starting from a high base in two of the three study areas. Thus, the incorporation of Pbo into the Strategy may provide another successful delaying tactic as long as it is not over exploited and that selection does not favour the alternative nerve insensitivity resistance mechanism. The relative frequency of the two resistance mechanisms is closely monitored. The second possible chemical countermeasure for oxidative metabolic detoxification is to alter the pyrethroid molecule so that it is no longer a suitable substrate for the monooxygenase enzyme/s. This is possible and a number of resistance breaking compounds have been identified (9). However, a number of problems have to be overcome before these compounds can be commercially exploited : they should be active at low rates; safe to mammals and the environment; photostable and of equivalent persistence to current standard pyrethroids. It is probably only a matter of time before all these requirements are met but even then, agrochemical companies will not be keen to commercialise these 'resistance breakers' because of the inability of these compounds to overcome the alternative nerve insensitivity mechanism. This concern is probably justified as they will undoubtedly be of greatest value in those (up until now) relatively few resistance situations managed by a closely regulated IRM Strategy, where the nerve insensitivity resistance mechanism can be effectively contained to low levels. However, this does not mean that this research direction should be neglected as there is significant international effort being directed into the implementation of IRM Strategies worldwide (10). It remains to be seen whether 'resistance breaking' pyrethroids or synergists can be used successfully over a long period. The controlled use of Pbo in the Australian IRM Strategy may shed some light on this question.

Ecological Countermeasures

Reliance solely on chemicals for control of insects has been shown to often lead to resistance problems (11). In reality IRM is complementary to the accepted principles of good IPM (Integrated Pest Management), where chemical controls are integrated with a range of non chemical control measures, both biological and cultural. The range of non chemical countermeasures employed in the Australian IRM Strategy are discussed below.

The Australian IRM Strategy aims to gain maximum benefit from natural enemies by relying on 'soft' chemicals early in the season. This also assists in reducing secondary pest outbreaks of such pests as mites, aphids and whitefly. Endosulfan has been the main early season 'soft' insecticide used in the past, but this is being strongly supported by increasing use of the bacterial pathogen *Bacillus thuringiensis*, the insect growth regulator, chlorfluazuron and the carbamate, thiodicarb. There is also an increasing trend to use these three 'soft' insecticides late in the season to replace the broad spectrum organophosphate insecticides which are used for dual *Heliothis* / mite control. If this trend continues and mites can be contained by better cultural management, alongwith the new specific miticides coming onto the market, then it is possible that levels of parasitism of late season larvae and pupae may increase. This would be particularly useful as it is these late season populations which are the most resistant and provide the overwintering pupal populations which carry over the resistance problem into the following season.

Improved agronomic practices aimed at setting an early cotton crop without sacrificing yield, have also been important in the Australian IRM Strategy. Early crops are actively pursued in Australia mainly because of the late *H. armigera* problem and the high cost of late season chemical control. A range of practices have been adopted to avoid the need for late season insecticides. These include: better irrigation management through monitoring of water use with neutron probe moisture meters; better nitrogen management through monitoring of N levels and matching fertiliser rates to the crop's specific need as well as the judicious use of plant growth regulators (such as mepiquat and chlormequat chloride), to control excessive vegetative growth. Excessive early season insect control and/or the prophylactic use of granular insecticides at sowing is not the recommended method to achieve earliness. Normal pest thresholds should be set and adhered to and earliness gained through early sowing and improved agronomic practices (as discussed above). Raingrown cotton allows much less flexibility but the principle is the same : manage the crop to produce an economic yield in as short a time as possible.

Avoidance of unfavourable cropping or rotation programmes is also a major part of the Australian IRM Strategy. Crops of corn or sunflowers which flower in December will coincide with the second *H. armigera* generation of the season. These then produce moths about six weeks later in mid January, just after the opening of the Stage 2 pyrethroid period. These crops act simply as nursery crops producing large numbers of *H. armigera* which may then colonise cotton crops. This has led to the recommendation for cotton growers to avoid growing these crops in predominantly cotton areas. In mixed cropping areas, cotton growers have no control over their neighbours' cropping programmes and may often have to deal with heavy infestations of *H. armigera* produced on neighbouring maize, sorghum or sunflower fields. In these situations, growers are strongly advised to avoid consecutive sprays of pyrethroids as resistance levels will be exacerbated by selection of moths before mating (9). Chickpeas are another crop favoured by *H. armigera* and cotton growers are advised against growing this early spring host in predominantly cotton areas. In fact, it is probably easier to manage insecticide resistance in a cotton monoculture ecosystem than in a diverse multi-cropping system. The same has been noted for *Heliothis* management in general (12). The exception to this may be late sown corn which could act as a trap crop diverting late season *H. armigera* populations off cotton and arresting them in pupal diapause where they can be easily destroyed by cultivation (see below).

Cotton stubble has been shown to be the major source of overwintering *H. armigera* pupae in the Namoi/Gwydir cotton growing areas of northern NSW (13). Because of the intensity of spraying on this crop these populations have low levels of parasitism (13) and are also the most resistant population of the season as they derive from eggs laid during the late season peak resistance period in early March (14, Fig 1). Thus the overwintering population under cotton constitutes the major source for the carryover of resistant *H. armigera* from one season to the next if left undisturbed. Cultivation of winter cotton stubble has been shown to be an effective means of killing these overwintering pupae (15,16) and growers are strongly recommended to sample for pupae under cotton stubble and to cultivate if necessary. The overwintering pupal stage is the weak link in the *H. armigera* life cycle as it remains vulnerable to simple 'resistance proof' physical control measures for almost six months. Despite this advice, some growers continue to direct drill or aerially sow winter crops into cotton stubble without cultivation or are unable or unwilling to cultivate because of wet weather or potential soil structural damage. In some seasons, up to 38% of the cotton area has been left effectively uncultivated over the winter (NWF & GPF, unpub. data). A possible answer to this problem is encouragement of the ichneumonid pupal parasite *Pterocormus promissorius* which is active over the winter period. Research is underway on the ecology of this parasite and ways to increase its beneficial impact (GPF, unpub. data) possibly even through augmentative release.

Finally, but certainly not least, host plant resistance (HPR) has been utilised wherever possible. *Heliothis* spp. are notoriously difficult pests to counter with HPR and cotton is no exception. Some progress has been made through conventional plant breeding techniques using morphological characters (17) but genetically engineered host plant resistance offers considerable scope for the future. For the present, the okra leaf shape remains the most useful HPR mechanism in cotton in Australia reducing the damage potential of mites in particular (18), but also offering some degree of *Heliothis* suppression (17). Okra leaf has also improved the penetration and hence efficiency of both chemical and biological sprays and will no doubt assist in suppressing potential future problems with whitefly.

CONCLUSION

The Australian IRM Strategy has successfully integrated both chemical and ecological countermeasures for control of resistant *H. armigera*. It is a working example of an IRM Strategy complementing good IPM practice. Successful implementation of ecological countermeasures is dependent on an equally intimate knowledge of both pest and host ecology. The Australian Strategy has been willingly adopted by all sectors of the cotton and field crop growing communities and has allowed the Australian cotton industry to continue and expand despite resistance. The Strategy has not overcome resistance but has proven to be a successful delaying tactic extending the useful life of the pyrethroids and buying time to allow the discovery, development and implementation of alternative control measures, both chemical and non-chemical. The Strategy has also focussed attention on the need to conserve cheap, safe and environmentally acceptable control measures. As such, it has engendered an awareness of the need to exploit new technologies in a controlled and responsible manner.

ACKNOWLEDGEMENT

The authors wish to thank the Australian Cotton Research & Development Corporation for significant financial contribution to the studies discussed in this paper.

REFERENCES

1.   Gunning, R.V., Easton, C.S., Greenup, L.R. and Edge, V.E., Pyrethroid resistance in *Heliothis armigera* (Hübner)(Lepidoptera:Noctuidae) in Australia. *J. Econ. Entomol.*, 1984, **77**, 1283-1287.

2.   Forrester, N.W., Designing, implementing and servicing an insecticide resistance management strategy. *Pestic. Sci.*, 1990, **28**, 167-179.

3.   Forrester, N.W., Strategy change for the 1990-91 insecticide resistance management strategy for *Heliothis*. NSW Agriculture & Fisheries *Advisory Note*, 1990, No. **8/90**. Agdex 622.

4.   Room, P.M., Calculations of temperature driven development by *Heliothis* spp. (Lepidoptera:Noctuidae) in the Namoi valley, New South Wales. *J. Aust. ent. Soc.*, 1983, **22**, 211-215.

5.   Mani, G.S., Evolution of resistance in the presence of two insecticides. *Genetics*, 1985, **109**, 761-783.

6.   Curtis, C.F., Theoretical models of the use of insecticide mixtures for the management of resistance. *Bull. ent. Res.*, 1985, **75**, 259-265.

7.   Comins, H.N., Tactics for resistance management using multiple pesticides. *Agric. Ecosys. Environ.*, 1986, **16**, 129-148.

8.   Daly, J.C., Fisk, J.H. and Forrester, N.W., Selective mortality in field trials between strains of *Heliothis armigera* (Lepidoptera:Noctuidae) resistant and susceptible to pyrethroids : functional dominance of resistance and age class. *J. Econ. Entomol.*, 1988, **81**, 1000-1007.

9.   Forrester, N.W., Management of pyrethroid and endosulfan resistance in *Heliothis* (= *Helicoverpa*) *armigera* (Hübner) in Australia. Ph.D.thesis (Queensland University) in prep.

10. Smale, B.C., International Organisation for Pest Resistance Management. *Pest Resistance Management*, 1990, **2(2)**, 8.

11. Georghiou, G.P., The magnitude of the resistance problem. In *Pesticide Resistance : Strategies and Tactics for Management*, eds. National Academy of Science, National Academy Press, Washington DC, 1986, pp. 14-43.

12. Fitt, G.P., The ecology of *Heliothis* species in relation to agroecosystems. *Annu. Rev. Entomol.*, 1989, **34**, 17-52.

13. Fitt, G.P. and Daly, J.C., Abundance of overwintering pupae and the spring generation of *Helicoverpa* spp. (Lepidoptera:Noctuidae) in northern NSW, Australia. *J. Econ. Entomol.*, 1990, **83**, 1827-1836.

14. Daly, J.C and Fitt, G.P., Resistance frequencies in overwintering pupae and the first spring generation of *Helicoverpa armigera* (Lepidoptera : Noctuidae) : selective mortality and immigration. *J. Econ. Entomol.*, 1990, **83**, 1682-1688.

15. Fitt, G.P. and Forrester, N.W., Overwintering of *Heliothis* : the importance of stubble cultivation. *The Aust. Cottongrower*, 1987, **8(4)**, 7-8.

16.   Fitt, G.P. and Daly, J.C., The overwintering foe : winter populations of *Heliothis* in cotton growing areas and the importance of stubble cultivation. Proceedings Australian Cotton Conference, Surfers Paradise, Queensland, 1988, 13-24.

17.   Thomson, N. J., Host plant resistance in cotton. *J. Aust. Instit. Agr. Sci.*, 1987, **53**, 262-270.

18.   Wilson, L. J. and Fitt, G. P., Varietal resistance to spider mites. *The Aust. Cottongrower*, 1987, **8(3)**, 8-10.

RESISTANCE TO PHENYLAMIDE FUNGICIDES: STRATEGIES AND THEIR EVALUATION

LOUISE R. COOKE

Plant Pathology Research Division, Department of Agriculture for Northern Ireland, Newforge Lane, Belfast, BT9 5PX, UK

ABSTRACT

Strategies for the management of phenylamide resistance may be evaluated by mathematical modelling, experimentation or study of performance in practice. Modelling and experimentation both predict that fungicide mixtures slow down, but do not prevent, the build-up of resistant pathogen strains. The strategy developed by the Fungicide Resistance Action Committee's Phenylamide Working Group advocates use of pre-packed mixtures of phenylamides with non-systemic residual fungicides, a limited number of applications per season and no curative use. Strategies based on this approach for *Phytophthora infestans* and *Plasmopara viticola* in Europe have not prevented the build-up of resistant strains. Phenylamide resistance in *Bremia lactucae* has been managed more successfully by using host-plant resistance in conjunction with fungicide use. Phenylamide-resistant *Peronospora tabacina* has only now appeared on commercial tobacco crops in the USA despite 10 years' use of metalaxyl alone. Where phenylamide resistance has been managed or avoided, factors other than the purely fungicide-based elements of the FRAC strategy are responsible. Successful strategies must take account of pathogen epidemiology as well as fungicide usage.

INTRODUCTION

Resistance to phenylamide fungicides was first detected in 1980 in *Pseudoperonospora cubensis* in Israel [1]. Subsequently, crop losses caused by phenylamide-resistant *Phytophthora infestans* in the Netherlands and the Republic of Ireland in 1980 [2,3] provided a major stimulus to the development of strategies to combat resistance. These have three objectives: 1. to prevent the appearance of phenylamide resistance in pathogens and areas where it is not present, 2. where resistant strains have been detected, to prevent their build-up, 3. to prevent crop loss and achieve a continuing contribution from phenylamides to disease control.

In an attempt to achieve these objectives, the phenylamide manufacturers, cooperating in a Fungicide Resistance Action Committee (FRAC) Working Group evolved a strategy [4] with six major elements:

1. Sale of phenylamides only as pre-packed mixtures with residual fungicides.
2. Use of the residual partner at three-quarters to full rate.
3. Application interval not to exceed 14 days.
4. A limit of between two and four phenylamide applications per crop per season.
5. No curative or eradicative use.
6. No soil application for foliar pathogens.

These elements seem to have been based on an empirical and intuitive approach, but since their initial proposal attempts have been made to validate them. In this paper, I shall review how this strategy has been adapted to different pathogens and to what extent it has been effective, using the major foliar pathogens, *Phytophthora infestans*, *Plasmopara viticola*, *Bremia lactucae* and *Peronospora tabacina* as examples. Strategy evaluation will be considered under three main headings, *viz.* mathematical modelling, experimentation and effectiveness in practice.

MATHEMATICAL MODELS

The use of modelling will be considered more fully in other chapters and I shall not attempt to deal with the relative merits of different mathematical models which have been developed to evaluate strategies for limiting fungicide resistance. Such models have predicted that use of combinations of fungicides with differing modes of action will slow down the build-up of resistance compared with use of the at-risk fungicide alone. Most have suggested that use of fungicide mixtures rather than alternation is likely to give a slower rate of increase [5,6,7,8]. All agree that whilst such strategies slow down the build-up of resistance, they will not prevent it.

These conclusions have not had a direct impact on anti-resistance strategies. The decision by phenylamide manufacturers to recommend mixtures rather than product alternation, was made for practical reasons rather than on the basis of mathematical models. The use of prepack mixtures gives greater assurance of user compliance [9].

In future, models of specific host-pathogen-fungicide combinations may provide information on areas of anti-resistance strategy which particularly need critical evaluation, notably the timing of phenylamide applications within the season [10], and the risks of "critical period" or curative application. However, as yet such models have not contributed to the practical implementation of anti-resistance strategies.

EXPERIMENTAL VALIDATION

Relatively few attempts appear to have been made to evaluate experimentally the benefits of different strategies for slowing the build-up of resistance. The reasons for this are probably two-fold [11]. First, to generate valid results with airborne pathogens, it is essential to isolate populations undergoing different treatment regimes to prevent interference between populations. This necessitates the use of separate growth rooms or plastic tunnels. Second, methods available for monitoring resistance levels are rather imprecise [12].

Experiments to date all appear to relate to *P. infestans* or *P. viticola*. Staub & Sozzi [9,13] carried out a series of experiments in growth rooms and showed that mixtures of metalaxyl+mancozeb produced a slower build-up of resistance than metalaxyl alone.

Samoucha & Gisi [14] also using growth rooms, reported that a three-way fungicide mixture (oxadixyl+cymoxanil+mancozeb) prevented the build-up of phenylamide-resistant strains of both *P. infestans* and *P. viticola*. Similarly, in walk-in plastic tunnels, Cohen & Samoucha [11] found that a two-way mixture (oxadixyl+mancozeb) only slightly reduced the build-up of phenylamide-resistant *P. infestans* compared with oxadixyl alone, but that three-way mixtures produced a much slower rate of build-up. The effectiveness of three-way mixtures was attributed to enhanced systemicity and persistence of cymoxanil in the presence of oxadixyl [15], and to synergistic interactions between the fungicides [16]. Although such mixtures have been introduced commercially, data on their influence on natural pathogen populations are apparently not available, perhaps because their use has not been widespread enough.

A major problem with the interpretation of the results of small-scale experiments on resistance management is their dependence on the particular isolates selected. There is substantial variation within natural pathogen populations in characters which may influence pathogenicity. Selection of, for example, a phenylamide-resistant isolate which is fitter than the phenylamide-sensitive one, as with *P. infestans* in Israel [17,11] may profoundly influence the outcome. Caution must be exercised in extrapolating results from small-scale experiments to the behaviour of country- or continent-wide pathogen populations.

MONITORING AND STUDY OF STRATEGIES IN PRACTICE

Potato Late Blight: UK, Republic of Ireland and the Netherlands

On a world-wide scale, control of *Phytophthora infestans* probably provides the largest market for phenylamides. It is also one of the most dangerous pathogens in its potential to cause devastating losses if control fails. Inevitably an appreciation of the penalties of failure has been a major consideration when developing strategies for use of phenylamides in its control.

The UK Strategy. In the UK, phenylamide-resistant *P. infestans* was first detected in 1981 [18,19]. Metalaxyl alone was never recommended for the control of potato blight, although the first phenylamide formulation contained only a "half rate" of mancozeb. The FRAC strategy on mixtures was adopted and the rates of residual fungicides in phenylamide formulations brought into line. However, UK phenylamide manufacturers continue to recommend a maximum of five applications per crop per season, rather than the FRAC limit of two to four, although in practice most growers who use phenylamides now apply three applications per season. In the UK, a major emphasis has been on use of phenylamides from the start of the season with a switch to a non-systemic fungicide when active crop growth ceases, and the recommendation against curative use has been very firmly advocated. Use of phenylamides on seed crops has never been restricted (except in the Channel Islands).

Republic of Ireland Strategy. In 1980, phenylamide resistance was detected in crops treated with metalaxyl alone in the Republic of Ireland [3] and no further use of phenylamides was permitted until 1985 [20]. From then on the strategy adopted has been similar to that in the UK with an emphasis on early season use of phenylamides and avoidance of curative treatment. However, a maximum of three phenylamide applications has been recommended from the start of the season, and that use on seed and early potatoes is not advocated [21].

The Netherlands Strategy. In the Netherlands, resistance was first detected in 1980 [2], again in crops treated with metalaxyl alone and phenylamide use was not advocated again until 1984. The strategy subsequently adopted has been reviewed [22,23]. In contrast to the UK and Ireland, phenylamide use is only permitted when conditions are favourable to the spread of blight, when up to two consecutive applications are allowed. At all other times non-phenylamide fungicides must be used. Use of phenylamides on seed potatoes is not permitted.

Results of Strategies. In each of these countries, phenylamide resistance has been monitored annually. Strategy success has been judged in terms of trends in the proportion of resistant strains detected and on the basis of product performance in practice. In addition in the UK and Ireland, field trials have been used to estimate the likely benefits of phenylamide treatment with the prevailing resistance levels.

In both the Netherlands and the Republic of Ireland, during the period when phenylamide use was suspended, the proportion of isolates containing resistant strains declined from *c.* 80% in 1980/1 to 5-10% in 1983/4 [20,22]. However, after 1986, in both Great Britain and Ireland, the proportion of isolates containing resistant strains increased dramatically to *c.* 80% [24,25,26]. The results of the different surveys parallel each other strikingly closely. In the Netherlands, the current situation is only slightly more satisfactory. The proportion of resistant strains has been maintained at 43%-46% in 1988-90 over the country as a whole but in the area of starch potato production, where there is considerable saving of seed, it is *c.* 80%. [23,27].

Assessment of whether the phenylamides still make a worthwhile contribution to blight control has given more equivocal results. In the Republic of Ireland, on the basis of trials over six years showing benefit from phenylamide treatment, three early season applications are currently recommended by the Agriculture and Food Development Authority [21]. In England and Wales, results of ADAS trials have indicated much reduced benefit from phenylamide treatments compared with the situation in the early 1980's [24] and their use has been questioned [28], whilst in Northern Ireland trials have shown that where infection is initiated by 100% phenylamide-resistant strains, control by metalaxyl+mancozeb is only equivalent to that by mancozeb alone [29]. In the Netherlands, phenylamide treatment is not recommended in the starch-producing region [23] and the future value of treatment elsewhere is uncertain.

The results from these countries all show a failure to prevent the build-up of resistant strains. They do not provide clear evidence on the relative merits of early season versus critical period application strategies. Indeed, the similarity in resistance trends suggests that the differences between strategies have been relatively unimportant compared with the unifying influence of the behaviour of *P. infestans*. In each case, a succession of seasons with weather favourable to the spread of blight produced an increase in the detection of phenylamide-resistant strains [23,25]. The somewhat greater success in delaying the build-up of resistance in the Netherlands is probably due to their prohibition on treatment of seed crops [25]. Where ware and seed potatoes are produced in geographically distinct areas, this effectively breaks the selection cycle. However, where successive generations are produced in the same area such a prohibition is ineffective.

Thus, the success of attempts to manage phenylamide resistance in *P. infestans* in Europe has been limited to slowing down, not preventing, resistance build-up and reducing the risk of control failure. Although phenylamides may continue to have a limited role in blight control, this is far from the rosy future which was predicted for them when they first appeared. Phenylamide-resistant *P. infestans* has recently been detected in the USA in Washington State and an adaptation of the FRAC strategy is being adopted in 1991 [30]. It remains to be seen whether this will prove more effective than in Europe.

Vine Downy Mildew in Europe

The situation with *Plasmopara viticola* appears very similar to that with *P. infestans*. Phenylamide-resistant strains were first detected in 1981 in France [31,32], although phenylamides had only been used commercially in mixtures with non-systemic fungicides. By 1983, they were detected in over 50% of Bordeaux and Loire valley vineyards [33] and in nearly 100% of isolates from Cognac [34]. Artificially-inoculated trials showed that phenylamides did not contribute to disease control when resistance levels were high [33]. These findings led to a strategy of using no more than three phenylamide applications in mixture with residual fungicides, and alternating these with other fungicides [33]. Curative treatments are not recommended, although growers may still use them in practice. Despite this, phenylamide-resistant *P. viticola* has now been detected throughout Western Europe except in Spain and Portugal [35]. Thus, as in *P. infestans*, management to limit resistance has not proved very successful.

Lettuce Downy Mildew in the UK

In the UK, use of phenylamides in an integrated control strategy for lettuce downy mildew, caused by *Bremia lactucae*, has been studied by Crute and his co-workers. This work has been fully described [36,37,38]; the outline below uses information from these papers and later unpublished reports.

Metalaxyl alone was introduced for the control of lettuce downy mildew in 1978. It gave spectacularly more effective control than existing non-systemic fungicides, and was immediately widely adopted. Curative treatment became standard practice, although this was contrary to the manufacturer's recommendation. In 1980, metalaxyl alone was replaced by metalaxyl+mancozeb. Up to 1983, control of *B. lactucae* remained excellent, although contrary to recommendations, residual stocks of metalaxyl alone continued in use. In late 1983, control of *B. lactucae* failed in an intensive lettuce-producing area and this was shown to be due to a phenylamide-resistant pathotype. By 1985, resistant isolates were obtained from sites in almost every lettuce-producing region of Great Britain; circumstantial evidence linked this spread to movement of transplants.

Host plant resistance also plays an important role in controlling *B. lactucae* and thirteen dominant downy mildew (*Dm*) resistance genes have been identified. Of the initial phenylamide-resistant isolates, all proved to be of a single virulence phenotype (the "NL10 type"), lacking virulence to *Dm11*, *Dm16* and *Dm18*. Resistant isolates proved as fit as sensitive ones.

A control strategy was developed which restricted metalaxyl+mancozeb treatment to cultivars with resistance gene *Dm11*. This *Dm* gene prevented infection by the phenylamide-resistant *B. lactucae*, whilst the metalaxyl controlled any pathotype virulent on *Dm11*. In 1987, a new phenylamide-resistant pathotype was detected in the UK. This proved to be virulent on *Dm11*, but avirulent on *Dm6*, *Dm16* and *Dm18*. It initially produced control failures, but was controlled by a new strategy of metalaxyl + *Dm16* or *Dm6+Dm11* or *Dm18*.

The basic strategy may be summarised as follows (after Crute [37]):
1. Use of phenylamide in mixture with residual fungicides.
2. Consider use of alternative fungicides (fosetyl-Al) in alternation.

3. Use good cultural practices.

4. Grow cultivars carrying *Dm* genes to which phenylamide-resistant *B. lactucae* is avirulent.

5. Continue use of phenylamides regardless of phenylamide resistance problems, particularly on cultivars with appropriate *Dm* genes.

6. Avoid growing cultivars with and without specified *Dm* genes in close proximity.

The *B. lactucae* control strategy has been monitored by annual surveys in which isolates have been tested for phenylamide resistance, virulence phenotype and sexual compatibility type. These data have also been related to performance of cultivar/fungicide combinations in controlling the pathogen in practice. The decline in the occurrence of the original "NL10" phenylamide-resistant pathotype, and the continuation of good control of the disease, both provide evidence of the overall success of the approach. At present, it appears that it should be possible to continue this strategy for the foreseeable future [39], complementing the role of metaxyl by selection of appropriate *Dm* genes, providing these genes can be incorporated into commercially acceptable lettuce cultivars.

Tobacco Blue Mold in the USA

The introduction of metalaxyl in 1980 led to the adoption of fungicide treatment in the majority of US tobacco fields for the control of blue mold (*Peronospora tabacina*), black shank (*Phytophthora parasitica*) and Pythium soft rot. Phenylamide-resistant isolates of *P. tabacina* were first detected in Central America in the early 1980's and are now so widespread in this region that metalaxyl is no longer an effective fungicide [40]. Trials in Mexico with metalaxyl+mancozeb showed that the metalaxyl did not contribute to control of a phenylamide-resistant population [41].

Despite this, metalaxyl alone has continued to be recommended in the US as a soil treatment at planting [42]. The rationale is that spray application is unlikely to achieve adequate doses of fungicide within foliage of this very leafy crop. The use of metalaxyl+mancozeb foliar sprays is also unacceptable because of export market concerns. The main tool for management of the *P. tabacina* population in North America is the Blue Mold Warning System. This provides growers with advance warning of the spread of the pathogen and is also used to monitor isolates for sensitivity to metalaxyl.

Surprisingly, up to 1991, phenylamide-resistant *P. tabacina* had not been isolated from commercial tobacco crops in the US. This is probably because the pathogen does not overwinter in N. America. However, during April-June 1991 isolates of phenylamide-resistant *P. tabacina* were obtained from outbreaks in commercial crops in Georgia, Florida and South Carolina [43]. The resistant strains may have moved from Mexico via Texas, where the wild tobacco, *Nicotiana repanda*, could act as a host, to the commercial production areas of the US, or it could have originated via Cuba-Florida or Cuba-Texas [44].

The lack of phenylamide-resistant *P. tabacina* in the US up to 1991 appears to have resulted from escape rather than management. As long as there is a short cycle of selection for resistance, and any resistant strains perish in the winter, it does not much matter what phenylamide usage strategy is adopted. However, if resistant strains develop which can persist over winter, or if the weather pattern shifts so that resistant strains are introduced each season, then the use of soil treatment will carry a high risk of producing widespread control failures. US growers and advisers will now have to determine whether this is more or less unacceptable than the use of fungicide mixtures.

PROBLEMS OF PHENYLAMIDE RESISTANCE MANAGEMENT IN PRACTICE

What can be learnt from the above examples? Do strategies based on use of phenylamides in mixture with residual fungicides, and on limitation of number of applications and avoidance of curative use, work?

The evidence above suggests that strategies based solely on these fungicide-linked criteria are not effective. A number of other key factors which contribute to strategy failure or success may be identified.

Sensitivity of the pathogen to climatic conditions. With pathogens such as *P. infestans* and *P. viticola*, strategies may appear to work as long as pathogen reproduction is limited by the prevailing weather, but fail when optimum conditions permit maximum rates of reproduction, and rapid spread of the resistant phenotype.

Cycle of selection for resistance. It is often impossible to prevent the primary inoculum which initiates each season's infection cycle from receiving prior exposure to phenylamides. Only when the selection cycle can be broken, is it feasible to prevent, rather than merely slow down, the build-up of resistance. Where seed and progeny potatoes are grown in the same region, the strains of *P. infestans* which initiate each season's epidemics may have been exposed to phenylamides the previous year. In contrast, the lack of overwinter survival of *P. tabacina* in the US breaks the selection cycle.

Fitness of phenylamide-resistant pathogen strains. Although some initially-selected phenylamide-resistant strains of *P. infestans* appeared less fit [45], current phenylamide-resistant *P. infestans*, *P. viticola* and *B. lactucae* seem to have a similar level of fitness to the wild-type. Thus a short break in phenylamide exposure within a season is unlikely to result in a significant decline in the proportion of resistant strains.

Inadequacy of other control measures. The fungicides available for use in conjunction with phenylamides are far from ideal. Non-systemic fungicides leave areas such as abaxial leaf surfaces and developing foliage unprotected, providing regions where phenylamide-resistant strains can be selected. The contribution of host plant resistance to control of *B. lactucae* has allowed a successful management strategy to be developed. In contrast, host plant resistance contributes much less to potato blight control; major gene resistance has proved useless because of the rate of development of new races of the pathogen [46].

Mutability of the pathogen. *P. infestans* appears to be an inherently more mutable pathogen than *B. lactucae*; in potatoes, major gene resistance to late blight has been abandoned because the pathogen quickly overcomes it, whereas in *B. lactucae* it is still an effective control strategy. Similarly, in *P. infestans* phenylamide resistance appears to have developed many times in different genetic backgrounds [47,48], whereas in *B. lactucae* it seems to be a rare event.

CONCLUSIONS

Mathematical modelling, experimentation and results in practice all show that whilst two-way phenylamide-residual fungicide mixtures delay the initial appearance of resistance, the respite may be quite short. Once the proportion of resistant individuals within the population reaches a threshold level, it increases rapidly if phenylamides continue in use, unless there is a break in the selection cycle. However, much stress is laid on using mixtures because apart from any impact on resistance management, they minimise the chance of control failure; this may also underlie the antipathy to curative use. Such an approach benefits the grower in the short term, since otherwise he might suffer crop loss. From the advisers' and manufacturers' point of view, it reduces the risk of litigation and for the manufacturer, it may also serve to disguise the reducing contribution of phenylamides to disease control. Thus it may be argued that the basic FRAC approach is a strategy to limit the risk of short-term loss rather than to manage resistance.

Could synergistic mixtures provide a better approach? It has been suggested that even with two-way phenylamide mixtures, synergy may operate [49]. However, with *P. infestans*, *P. viticola* and *P. tabacina*, two-way metalaxyl+mancozeb mixtures give control of highly resistant populations only equivalent to that achieved by mancozeb alone [29,33,41]. With three-way mixtures, although striking effects have been demonstrated experimentally, their role in practical management of a large-scale population has yet to be demonstrated.

Where effective management strategies for phenylamide resistance have been developed, they have not relied solely on the basic FRAC guidelines, but have made use of at least one other factor to break the selection cycle. In the future, a full appreciation of the epidemiology of pathogens will be vital to developing integrated approaches to handling fungicide resistance. Strategies geared to specific pathogen-fungicide combinations are essential: strategies based purely on manipulation of fungicide use without regard to the biology of the pathogen will fail.

ACKNOWLEDGEMENTS

I am most grateful to the many people who have assisted me with information and comments; particularly Mr N.J. Bradshaw, ADAS, UK, Dr I.R. Crute, Horticultural Research International, UK, Mr L.J. Dowley, Teagasc, Republic of Ireland, Dr L.C. Davidse, Royal Sluis, the Netherlands, Dr K.L. Deahl, USDA, Maryland, USA, Professor W.E. Fry, Cornell University, USA, Mr A.J. Leadbeater, Ciba-Geigy Agrochemicals, UK, Professor M.R. Siegel, University of Kentucky, USA. I also thank my colleagues for their help in studies of *P. infestans* in Northern Ireland and in preparation of this paper.

REFERENCES

1. Reuveni, M., Eyal, H., and Cohen, Y., Development of resistance to metalaxyl in *Pseudoperonospora cubensis*. Plant Disease, 1980, **64**, 1108-9.

2. Davidse, L.C., Looijen, D., Turkensteen, L.J. and van der Wal, D., Occurrence of metalaxyl-resistant strains of potato blight in Dutch potato fields. Netherlands Journal of Plant Pathology, 1981, **87**, 65-8.

3. Dowley, L.J. and O'Sullivan, E., Metalaxyl-resistant strains of *Phytophthora infestans* (Mont.) de Bary in Ireland. Potato Research, 1981, **24**, 417-21.

4. Urech, P.A. and Staub, T., The resistance strategy for acylalanine fungicides. EPPO Bulletin, 1985, **15**, 539-43.

5. Kable, P.F. and Jeffery, H., Selection for tolerance in organisms exposed to sprays of biocide mixtures: a theoretical model. Phytopathology, 1980, **70**, 8-12.

6. Delp, C.J., Coping with resistance to plant disease. Plant Disease, 1980, **64**, 652-7.

7. Skylakakis, G., Effects of alternating and mixing pesticides on the buildup of fungal resistance. Phytopathology, 1981, **71**, 1119-20.

8. Josepovits, G. and Dobrovolszky, A., A novel mathematical approach to the prevention of fungicide resistance. Pesticide Science, 1985, **16**, 17-22.

9. Staub, T. and Sozzi, D., Fungicide resistance. Plant Disease, 1984, **68**, 1026-31.

10. Doster, M.A. and Fry, W.E., Evaluation by computer simulation of strategies to time metalaxyl applications for improved control of potato late blight. Crop Protection, 1991, **10**, 209-14.

11. Cohen, Y. and Samoucha, Y., Competition between oxadixyl-sensitive and -resistant field isolates of *Phytophthora infestans* on fungicide-treated potato crops. Crop Protection, 1990, **9**, 15-20.

12. Sozzi, D. and Staub, T., Accuracy of methods to monitor sensitivity of *Phytophthora infestans* to phenylamide fungicides. Plant Disease, 1987, **71**, 422-5.

13. Staub, T. and Sozzi, D., Recent practical experiences with fungicide resistance. Proceedings 10th International Congress of Plant Protection, 1983, **2**, 591-8.

14. Samoucha, Y. and Gisi, U., Use of two- and three-way mixtures to prevent buildup of resistance to phenylamide fungicides in *Phytophthora* and *Plasmopara*. Phytopathology, 1987, **77**, 1405-9.

15. Samoucha, Y. and Gisi, U., Systemicity and persistence of cymoxanil in mixture with oxadixyl against *Phytophthora* and *Plasmopara*. Crop Protection, 1987, **6**, 393-8.

16. Gisi, U., Synergism between fungicides for the control of *Phytophthora*. In *Phytophthora*, eds J.A. Lucas, R.C. Shattock, D.S. Shaw and L.R. Cooke, Cambridge University Press, Cambridge, 1991, pp. 361-372.

17. Kadish, D. and Cohen, Y. Competition between metalaxyl-sensitive and metalaxyl-resistant isolates of *Phytophthora infestans* in the absence of metalaxyl. Plant Pathology, 1988, **37**, 558-64.

18. Cooke, L.R., Resistance to metalaxyl in *Phytophthora infestans* in Northern Ireland. Proceedings British Crop Protection Conference - Pests and Diseases, 1981, **2**, 641-9.

19. Holmes, S.J., The incidence of metalaxyl-insensitivity in late blight of potatoes in the West of Scotland. Proceedings Crop Protection in Northern Britain, 1984, 108-13.

20. Dowley, L.J. and O'Sullivan, E., Monitoring metalaxyl resistance in populations of *Phytophthora infestans*. Potato Research, 1985, **28**, 531-4.

21. Dowley, L.J., Potato blight control: a six-year review of systemic/protectant mixtures. Oak Park Research Centre Report, Carlow, Ireland, 1991, 8 pp.

22. Davidse, L.C., Henken, J., van Dalen, A., Jespers, A.B.K. and Mantel, B.C., Nine years of practical experience with phenylamide resistance in *Phytophthora infestans* in the Netherlands. Netherlands Journal of Plant Pathology, 1989, **95** (Supplement 1), 197-213.

23. Davidse, L.C., van den Berg-Velthuis, G.C.M., Mantel, B.C. and Jespers, A.B.K., Phenylamides and *Phytophthora*. In *Phytophthora*, eds J.A. Lucas, R.C. Shattock, D.S. Shaw and L.R. Cooke, Cambridge University Press, Cambridge, 1991, pp. 349-60.

24. Cock, L.J., Efficacy of phenylamide in potato blight fungicides - England and Wales. ADAS Report, Trawsgoed, Wales, 1990.

25. Cooke, L.R., Current problems in the chemical control of late blight: the Northern Ireland experience. In *Phytophthora*, eds J.A. Lucas, R.C. Shattock, D.S. Shaw and L.R. Cooke, Cambridge University Press, Cambridge, 1991, pp. 337-48.

26. Dowley, L.J., O'Sullivan, E., Changes in the distribution of metalaxyl-resistant *Phytophthora infestans* (Mont.) de Bary in Ireland 1985-89. Potato Research, **34**, 67-9.

27. Davidse, L.C. Personal communication, 1991.

28. Anon., Spray strategy for potato blight. Farmers Weekly, Fungicides (ADAS) Supplement, 1 February 1991, p.14.

29. Cooke, L.R. and Little, G., The control of phenylamide-resistant potato blight infection by fungicide formulations. Tests of Agrochemicals and Cultivars No. 10, Annals of Applied Biology, 1989, **110** (supplement), 174-5.

30. Deahl, K.L. Personal communication, 1991.

31. Staub, T, and Sozzi, D. Résistance au métalaxyl en pratique at les conséquences pour son utilisation. Phytiatrie-Phytopharmacie, 1981, **30**, 283-91.

32. Clerjeau, M. and Simone, J., Apparition en France de souches de Mildiou (*Plasmopara viticola*) résistantes aux fongicides de la famille des anilide (métalaxyl, milfurame). Le Progrès Agricole et Viticole, 1982, **99**, 59-61.

33. Clerjeau, M., Moreau, C., Piganeau, B. and Malato, G., resistance of *Plasmopara viticola* to anilide fungicides: evaluation of the problem in France. Meded. Fac. Landbouww. Rijksuniv. Gent., 1984, **49**, 179-84.

34. Staub, T. and Diriwaechter, G., Status and handling of fungicide resistance in pathogens of grapevine. Proceedings British Crop Protection Conference - Pests and Diseases, 1986, **2**, 771-80.

35. Leadbeater, A.J., personal communication, 1991.

36. Crute, I.R., Norwood, J.M. and Gordon, P.L., The occurrence, characteristics and distribution in the United Kingdom of resistance to phenylamide fungicides in *Bremia lactucae* (lettuce downy mildew). Plant Pathology, 1987, **36**, 297-315.

37. Crute, I.R., Lettuce downy mildew: a case study in integrated control. In Plant Disease Epidemiology, Volume 2, Genetics, Resistance and Management, eds Leonard, K.J. and Fry, W.E., McGraw-Hill, New York, 1989, pp.30-53.

38. Crute, I.R. and Harrison, J.M., Studies on the inheritance of resistance to metalaxyl in *Bremia lactucae* and on the stability and fitness of field isolates. Plant Pathology, 1988, **37**, 231-50.

39. Crute, I.R., personal communication, 1991.

40. Todd, F., The blue mold situation in Central America and Caribbean region. Blue Mold Symposium III. 30th Tobacco Workers Conference, 1983, pp. 14-9.

41. Wiglesworth, M.D., Reuveni, M., Nesmith, W.C., Siegel, M.R., Kuc, J. and Juarez, J., Resistance of *Peronospora tabacina* to metalaxyl in Texas and Mexico. Plant Disease, 1988, **72**, 964-7.

42. Nesmith, W. C., Fungicide resistance in tobacco production. In Fungicide resistance in the United States, ed. P.L. Sanders, National Agricultural Pesticide Impact Assessment Program (NAPIAP), U.S. Department of Agriculture, Washington D.C., 1991, pp. 60-2.

43. Nesmith, W. C., Current Blue Mold Status, Reports 91: 3,4,5,6, April-June 1991.

44. Siegel, M.R., personal communication, 1991.

45. Dowley, L.J., Factors affecting the survival of metalaxyl-resistant strains of *Phytophthora infestans* (Mont.) de Bary in Ireland. Potato Research, 1987, **30**, 473-5.

46. Dowley, L.J., O'Sullivan, E. and Kehoe, H.W., Development and evaluation of blight resistant potato cultivars. In *Phytophthora*, eds J.A. Lucas, R.C. Shattock, D.S. Shaw and L.R. Cooke, Cambridge University Press, Cambridge, 1991, pp. 373-82.

47. O'Sullivan, E. and Dowley, L.J., Physiological specialisation in strains of *Phytophthora infestans* sensitive and resistant to metalaxyl. Irish Journal of Agricultural Research, 1983, **22**, 105-7.

48. Fry, W.E., Drenth, A., Spielman, L.J., Mantel, B.C., Davidse, L.C. and Goodwin, S.B., Population genetics of *Phytophthora infestans* in the Netherlands. Genetics, in press.

49. Gisi, U., Binder, H. and Rimbach, E., Synergistic interactions of fungicides with differing modes of action. Transactions of the British Mycological Society, 1985, **85**, 299-306.

THE INTERNATIONAL ORGANIZATION FOR RESISTANT PEST MANAGEMENT (IRPM): A FRESH COLLABORATIVE APPROACH

Bernard C. Smale
U.S. Environmental Protection Agency
401 M Street, S.W. (H7506C)
Washington, D.C. 20460, U.S.A.

ABSTRACT

The goal of the International Organization for Resistant Pest Management (IRPM) is to provide an international forum to promote the concept of pest resistance management within the context of Integrated Pest Management (IPM) systems, and to facilitate implementation programs in industrial and develop- ing nations and the emerging democracies. A General Assembly of IRPM, the First International Resistant Pest Management Congress, will be held in 1992 in Washington, D.C. The Con- gress will bring together key representatives of the institu- tions in the public and private sectors that are critical to implementing resistant pest management programs. This body will identify practical approaches to encourage and coordi- nate internationally the implementation of local resistance management programs and help to establish a continuing and expanding communication network world-wide. Programs which have been funded and are under development for implementation include Resistance Management of Apple Insects and Fire Blight in Mexico; Resistance Management of Apple Insects, Diseases and Weeds in Poland; and Resistance Management of Heliothis on Cotton in India.

BACKGROUND

Resistance to a broad range of biological and chemical mater- ials is progressively undermining their use to control agri- cultural pests and disease vectors world-wide. Although resistance is a consequence of basic evolutionary processes that occur among all classes of pests, including insects, pathogens, and weeds, its development has been accelerated by excessive dependence upon single pest control tactics. Such selection pressure, if left to continue without the interven- tion of comprehensive resistance management efforts, will eventually limit the usefulness of many present and future pest control materials. Consequently, it could handicap efforts to promote integrated pest management (IPM) approaches. Dr. John Perfect, of the Natural Resources Institute, UK, clearly linked the role of resistance management to IPM in his presentation at the Malaysia Seminar on Pest Management and the Environment in the Year 2000. "It is worth emphasizing that resistance management is likely to be the key to successful IPM in the future; the concept must embrace both resistance of pests to pesticides and resistance of crop plants to pests."[1] Both of these issues are included in the foundation principles of IRPM.

Comprehensive strategies for managing resistance, either to minimize its occurrence or mitigate its impact, have been devised and are being implemented in a number of problem areas around the world. The information and experiences available on this subject now need to be more widely converted into practical local programs, as part of IRPM efforts. Widespread cooperation is crucial because individual resistance management programs may not work effectively if tactics are not coordinated within the geographical range of the pest. Furthermore, much can be learned by sharing experiences among nations. Therefore, an international scope is necessary for the success of resistance management efforts.

OBJECTIVE OF IRPM

The goal of the International Organization for Resistant Pest Management (IRPM) is to provide an international forum to promote the concept of pest resistance management within the context of Integrated Pest Management (IPM) systems, and to facilitate implementation programs in industrial and developing nations and the emerging democracies.

A General Assembly of IRPM, the First International Resistant Pest Management Congress, will be held in 1993 in Washington, D.C. The Congress will bring together key representatives of the institutions in the public and private sectors that are critical to implementing resistant pest management programs. This body will identify practical approaches to encourage and coordinate internationally the implementation of local resistance management programs and help to establish a continuing and expanding communication network world-wide.

The IRPM has an agreement with the Agricultural Research Institute (ARI) to execute the physical and administrative affairs of the Congress. ARI, formerly with the Board on Agriculture of the National Academy of Sciences, is a consultative, non-profit institution that brings together the agricultural interests of governments, academia and industry and serves as a forum to generate cooperative action among its member institutions. Contributions to the Congress will be managed by ARI and subject to standard auditing procedures by a certified public accountant.

RATIONALE

It is the intent of IRPM to facilitate implementation programs by promoting a high degree of coordination with international development agencies, research and extension organizations in industrial and developing nations, and international organizations engaged in pesticide and IPM training
programs. Global networking and data base development are
key components of IRPM's strategy to transfer resistance
management technologies as broadly as possible. Since IRPM
programs are highly site-specific, successful implementation
hinges on farmer understanding and acceptance of the problem-
solving approach to resistant pest management. Therefore, a
high degree of farmer involvement and in-country coordination
among policy, research, and information transfer infra-
structures is critical.
     In order to leverage limited resources and avoid dupli-
cation of effort, close cooperation with FAO, WHO, and other
international organizations will allow IRPM to augment programs sponsored by these institutions by focusing international training efforts on pest resistance management within
the context of IPM systems. IRPM's commitment is projected
to be long-term with adequate provision for follow-up and
tracking of implementation efforts to assure program sustain-
ability.

WORKING GROUP CHAIRPERSONS

To achieve the IRPM goals, five working groups were established and the Planning Committee selected the following specialists in resistance management as Co-Chairs:

Insect Resistance Management
1.   Dr. G. Jackson (IRAC/GIFAP, UK)
2.   Dr. R. Frisbie (Texas A&M, US)

Plant Pathogen Resistance Management
1.   Dr. G. Lorenz (BASF)
2.   Dr. J. Northover (Agriculture Canada)

Weed Resistance Management
1.   Dr. H. LeBaron (HRAC, US)
2.   Dr. R. Gressel (Weizmann Institute of Science, Israel)

Implementation Constraints
1.   Dr. N. van der Graaff (FAO, Rome, Italy)
2.   Dr. L. Hawkins (California Department of Food and
     Agriculture, US)

Data Management and Communications
1.   Dr. S. Gage (Michigan State University, US)
2.   Dr. M. Whalon (Michigan State University, US)

The charge to each of the three Technical Working Groups is to:

- Consider the background and scope of the problem.
- Consider existing resistance management strategies.
- Consider transferable resistance management strategies.
- Develop multi-disciplinary production systems approaches across all pest classes.
- Develop detailed proposals for local implementation, taking into account such factors as current production practices; the nature and severity of the potential resistance problem; technologies appropriate to manage the problem; prevailing social, political, economic, and biological conditions; and any research necessary for successful implementation.
- Identify and recommend personnel to be involved in project implementation.
- Develop a timetable and construct a budget for project implementation.

Insect Resistance Management Working Group

This is comprised of an Insecticides/Acaricides Steering Committee and four Task Groups on Cotton, Vegetables, Tree Crops, and Public Health Pests, and held its first meeting the summer of 1990. The Chairperson of each Task Group (listed below) sits on the Steering Committee.

Cotton Task Group -- Mr. Neil Forrester
Vegetable Crop Task Group -- Vacant
Tree Fruit Task Group -- Dr. Brian Croft
Public Health Vector Task Group -- Dr. George Georghiou

The Cotton Task Group has proposed assembling multi-disciplinary investigatory teams to develop a complete profile of the candidate countries, in order to identify significant opportunities for, and constraints to, implementation. Three to four person investigatory teams will include an entomologist, agronomist, agrichemical industry representative, and a sociologist or economist. Agricultural, environmental, socio/political, market and economic factors will be addressed in the profile. Networks will be developed within government, international development organizations, research entities, extension personnel, the pesticide industry, and farmers. Teams will be assembled to address resistance management in:
- West Africa and Madagascar
- Thailand
- Latin America
- India/Pakistan/Soviet Union
- China

The first meeting of the Tree Fruit Task Group was held September 11-13, 1990 in Portland, Oregon. The group has proposed implementation projects for mite resistance management on apples in Mexico; for mite and disease resistance on apples in Poland; and for mite and disease resistance on citrus in Brazil.

The Public Health Vector Task Group, under Dr. Georghiou's leadership, held its first meeting in Geneva, Switzerland in March 1991 in the offices of WHO.

Dr. Edward Cheng, of the Taiwan Agricultural Research Institute, as past chairman of the Vegetable Crop Task Group did an excellent job of organizing and guiding the Group and convening their first meeting. Unfortunately, personal reasons require that he resign his position. The Vegetable Crop Task Group and other participants in IRPM thank him for his significant contribution to this effort and hopefully welcome his future contributions. The Insecticides/Acaricides Steering Committee is in the process of selecting a new chairperson.

Plant Pathogen Resistance Management Working Group

The Plant Pathogen Resistance Management Working Group held its first meeting in November 1990 in Brighton, England. This Working Group is targeting disease resistance management programs on cereals in Western Europe, Russia and Brazil; rice in the Philippines, and apples in Poland and Mexico. This latter program would be conducted jointly with the Insect Resistance Management Working Group.

Weed Resistance Management Working Group

The first meeting of the Weed Resistance Management Working Group was to be held in March 1991 in London, but was cancelled for security reasons associated with the Persian Gulf war. The meeting was rescheduled and held July 10-12, 1991 at Gatwick, UK.

Implementation Constraints Working Group

The Constraints Working Group is charged with considering government, agricultural, sociological, political, economic, ecological and regulatory constraints to implementing the resistance management programs on a broad scale, as well as recommendations for circumventing those constraints. IRPM's long-term goals of establishing effective resistance management programs as components of integrated pest management (IPM) and sustainable agricultural programs dictate totally new approaches to technology transfer and grower training. This Working Group will develop strategies for broad scale, efficient transfer of resistance management organization, considering such options as established networks of international volunteer organizations.

Data Management and Communications Working Group

The Communications Working Group is charged with establishing a global data base on pest resistance management. Such a data base is expected to play a key role in global networking between growers, researchers, extension, and plant protection specialists in developing and industrialized nations.

CURRENT ACTIVITIES AND PROGRESS

During the past 18 months or so since the official formation of IRPM, we have made great strides toward our stated objectives of providing an international forum (1) to promote the concept of resistant pest management within the context of IPM systems and (2) to identify and facilitate implementation of resistant pest management programs in industrial and developing nations and emerging democracies.

Leadership of the Technical Working Groups by Drs. Jackson and Frisbie (Insecticides/Acaricides); Drs. Lorenz and Northover (Pathogens); and Drs. Gressel and LeBaron (Weeds) has been outstanding. Their choices of working group members and approaches to program management provided the technical expertise, opportunity and incentive for sound project development.

The agrichemical industry and USEPA contributed $135,000 to IRPM. USAID provided $25,000 for development of the Mexican apple project with additional funds ($35,000) for the project forthcoming from the Organization of American States (OAS) and UNIFRUT, the apple growers group. Funding for the India (cotton) and Poland (apple) projects is provided through OICD/USDA under US Public Law 480. The expanded project development meetings in-country with Indian (1 week) and Polish (2 weeks) specialists and government staff are, because of their duration, estimated to cost about $65,000 in PL-480 funds. Total receipts, both cash and PL-480 funding, to date are approximately $260,000.

Disbursements of approximately $115,000 were required for travel and expenses of the 90 members of the three technical working groups (five meetings) and the Charter Working Group (two meetings). Administrative costs incurred by the Agriculture Research Institute total $15,000. Because of the common aims of WRCC-60 and IRPM, the executive committee has, at the request of WRCC-60, become a supporting member and has provided $4,000 to aid in publication of the newsletter.

REPORT ON ACTIVITIES OF THE TECHNICAL
WORKING GROUPS

The Tree Fruit Task Group (TFTG) of the Insecticides/ Acaricides Steering Committee and Pathogen Resistance Working Group (PRWG). Funding provided by USAID, and UNIFRUT.

A. Mexico: Resistance Management of Apple Insects and Fire Blight. Drs. Brian Croft and John Northover.
The TFTG and PRWG will jointly convene a working research/- education meeting in November of 1991 that will include rep- resentatives from several key groups concerned with manage- ment of resistant pests of tree fruits in Mexico. We will also involve other key leaders in this field of work from countries of Central and South America. The initial meeting group would include:
1. Officials of the grower group, UNIFRUT, which has headquarters in Chihuahua City and maintains a research sta- tion near the center of the fruit region.
This group will provide resources to support the project now that USAID support has been committed. From their facility at Cuauhtemoc, UNIFRUT provides pest monitoring, tree fruit nutrition analysis, several biological control agents for release and other related IPM services.
2. Representatives from the Mexican National Institute of Agriculture and Forestry Research (INIFAP). These scien- tists are anxious to expand their ability to deliver technol- ogy to growers. They have a field station in the nearby val- ley that could be used for demonstration programs.
3. One or two key leaders in this field from Central or South America will be invited to the meeting to establish liaisons with other international programs. A good example of such a person is Dr. Roberto Gonzales from Santiago Uni- versity in Chile. He worked for many years as an interna- tional specialist in agricultural research at FAO in Rome and has been a leader in tree fruit pest control in South America for many years now.
4. A small team of IRPM scientists who represent speci- alists with expertise in the following areas of resistance pest management (these areas were identified as priorities in earlier discussions in Mexico): a) codling moth resistance monitoring and management, b) pesticide selectivity evalua- tion and releases of pesticide-resistant populations of the beneficial wasp, _Trichogramma_ _prateosum_, c) resistance moni- toring and management of acaricides in spider mites, d) pes- ticide selectivity and release of insecticide-resistant pre- datory mites, e) resistance monitoring and management of insecticides in the wooly apple aphid, _Eriosoma_ _lanigerum_, and f) resistance monitoring of streptomycin and resistance management in fire blight disease.
In our meeting we will identify research/implementation teams, identify sites of work and design experiments and edu- cational programs to implement programs of resistance manage- ment for the pest complex groups cited above. Meeting agenda

development and coordination will be handled by Dr. Brian Croft, of Oregon University and Dr. Carlos Garcia from INIFAP. Dr. Garcia will act as a liaison between the people from the UNIFRUT and INIFAP organizations. At the meeting, small teams having representatives from both INIFAP and UNIFRUT would be identified who will carry out the proposed program.

B. Brazil: Resistant Management of Two Mite Pests and One Fungal Pathogen on Citrus. Dr. Tim Dennehy. Funding is being negotiated and will probably be available by end of 1991.
IRPM, in cooperation with the established citrus IPM program in the State of Sao Paulo, plans to establish and implement a provisional resistance management strategy targeting two mite pests and one fungal pathogen. The resistance management strategy will involve rotations of different classes of acaricides and fungicides and will be implemented via a multi-tactic IPM program that uses chemicals only when economically justified and maximizes biological control ("soft" insecticides, conservation of Hirsutella thompsoni), monitoring key pests and use of reasonable thresholds.

Within the large implementation project, we will conduct large-plot, replicated evaluations of the benefit of the resistance management strategy (i.e., chemical rotations plus other integrated management techniques employed). Definitive evaluation of the benefit of the resistance management strategy will be made by bio-assaying the changes in frequency of resistant pests. Evaluation trials will be conducted at a subset of the locations where the larger implementation program is being conducted.

Pests Targeted for Resistance Management are:

Citrus leprosis mite	Brevipalpus phoenicis
Citrus rust mite	Phyllocoptruta oleivora
Citrus scab	Elsinoe australis, E. fawcetti

C. Poland: Resistance Management of Apple Insects and Diseases. Drs. Gisela Lorenz and David Pree. Funding provided by Poland under the authority of US Public Law 480, as managed by OICD/USDA.
The TFTG and PRWG will jointly convene a two to three week working research education meeting in Poland with key representatives of the government, academic, and grower communities.

The first phase of the September 23-27 visit of Drs. David Giles, Gisela Lorenz, David Pree and Wayne Wilcox of IRPM with Drs. Kropcznynska, Bielenin and other Polish scientists and growers will entail a three to four day field trip to various apple production areas to observe first-hand the insect and disease problems and select potential implementation sites. The second phase will involve development,

by the IRPM and Polish scientists, of a detailed resistance
implementation proposal.

**The Vegetable Crop Task Group of the Insecticides/Acaricides
Steering Committee.**

**Central America-Caribbean Resistance Management of Diamond-
back Moth on Crucifers. Drs. Keith Andrews, Janice Reed,
Ronald Estrada and Jeff Waage. Project not funded.**
Insect pests and diseases threaten cabbage and broccoli production throughout the Central American and Caribbean countries. Resistance is a major problem which complicates the
availability and value of plant protection materials. Pilot
extension programs have shown that several nonpesticide
alternatives are highly cost-effective means to reduce insecticide use. Pesticide use in crucifers can be rationalized
without jeopardizing production. Public concern, grower des-
peration, industry support, the political will, and technological capabilities all exist. A multi-national crucifer
IPM outreach program aimed at mitigating resistance problems
would have a high probability of success, and would create
momentum for future regional IPM efforts in vegetables.

Farmers and technicians who learn pesticide resistance
management procedures in crucifers will transfer them to
other horticultural crops. Developed country consumers and
importers will benefit from produce which is in conformance
with established tolerances.

The program, with funding currently under negotiation,
is conceived as a five-year long project with a ten-year
horizon. A networking arrangement will link the isolated,
underfunded implementation efforts underway in Central America and the Caribbean. Honduras will lead the effort with
Guatemala, Jamaica and Trinidad-Tobago as key collaborating
countries. These countries have all made progress in
researching and implementing certain components of crucifer
IPM.

Resistance management programs will make use of different combinations of the following alternatives: reduced use
of synthetic insecticides; rotations of synthetics, botani-
cals and microbials; mosaic spraying; use of synergists; use
of complementary biological and cultural controls. An internationally recognized specialist will monitor resistance levels in all implementation sites using appropriate techniques.
The outreach programs will be living laboratories in which
large numbers of technology transfer specialists from other
Central American and Caribbean countries will receive in-ser-
vice training and obtain validated training materials.

The Cotton Working Group of the Insecticides/Acaricides Steering Committee.

India: Resistance Management of Heliothis on Cotton. Dr. Neil Forrester. Funding provided by India under the authority of US Public Law 480, as managed by OICD/USDA.
A group meeting sponsored by the Indian Council of Agricultural Research (ICAR), International Organization of Resistance Pest Management (IRPM), and the Far Eastern Regional Research Office (FERRO) of the United States Department of Agriculture will be held in Hyderabad from October 14-18, 1991.

Insecticide resistance has become a major limiting factor to economic cotton production world-wide. Since synthetic organic insecticides were first used, insect pests attacking cotton have developed resistance to virtually all classes of insecticides. The development of pyrethroid resistance by Heliothis spp. has caused great concern and economic hardship in major cotton-producing areas of Asia, Australia, Central America, USA and USSR. In addition to Heliothis spp., pink bollworm (Pectinophora gossypiella), white fly (Bemisia spp.), aphids, spider mites (Tetranychus spp.) and other insects have become resistant to a wide range of insecticides/acaricides.

In India, in the last few years, the outbreaks of Heliothis in Andhra Pradesh, Tamil Nadu and more recently in Punjab and Haryana have caused great alarm. A very high degree of resistance to pyrethroids has been found in Heliothis populations in all the major cotton growing areas of the country. In addition, resistance has also been reported in Heliothis on pigeonpea and chickpea. A sound resistance management strategy is needed to ensure the long-term effectiveness of all classes of pesticides, especially synthetic pyrethroids.

The main objective of this meeting is to facilitate the development of a resistance management program for India. The meeting would address itself to the following:

1. Determine the nature and extent of the resistance problems in India.

2. Review cotton culture and existing methods of pest management within India.

3. Identify existing technologies from around the world that may serve to guide resistance management research programs.

4. Under the guidance of representatives from India, identify regions within the country where pilot research programs may be implemented.

5. Develop research priorities and a timetable of activities for the pilot programs.

6. Determine opportunities and constraints for the implementation of resistance management research results through a series of demonstrations.

Five members of the Working Group (Drs. Forrester, Matthews, LeRumeur, Frisbie, Thomas and Smale [IRPM Executive

Director]) were invited to attend a USDA-sponsored meeting on resistance problems in Indian cotton last February (funded via the Office of International Cooperation and Development, Far Eastern Regional Research Office). However, the Gulf War intervened and this has now been postponed to October 14-18, 1991. It is proposed to hold a joint three-day workshop with Indian researchers at Hyderabad, followed by a brief field visit to the cotton belt of Guntur/Prakasham districts in Andhra Pradesh. The main objective of the meeting is to facilitate the development of a resistance management program for India.

EXECUTIVE AND HOST NATION PLANNING COMMITTEES

Dr. Bernard C. Smale*
Executive Director, IRPM
USEPA

Dr. Keith Brent*
Chairman of the Congress
Long Ashton Res. Stn.
(UK)

Mr. Ken Byrd
US Peace Corps

Mr. Wm. Stanwood Cath*
Treasurer, IRPM
Agricultural Research
Institute

Dr. Charles Delp
Consultant

Dr. Stanford W. Fertig
USDA/ARS

Dr. Ray Frisbie*
Texas A&M

Dr. William Furtick
USAID

Dr. Stuart Gage*
Michigan State University

Dr. Jonathan Gressel*
Weizmann Institute of
Science (Israel)

Ms. Faith L. Halter, J.D.
USEPA

Dr. Lyn Hawkins*
California Dept. of Food
and Agriculture

Dr. Fumikiko Hayashi
USEPA

Dr. Richard A. Herrett*
ICI Americas Inc.

Ms. Maureen K. Hinkle
National Audubon Society

Dr. William L. Hollis*
Secretary, IRPM
USDA/ARS

Dr. Polly Hoppin*
The Conservation Foundation

Dr. Diana Horne
USEPA

Dr. Geoffrey J. Jackson*
Insecticide and Fungicide
Resistance Action Committee (UK)

Mr. Edwin L. Johnson
USEPA

Dr. Agi Kiss
World Bank

Dr. Homer M. LeBaron*
Ciba-Geigy Corporation

Dr. Gisela Lorenz*
BASF Ag

Dr. Julius J. Menn
USDA/ARS

Dr. John Northover*
Agriculture Canada

Dr. Walter T. Reed
DuPont Co.

Dr. T. Saito*
Nagoya University (Japan)

Dr. Ashock Seth*
World Bank

Dr. Allen L. Steinhauer
University of Maryland
Consortium for Interna-
tional Crop Protection

Dr. N. van der Graaff*
FAO (Italy)

Dr. Benjamin H. Waite*
USAID

Dr. Mark E. Whalon*
Pesticide Research Center
Michigan State University

* Executive Committee of the Congress

REFERENCES

1. Perfect, J., IPM in 2000 AD. In Pest Management and the
Environment in the Year 2000. Proceedings of a seminar held
in Kuala Lumpur, Malaysia, May 7-8, 1991, Agricultural Institute of Malaysia, 1992, in press.

WHAT DO WE REALLY KNOW ABOUT
MANAGEMENT OF INSECTICIDE RESISTANCE?

BRUCE E. TABASHNIK
Department of Entomology
University of Hawaii
Honolulu, Hawaii 96822 U.S.A.

JAY A. ROSENHEIM
Department of Entomology
University of California
Davis, California 95616 U.S.A.

MICHAEL A. CAPRIO
Department of Entomology
University of California
Berkeley, California 94720 U.S.A.

ABSTRACT

Evidence is presented that challenges common beliefs about
the roles of generation time and dispersal in evolution of
insecticide resistance. Previously reported empirical and
theoretical analyses have concluded that resistance develops
faster as the number of generations per year increases.
Analysis of data for 682 species of North American arthropod
pests shows no direct relationship between generation time
and resistance development. Reevaluation of a relatively
simple analytical model suggests that the rate of resistance
development is independent of generation time. Simulations
show that interactions with genetic, ecological, and
operational factors can cause variable relationships between
generation time and resistance development. Dispersal of
susceptible individuals from untreated to treated habitats
has been viewed as a potentially important influence that
retards resistance development. Analysis of allozyme
variation and geographical variation in resistance in
diamondback moth suggests that gene flow in this mobile
insect is too low to substantially delay resistance.
Conversely, gene flow is sufficiently high to introduce
alleles for resistance into susceptible populations from
resistant populations. Thus, the intermediate levels of gene
flow that typically occur among field populations may
promote, rather than retard, resistance development.

INTRODUCTION

The goals of resistance management are to slow the evolution of pesticide resistance in pests and to promote it in beneficial organisms. To manage resistance, we must first understand how the rate of resistance development in field populations is affected by various factors. Ideally, this knowledge can then be applied to achieve the desired results.

Our understanding of how various parameters influence the rate of resistance development has come primarily from three investigative approaches: modeling, laboratory experiments, and field studies. Each of these approaches has yielded important insights, yet each has serious limitations.

Models tell us what is possible, and what will happen if certain conditions are met. In most cases, however, many of the assumptions in models are not based on data. Rarely are predictions from models tested against observations from the field [1, 2].

Rates of resistance development have been measured in numerous laboratory selection experiments. Key differences between conditions in the laboratory and field, however, suggest that extrapolation of management strategies from laboratory results to field populations may not be valid [3]. In general, laboratory populations are relatively small. If alleles conferring resistance are rare in field populations, they may be absent in laboratory populations. Further, the spatial and temporal variation in pesticide dose that is characteristic of field conditions may be greatly reduced or absent in the laboratory.

In principle, the best way to test tactics for resistance management is to conduct controlled, replicated field experiments [4]. Yet, the utility of this approach is limited by practical constraints. Field experiments can worsen resistance problems by increasing resistance levels in field populations of pests [1]. Other drawbacks include the long time needed for resistance experiments in the field (perhaps several years to a decade or more), the difficulty in isolating treatments spatially so that gene flow doesn't obscure differences between treatments, and the expense and logistics of obtaining sufficient replication. Review of several resistance management programs in cotton shows that without experimental controls, success or failure of a program cannot be reliably attributed to any particular tactic [5].

Because of the limitations described above, our understanding of how most factors affect rates of resistance development in the field is limited, even though much information is available regarding resistance mechanisms, genetics, and spatial patterns [6-9]. We do know that resistance development is promoted by frequent insecticide use [4, 10]. Some of the more complex "principles" upon which insecticide resistance management is founded may be, at best, not rigorously proven, and at worst, incorrect. To illustrate this point, we present evidence here that challenges widely accepted views about the role of generation time and migration in resistance development.

GENERATION TIME

The conventional wisdom has been that generation time is one of the few factors that exerts a strong and consistent influence on the rate of evolution of resistance. Previously reported empirical and theoretical analyses have concluded that resistance develops faster as the number of generations per year increases [2, 11-15]. Here we review a new empirical analysis that challenges this conclusion [16]. We examine previous theoretical arguments and show that they fail to demonstrate a consistent relationship between generation time and resistance development [16]. We also present results of computer simulations showing a highly variable and complex relationship between generation turnover and resistance development [17].

The new empirical analysis was based on data for 682 species of arthropod pests from North America. Davidson and Lyon's [18] extensive list of pest species was the foundation for the data base. A resistance score was calculated for each species by counting the number of insecticide or acaricide classes to which at least one North American field population had been reported as resistant, using the resistance documentation list and pesticide classifications of Georghiou [19]. Nineteen percent of the 682 species had resistance scores >0. Estimates of the number of generations per year for each species were compiled from the literature.

Regression analysis showed that, in general, the number of generations per year had little effect on resistance development. Considering all 682 pest species, variation in generations per year accounted for only 1% of the variation in resistance score.

Pest severity, one of several other independent variables considered in the analysis, showed a consistent positive effect on resistance score (r^2=0.14, P<0.001). Generations per year and pest severity were almost completely uncorrelated (r^2=0.026). Thus, species with more generations yearly were not generally more severe pests, and pest severity did not hide the effect of generations per year in the analysis.

Species with intermediate numbers of generations per year generally had high resistance scores, whereas those with <5 generations or >12 generations per year had somewhat lower scores. Polynomial regression showed that the relationship between generations per year and resistance development was non-linear and highly variable (r^2=0.06, P<0.001 for linear, quadratic, and cubic terms.).

Consideration of eight subsets of pests within the data base (agricultural pests, stored products pests, pests of human and animal health, key pests, cotton pests (Fig. 1), grain and corn pests, apple and pear pests, solanaceous crop

pests) also showed no consistent relationship between
generation turnover and resistance development. In the one
exceptional case, however, generations per year explained a
significant portion of the variation in resistance
development for pests of apple and pear ($r^2=0.09$, P=0.02,
n=56 species). This exception is notable because the rate of
development of resistance to azinphosmethyl for 12 pests and
12 natural enemies in North American apple orchards was
directly proportional to generations per year [13]. Despite
being derived from different measures of resistance
development and different samples of arthropod species, both
analyses suggest a positive effect of generations per year on
resistance development for apple arthropods. None of the
apple or pear pests had >12 generations per year. The
positive effect observed may therefore simply reflect the
increasing trend in rates of resistance development for
species with low to intermediate numbers of generations per
year (see above).

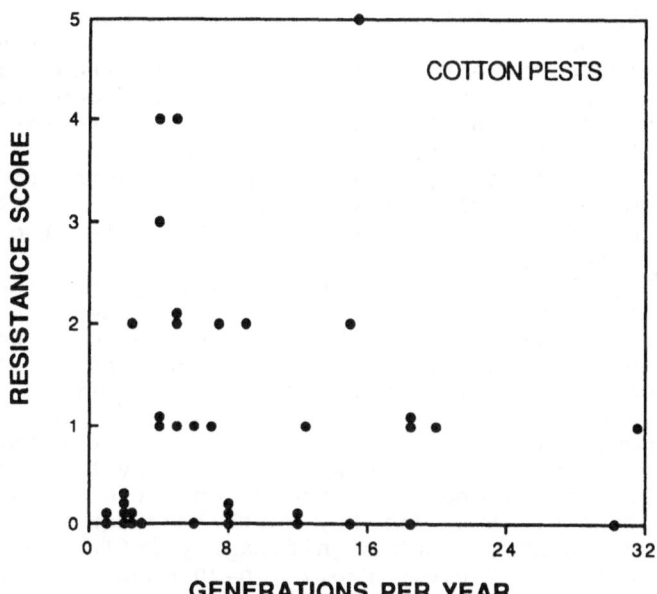

FIGURE 1. Relationship between generations per year and
resistance for North American cotton pests. Resistance score
is the number of classes of insecticides and acaricides to
which resistance has been documented. For the 8 cases in
which 2 to 4 species have identical coordinates, points are
stacked vertically to enhance visualization of the data. The
actual resistance score for all points in a stack is the
whole number represented by the point at the bottom of the
stack. $r^2 = 0.000$, n= 38 species, P > 0.9 [16]

Previously reported empirical evidence supporting a positive relationship between generations per year and resistance evolution is of limited scope. Georghiou [12] reported a linear relationship between rate of development of resistance to aldrin or dieldrin and generations per year for seven soil-dwelling insect species. Generations per year and time for field resistance development was specified for only three arthropod species by May & Dobson [14].

The analytical model of May and Dobson [14] provides one of the most lucid expressions of the dominant view that the time for resistance development is linearly related to generation time. The key assumption of their model (and others that reach the same conclusion; see [20]) is that the selection intensity per generation is the same for species with different generation times. This assumption may be valid for laboratory selection experiments, but it does not generally hold in the field, as demonstrated by the following example.

Consider two pests that occupy the same grower's field: Species A has one generation per year, whereas species B has two. The two species are alike in all other aspects, including their susceptibility to insecticide. The grower treats his field twice yearly. Species A is treated twice per generation, while B is treated only once per generation. Thus, the selection intensity per generation is greater for the species with longer generation time. The selection intensity per year, however, is independent of generation time--both species are selected twice yearly. Thus, all else being equal, the rate of resistance development is also independent of generation time.

Comins [11] also reasoned that if pesticide treatments imposed a constant additional daily mortality, then the rate of selection for resistance would be independent of a pest's generation time. He further assumed, however, that pests with shorter generation times were treated more often. This assumption may apply in a few cases, but there is no causal link between generation time and treatment frequency for the vast majority of pests which are not the primary targets of pesticide applications, and therefore do not affect the timing of treatments. Even for key North American pests, resistance development was not significantly influenced by generations per year (r^2=0.01, P=0.72, n=70 species, [16]).

Although reevaluation of simple, heuristic models suggests that the rate of evolution of resistance is independent of generation time, simulations show that generation time can interact with other factors to influence resistance development [16]. For example, if the criterion for resistance includes not only a genetic component (e.g., R allele frequecy >0.50), but also a population density component (e.g., economic threshold of pest exceeded), then species with shorter generation times will attain resistance faster. This occurs because their population density increases more rapidly; the rate of change in allele frequency is still similar across species.

TABLE 1
Summary of effects of number of generations per year (GPY) on
the rate of resistance development under various conditions
[9]

Conditions	Effect of GPY on rate of resistance development[a]
Direct effect of GPY	0
Population density criterion	+
Fitness cost	-
Frequent treatments, high dose	-
Life stage refuges[b]	+
Gene flow[c]	+

[a] 0 = no effect, + = resistance evolves faster as GPY increases, - = resistance evolves slower as GPY increases

[b] assumes one or more life stages (e.g., pupa) are not exposed or not susceptible to pesticide

[c] assumes fixed daily rate of immigration of susceptibles

In some cases, a fitness cost is associated with resistance, so that the frequency of the R allele declines in the absence of treatments [21]. If the fitness cost of resistance per generation is independent of generation time, then species with more generations per year will be affected more by this cost. Thus, the rate of resistance development will decline as generations per year increases. Because the relationship between generations per year and rate of resistance development can be positive or negative (Table 1), certain combinations of assumptions can produce non-linear relationships (Fig. 2) that mirror the pattern observed in the empirical data for North American arthropod pests [16].

In summary, the conventional wisdom that species with more generations per year develop resistance more rapidly was not supported by analysis of an extensive data base nor by reevaluation of theoretical models. In the simplest case, resistance development is independent of generation time. Interactions with other factors, however, can cause positive, negative, and non-linear relationships between generation time and resistance development.

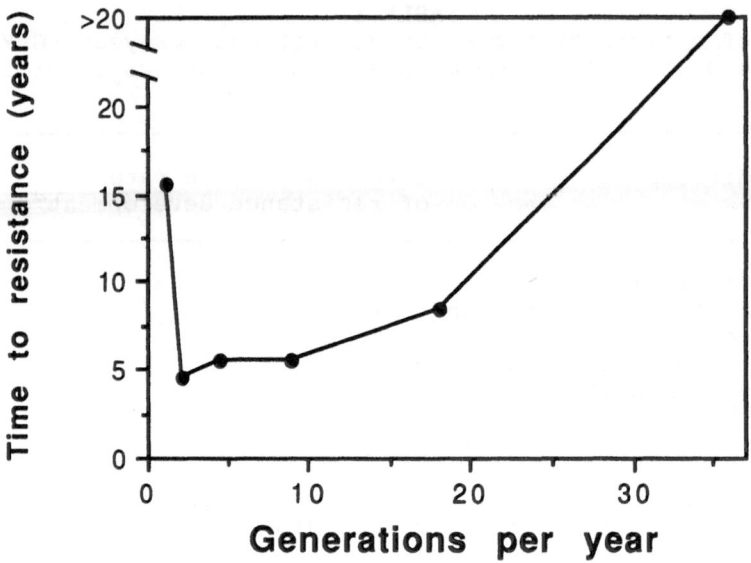

FIGURE 2. Example of a multiple-factor interaction that
results in a U-shaped relationship between generations per
year and the time to evolve resistance. Simulation
assumptions: one spray per year, a moderate fitness cost of
resistance in the absence of pesticide. Criterion for
resistance: R allele frequency > 0.50 *and* economic threshold
of pest exceeded [9]

DISPERSAL

Many authors have depicted dispersal as a factor that can
slow the evolution of insecticide resistance in the field.
Modeling studies have shown that immigration of susceptible
individuals into treated areas retards resistance development
by increasing the frequency of susceptible alleles in a
treated population [20, 22-25]. Although most considerations
of tactics for resistance management have stressed this
effect of dispersal (e.g., 6), increased dispersal can also
promote resistance development [26, 27]. Indeed, theoretical
analyses and empirical estimates of gene flow suggest that
dispersal may generally enhance rather than retard resistance
development in field populations.
     Theoretical analyses that emphasize the potential for
dispersal to slow resistance development focus on movement
from untreated to treated areas, but they generally ignore
movement in the opposite direction. In one of the first
modeling studies of resistance, however, Comins [22] analyzed
a system with a treated area, an untreated area, and movement
between the two areas in *both* directions. In this type of
system, immigration of susceptible individuals into the
untreated area slows resistance development, but emigration
of resistant individuals from the treated area speeds

resistance development in the untreated area. If treated and
untreated areas have similar population sizes and the
proportion of individuals migrating is equal for the two
areas, then intermediate levels of migration are optimal for
delaying resistance.

Stochastic computer simulations showed that interactions
among gene flow, population size, initial genetic variation,
and environmental heterogeneity affected resistance
development in subdivided, finite populations [28]. The
simulations were based on empirical data from the diamondback
moth, Plutella xylostella, a mobile insect that is notorious
for its ability to evolve insecticide resistance [29, 30].

The simulations showed that if alleles for resistance
are sufficiently rare, then gene flow generally will promote
resistance development. If resistance alleles are present in
relatively few finite subpopulations, evolution of resistance
in subpopulations that lack alleles for resistance cannot
proceed until alleles for resistance are generated by
mutation in situ or are introduced by gene flow. The
tendency for dispersal to speed resistance development
becomes more pronounced as the initial frequency of the
resistance allele declines. Increases in gene flow from very
low levels (0.025% to 2.5% of each population exchanged per
generation) increased the rate of resistance development in
simulations when the initial frequency of the resistance
allele was 2×10^{-5}. With an initial resistance allele
frequency of 2×10^{-2}, however, gene flow had no significant
effect on resistance development.

Simulations of up to 49 subpopulations, including
approximately equal numbers of treated and untreated
subpopulations, showed that high rates of gene flow (>10% of
each population exchanged per generation) can slow resistance
development, but only at the expense of greatly increasing
the resistance allele frequency in untreated subpopulations.

The simulations suggest, therefore, that depending on
the magnitude of dispersal and initial levels of genetic
variation, increases in dispersal can have positive, neutral,
or negative effects on resistance development. To predict
the influence of changes in the extent of dispersal, one must
estimate initial genetic variation and gene flow in field
populations.

Lacking direct evidence, researchers have assumed that
initial resistance allele frequencies range from 10^{-13} to
10^{-2} [3]. Measurements of genetic variation among insect
populations suggest that gene flow rarely exceeds 10% per
generation [31, 32]. Analysis of allozyme variation among
diamondback moth populations suggests that gene flow is less
than 1% per generation [33]. Thus, gene flow in diamondback
moth is high enough to spread alleles for resistance when
they are rare, but it is too low to counteract selection for
resistance by markedly reducing the frequency of resistance
alleles in treated populations.

Precise predictions about the effect of dispersal on
resistance development require accurate estimates of several
parameters including selection intensity, initial genetic

variation, and population size. Nonetheless, we suggest that in most cases, gene flow among field populations is rarely sufficient to substantially slow resistance development by introducing susceptible individuals into treated areas. This argument is supported by the highly localized patterns of resistance development that occur even in relatively mobile insects such as diamondback moth and pear psylla [4, 30, 34]. Theoretical analyses show that increased dispersal can promote the spread of resistance when alleles for resistance are rare. This idea is supported by evidence that resistance to organophosphate insecticides in Culex pipiens is conferred by amplified esterase B2 genes that originated from an initial amplification event and subsequently spread throughout the world by migration [35, and Raymond and Pasteur in this volume].

CONCLUSION

We have presented evidence that challenges widespread beliefs about the influences of generation time and dispersal on resistance development.

A reevaluation of available empirical and theoretical analyses suggests that the rate of resistance development does not increase linearly as the number of generations per year increases. Among North American pests, species with intermediate numbers of generations per year had the highest resistance scores. Simulations showed that various factors can interact to produce a variable and non-linear relationship between generation time and resistance development.

The potential role of immigration by susceptibles in impeding resistance development has been discussed often. Estimates of gene flow from field populations, however, suggest that dispersal is not generally high enough to substantially slow the response to selection in treated areas. Conversely, the most significant effect of dispersal may be to spread alleles for resistance from resistant to susceptible populations, thereby promoting resistance development.

In summary, the effects on resistance development of two key factors, generation time and dispersal, are more complex and perhaps qualitatively different than what has been widely believed. If our assumptions about such apparently simple factors are uncertain or incorrect, then how well can we answer more difficult questions about resistance management? Although certain tactics may be successful with a particular pest and pesticide in a particular location, our ability to generalize about management of insecticide resistance is still very limited. There is only one general principle of resistance management in which we have the utmost confidence: reducing pesticide use will slow resistance development.

REFERENCES

1. Taylor, C.E., Evolution of resistance to insecticides: the role of mathematical models and computer simulations. In Pest Resistance to Pesticides, eds. G.P. Georghiou and T. Saito, Plenum, New York, 1983, pp. 163-73.

2. Tabashnik, B.E., Modeling and evaluation of resistance management tactics. In Pesticide Resistance in Arthropods, eds. R.T. Roush and B.E. Tabashnik, Chapman and Hall, New York, 1990, pp. 153-82.

3. Roush, R.T. and McKenzie, J.A., Ecological genetics of insecticide and acaricide resistance. Annu. Rev. Entomol., 1987, 32, 361-80.

4. Tabashnik, B.E., Croft, B.A., and Rosenheim, J.A., Spatial scale of fenvalerate resistance in pear psylla (Homoptera: Psyllidae) and its relationship to treatment history. J. Econ. Entomol., 1990a, 83, 1177-83.

5. Sawicki, R.M. and Denholm, I., Management of resistance to pesticides in cotton pests. Trop. Pest Manag., 1987, 33, 262-72.

6. National Research Council, Pesticide Resistance: Strategies and Tactics for Management, National Academy of Sciences, Washington, D.C., 1986.

7. Ford, M.G., Hollomon, D.W., Khambay, B.P.S., and Sawicki, R.M., eds., Biological and Chemical Approaches to Combating Resistance to Xenobiotics, Ellis Horwood, Chichester, 1987.

8. Green, M.B., LeBaron, H.M., and Moberg, W.K., eds., Managing Resistance to Agrochemicals From Fundamental Research to Practical Strategies,, American Chemical Society, Washington, D.C., 1990.

9. Roush, R.T. and Tabashnik, B.E., eds., Pesticide Resistance in Arthropods, Chapman and Hall, New York, 1990.

10. Rosenheim, J.A. and Hoy, M.A., Intraspecific variation in levels of pesticide resistance in field populations of a parasitoid, Aphytis melinus (Hymenoptera: Aphelinidae): the role of past selection pressures. J. Econ. Entomol., 1986, 79, 1161-73.

11. Comins, H.N., The management of pesticide resistance: models. In Genetics in Relation to Insect Management, eds. M.A. Hoy and J.J. McKelvey, Jr., Rockefeller Foundation, New York, 1979, pp. 55-69.

12. Georghiou, G.P., Insecticide resistance and prospects for its management. Residue Reviews, 1980, 76, 131-45.

13. Tabashnik, B.E. and Croft, B.A., Evolution of pesticide resistance in apple pests and their natural enemies. Entomophaga, 1985, 30, 37-49.

14. May, R.M., and Dobson, A.P., Population dynamics and the rate of evolution of pesticide resistance. In Pesticide Resistance: Strategies and Tactics for Management, National Academy of Sciences, Washington, D.C., 1986, pp. 170-93.

15. Hartl, D.L., A Primer of Population Genetics, 2nd ed., Sinauer Associates, Sunderland, Massachusetts, 1988.

16. Rosenheim, J.A. and Tabashnik, B.E., Influence of generation time on the rate of response to selection. Am. Nat., 1991, 137, 527-41.

17. Rosenheim, J.A., and Tabashnik, B.E., Evolution of pesticide resistance: interactions between generation time and genetic, ecological, and operational factors. J. Econ. Entomol., 1990, 83, 1184-93.

18. Davidson, R.H. and Lyon, W.F., Insect Pests of Farm, Garden, and Orchard, 8th ed., Wiley, New York, 1987.

19. Georghiou, G.P., The Occurrence of Resistance to Pesticides in Arthropods, FAO, Rome, 1981.

20. Tabashnik, B.E., and Croft, B.A., Managing pesticide resistance in crop-arthropod complexes: interactions between biological and operational factors. Environ. Entomol., 1982, 11, 1137-44.

21. Roush, R.T. and Daly, J.C., The role of population genetics in resistance research and management. In Pesticide Resistance in Arthropods, eds. R.T. Roush and B.E. Tabashnik, Chapman and Hall, New York and London, 1990.

22. Comins, H.N., The development of insecticide resistance in the presence of immigration. J. Theor. Biol., 1977, 64, 177-97.

23. Georghiou, G.P. and Taylor, C.E., Genetic and biological influences in the evolution of insecticide resistance. J. Econ. Entomol., 1977, 70, 319-23.

24. Curtis, C.F., Cook, L.M., and Wood, R.J., Selection for and against insecticide resistance and possible methods of inhibiting the evolution of resistance in mosquitoes. Ecol. Entomol., 1978, 3, 273-87.

25. Taylor, C.E. and Georghiou, G.P., Suppression of insecticide resistance by alteration of gene dominance and migration. J. Econ. Entomol., 1979, **72**, 105-9.

26. Maudlin, I., Green, C.H., and Barlow, F., The potential for insecticide resistance in Glossina (Diptera: Glossinidae)--an investigation by computer simulation and chemical analysis. Bull. Entomol. Res., **71**, 691-702.

27. Uyenoyama, M.K., Pleiotropy and the evolution of genetic systems conferring resistance to pesticides. In Pesticide Resistance: Strategies and Tactics for Management, National Academy of Sciences, Washington D.C., 1986, pp. 207-21.

28. Caprio, M.A. and Tabashnik, B.E., Gene flow accelerates local adaptation among finite populations: simulating the evolution of insecticide resistance. Submitted to: J. Econ. Entomol.

29. Talekar, N.S., ed., Diamondback Moth Management: Proceedings of the First International Workshop, Asian Vegetable Research and Development Center, Shanhua, Taiwan, 1986.

30. Tabashnik, B.E., Cushing, N.L., Finson, N., and Johnson, M.W., Field development of resistance to Bacillus thuringiensis in diamondback moth (Lepidoptera: Plutellidae). J. Econ. Entomol., 1990, **83**, 1671-76.

31. Pashley, D.P., Johnson, S.J., and Sparks, A.N., Genetic population structure of migratory moths: the fall armyworm (Lepidoptera: Noctuidae). Ann. Entomol. Soc. Am., 1985, **78**, 756-62.

32. Unruh, T.R., Genetic structure among 18 West coast pear psylla populations: implications for the evolution of resistance. Am. Entomol., Spring 1990, pp. 37-43.

33. Caprio, M.A., Gene flow as a factor in the evolution of insecticide resistance. Ph.D. Dissertation, University of Hawaii, Honolulu, 1990.

34. Tabashnik, B.E., Cushing, N.L., and Johnson, M.W., Diamondback moth (Lepidoptera: Plutellidae) resistance to insecticides in Hawaii: intra-island variation and cross-resistance. J. Econ. Entomol., 1987, **80**, 1091-99.

35. Raymond, M., Callaghan, A., Philippe, F., and Pasteur, N., Worldwide migration of amplified insecticide resistance genes in mosquitoes. Nature, 1991, **350**, 151-53.

HOW DOES SPATIAL STRUCTURE IN POPULATIONS AFFECT THE SPREAD OF FUNGICIDE RESISTANCE?

M. W. SHAW
Department of Agricultural Botany, School of Plant Sciences,
University of Reading, 2, Earley Gate, Whiteknights, P.O. Box 239
Reading RG6 2AU, UK

ABSTRACT

Spatial structure is imposed on pathogen populations both by the disposition of the host in space and by the pattern and scale of pathogen dispersal. Such structure influences both the initial genetic composition of a population to which fungicide is applied and the dynamics of any genetic change thereafter. First, consider application of fungicide to a small plot within a large untreated area. If plot size is small relative to gene flow distances, selection will be much weakened by immigration of susceptible individuals. If plot size is large relative to gene flow distances, selection will be effective, but underlying population differentiation will be large, so replicate experiments are likely to have a very large variance and a far from normal distribution of results. Next, consider uniform application over a large crop area. If resistance genes are common spatial structure can be ignored. If they are rare or very rare, resistance will spread from wherever it first becomes established. The way this spread occurs will differ according to the details of fungal dispersal biology. Last, consider mosaic treatments, in which parts of the range of a pathogen are treated, others untreated. This can produce substantial changes in the outcome of selection if treated patches are no larger than characteristic dispersal distances of the pathogen. Thus one of the most important features of a fungal pathogen is its dispersal scale; unfortunately several factors make it difficult to estimate this.

INTRODUCTION

Gene flow links individuals living in different places. Pathogens necessarily occur only where their host does, and so the spatial structure of host populations must be reflected in the spatial structure of the pathogen populations. This will not be true below a certain spatial length scale, set by the dispersal patterns of the pathogen. Two features describe this pattern ('contact distribution' in the terminology of [1]). One is the shape of the dispersal pattern: is it oriented in a certain direction? Are long-distance migrants common relative to short-distance? The other is the average distance that a new infection occurs away from its parent. This average cannot be a conventional mean, since then offspring moving in one direction would cancel out offspring moving in the opposite direction. Dispersal distance is therefore usually described by its root-mean-square or mean-square (σ^2). This measure has a simple interpretation: an individuals parent(s) probably lived somewhere within a circle of area σ^2 centred on the individuals

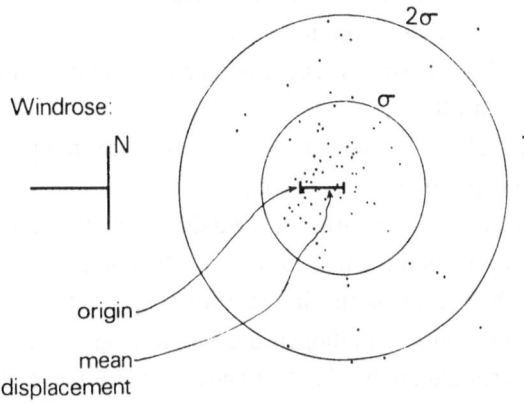

Figure 1.. The scatter of offspring of a single central lesion. See text for details.

present location (Figure 1). If there is a preferred direction of movement because of, for example, wind, a true mean displacement also occurs. In Figure 1, random offspring positions have been generated by superimposing two processes: directed travel by wind, which comes from a random direction but prevails from the West; and turbulent scatter, which is assumed to operate in any direction. For both processes the distance offspring move is a random number generated from an exponential distribution. Clearly, despite the strong influence of prevailing wind, a description in terms of a mean displacement and a scatter is not unreasonable.

This spatial scale set by dispersal mechanisms has consequences for both the initial genetic structure of a pathogen population to which fungicide is applied, and for the way resistance spreads once it arises. It is useful to consider two extreme cases, one in which pathogen dispersal distances are large compared with agronomically similar host areas, and the other in which dispersal distances are much smaller. If spatial scales are long it will be hard to do experiments which accurately depict what will happen when a fungicide is applied over a large area; when resistance arises it will appear to do so simultaneously over large areas. If spatial scales are short, experiments are liable to have irreproducible consequences, but resistance will tend to appear sporadically and may therefore be containable. All this clearly makes it very important to have a good understanding of dispersal patterns; but this is less easy than it seems. The rest of the chapter will discuss each of these aspects in turn.

DISPERSAL SCALE LONG

Small-scale field experiments

A common experiment has been to expose large field plots to fungicide and evaluate the effect of various mixtures, rates or timings on the development of fungicide resistance, or to determine whether sensitivity changes occur. Such experiments have been reported, for example, in [2], [3], or [4].

It is possible to calculate how the level of resistance in a population will change over the course of an experiment using a simple model of a small population receiving immigrant individuals from a much larger surrounding one. If a resistant form is present everywhere at the start of the growing season but the chemical is applied only in the experimental area, while over both the surroundings and the experimental area, conditions for the growth of the pathogen are similar, then the mathematical model given in the appendix should apply. Table 1 shows the predicted changes in resistance frequency at different levels of immigration, along with the effects on control.

The interesting thing to note is that small amounts of immigration affect the evolution of resistance much more than they affect the degree of control. Also, as immigration increases, so decreasing the rate at which resistance evolves, the control seen worsens, because in effect, the experimental plot is flooded with inoculum. In this case, therefore, monitoring resistance by performance would be extremely misleading. The overall conclusion is that interplot interference may be apparently negligible while experimental results on the evolution of resistance are still seriously contaminated by immigration.

Large-scale use of fungicide

Because we are assuming that dispersal distances are comparable to the size of the region within which fungicides are to be used, spatial variance in gene frequencies or quantitatively inherited phenotypic characteristics will be slight; that is, different parts of the region will be genetically uniform. In this case, the whole area treated will behave as a single population, and theory and experiments on the development of resistance in homogeneous populations apply. If variability is already present, resistance will rise steadily throughout the population. If the variability is based on a single gene, this increase will be logistic, leading to an apparent very sudden increase in resistance after fungicides have been applied for some time [5]; if the variability is polygenic the increase will be more nearly linear in time [6].

TABLE 1

Resistant frequency and extent of control by fungicide after 100 days in a small plot subject to immigration from an initially genetically identical surrounding area, as a function of immigration and the selection pressure.

Proportion of immigrants		Rate of growth of population		
		0.01	0.03	0.05
0%	Resistance*	3	17	60
	Control@	63	94	98
0.1%	Resistance	3	12	30
	Control	60	92	97
1%	Resistance	2	5	9
	Control	49	81	89
5%	Resistance	2	4	6
	Control	41	72	83

*: Percentage frequency of resistance in the treated plot
@: Percentage control, ie 100 x (untreated - treated)/untreated populations

DISPERSAL SCALE SHORT

Spatial variability in pathogen genetic composition

If two separate populations are sufficiently far apart, they will evolve more or less independently because they exchange individuals rarely and gene frequencies at one location cannot influence those at another. This observation can be generalised to populations which are not completely isolated, but linked by intervening populations which exchange individuals in every generation. Strong selection can maintain different gene frequencies at locations between which there is abundant genetic exchange [7]. However, from the point of view of the evolution of fungicide resistance, it is interesting to look at the case of unselected genes, which forms a limiting case: populations may be more different than suggested by the analysis, because of local selective factors, but they cannot be more similar.

For neutral genes, gene frequencies at any location tend to fluctuate by chance, simply because the local population is finite. Some multiple of σ^2 determines the area from within which the parent(s) of a new-born individual almost certainly came; its grandparents will have come from within an area twice this, its great-grandparent(s) from within an area three times it, and so on. Populations sufficiently far away will not have contributed any genes for a very long time, during which the gene frequency may have drifted considerably. Thus σ^2 allows the conversion of a distance into a time - the number of generations required for dispersal to cover the distance. Populations must then be evolutionarily independent if they are further apart than the characteristic time for drift to change allele frequencies substantially. Each population may then have very different allele frequencies at many loci. This argument can be quantified in several more or less similar ways. Two cases are of most interest in the current context: a single gene recurrently mutating to an alternative (e.g. resistant) form, and the case of a quantitatively varying trait.

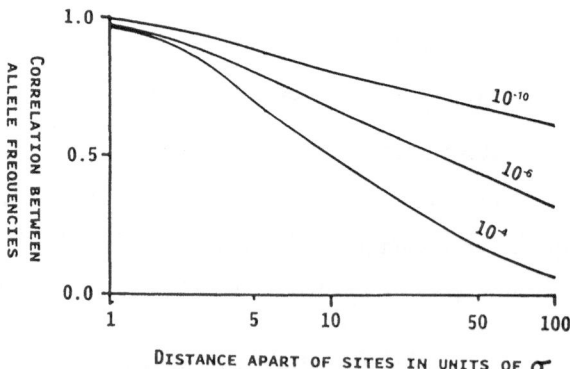

Figure 2. The decline in correlation between allele frequencies in two places as the distance apart increases, at three mutation frequencies.

Figure 2 shows how the correlation between gene frequencies falls away with distance in a population continuously distributed over a very large habitat, calculated from the model in [8]. In this model a single locus has two alleles which mutate to each other at a certain rate. The form of the curve depends only on the mutation frequency. However, the correlations are all relative to the limiting value for very close populations, and this depends on the population within any area σ^2. If population density is low, local populations are inbred and closely related, so that the correlation starts high but declines to a much lower figure; if population density is high local populations are more typical of the whole and small groups taken from one locality will be less correlated. For polygenically controlled traits, the correlation between the values of the traits exhibited by individuals against the distance apart of the individuals has a similar form to that of Figure 2 [9].

Consequently experiments at locations more than about 50σ apart are exposed to pathogens which are very likely to have different initial genetic compositions. A resistance allele, or one which can easily mutate to resistance may be absent or rare at some sites but relatively common at others. Equally, the fitness of a resistant allele in the absence of fungicide may be very different in the two populations. If a variant is actually absent throughout the range of a species, mutation to the variant form will occur after a very variable time-span at each location; but because dispersal is slow, the mutation will not move rapidly between localities. The net effect is that the results of experiments may be intrinsically very variable. Local replication may improve the accuracy of studies on the rate of selection without improving ability to predict resistance build-up at other locations. Table 2 shows an example of this variability from experiments done by ICI [10].

TABLE 2.

Plot to plot variation in response to selection for dual resistance to a benzimidazole-phenylcarbamate (diethofencarb) mixture. The pathogen was *Botrytis cinerea* growing on grapes at a trial site in the Loire valley, France. 200 isolates were tested in each plot, each year.

Plot	Frequency of dual resistance (%)	
	1986	1987
1	0	17.5
2	0	0
3	0	0

Movement of resistance outwards from a single origin

Where resistance arises initially at one location within a large uniformly treated region resistant forms are at an advantage over a wide range. Dispersal will have two effects: the increase in resistance at the site of origin will be diluted by immigrants, as described above, and the resistant genotype will move to locations away from its origin, and begin to increase in frequency there. If the dispersal distribution obeys certain conditions, a travelling wave of resistance will be set up, moving outwards from the origin into the surrounding population (Figure 3) as described in e.g. [11] or, in a different context, [1]. If s denotes the selective advantage of the resistant over the sensitive form, the speed of this wave will be proportional to $\sigma s^{\frac{1}{2}}$, and its width proportional to $\sigma/s^{\frac{1}{2}}$ but the constants of proportionality will differ according to the details of the dispersal distribution. These relations are intuitively reasonable: the stronger selection and the greater dispersal is, the faster the wave moves, and the faster it moves, the narrower it is.

However, this theory may have some defects. It is based on an approximation for which it is necessary to assume that the variance in dispersal, σ^2, is finite. Although this assumption must by definition be true in a finite world, a better model of pathogen behaviour may in fact be obtained by dropping it, because the commonly used power-law formulae which describe the transport of fungal spores [13] only have a finite variance if the exponent of the power law is greater than 2. The usual values fitted to experimental data-sets, and predicted from turbulent diffusion theory [14], are substantially less than 2. The physical meaning of this is that at any location immigration from all distances has to be considered; in the usual theory immigration from far away can be neglected. It is known that if long distance migrants are moderately common, wave speeds are dramatically increased [12]. Unpublished simulations suggest that the overall effect of dispersal patterns probably typical of plant pathogens is to be to produce a continuously accelerating and flattening wave of advance, as the resistant area grows and makes an important input of spores to locations further and further away. Thus, at least for diseases with windblown spores, the pattern of spread would be predicted to be a long period of waiting until a successful mutant is produced, then a period of apparent stasis or slow expansion which accelerates explosively to cover the whole region.

Some data are not inconsistent with the idea of an accelerating wave. Figure 3 shows the geographic pattern of spread of metalaxyl resistance in *Bremia lactucae* on lettuce, using the data in [15]. The isopaths shown on this diagram were constructed by joining the mid-points of the counties furthest from the outbreak origin from which resistant isolates had been received by the end of the time-period to which the isopath refers. It should be pointed out that in this case the authors did not implicate wind as the major cause of dispersal, but rather mass transport of lettuce seedlings between growing centres. Nonetheless, the overall patterns broadly fit the prediction, and the exact mechanism of dispersal is not relevant to the model, provided it can be described

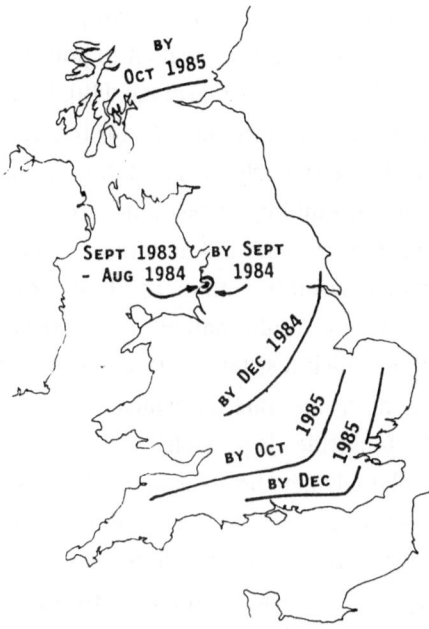

Figure 3. The advance of metalaxyl resistance in *Bremia lactucae* in the UK during the 1980's. The positions of the advancing front at various times were constructed from the data in [15] as described in the text.

statistically in the same way.

MOSAIC TREATMENT PATTERNS

A number of authors have discussed the use of spatial mosaics to reduce the rate at which insecticide resistance develops e.g. [16], [17], [18]; the same strategy is obviously applicable to fungicides, if it can be organised. The organisational aspect is important: an individual farmer can clearly use certain fungicides on some fields and not on others, and it is possible that the agrochemical industry might agree to restrict certain products, or combinations, to certain areas through dealer networks, although this would be hard to enforce. It seems improbable that agreement could be reached on spatial scales between these, because voluntary agreement between farmers or groups of farmers over very large areas would be required.

Therefore, the sort of mosaic which it is sensible to consider is either one in which adjacent fields are treated differently, with a typical patch size of c. 10 ha, or one in which whole regions receive different treatments, with patch sizes of perhaps thousands of km². In each case, there are two possible relations between patch size and dispersal scale: dispersal greater than patch size or patch size greater then dispersal.

If dispersal is comparable to or greater than patch size, and if some areas are simply unsprayed the strategy is simply equivalent to poor spray coverage, which is well established to slow the evolution of resistance: e.g. [19], [20]. However, if the alternate areas are sprayed with another chemical, the effect is to ensure that individuals resistant to either or both chemicals are strongly favoured, but no individual is ever exposed to both at once. This will favour build-up of resistance to each chemical separately, about half as fast as otherwise; then by a second mutation or recombination, the production of doubly resistant forms, favoured in both areas. In this case movement of individuals and their offspring is equivalent to alternating the fungicides very rapidly in time, and any reduction in the rate at which resistance evolves is likely to be very slight: [17], [18], [19]. Also, no reduction in the rate at which double resistance evolves due to interactions between the fungicides is possible.

If the patches are largely genetically isolated, because they are very much larger than the characteristic dispersal scale, evolution will proceed more or less independently in each patch, and resistance in each patch will develop more or less as fast as it would have done if the entire species range were treated. However, the evolution of combined resistance to two or more chemicals will be accelerated, because, at the boundaries of regions, each region will receive a continual small input of propagules resistant to the other chemical, allowing recombination or additional mutations to take place. Such doubly resistant forms will have a slight selective advantage, dependent on how many of their offspring find themselves in regions treated with a fungicide different from the parent. If isolation is reduced, the position will gradually become more similar to the previous scenario.

Thus, provided patch size is small relative to dispersal distance, development of single resistances should be slightly retarded by mosaic application of fungicides, but at the perhaps serious cost of substantially increasing the rate at which doubly resistant forms are selected. There would be no great advantage over alternation of the chemicals, which might be substantially easier to organise. (This analysis agrees in substance with that in [16] and [17], allowing for differences in biology between fungi and insects).

ESTIMATING DISPERSAL DISTANCES

To plan an experiment or a management strategy effectively, a reasonable estimate of σ^2 is needed. Characteristic lengths for spore dispersal gradients are available for many diseases [13]; however, there is evidence that these may not be very good guides. This evidence comes from two sources. First, for a wide range of organisms, physical measurements of movement seem to provide serious underestimates of gene flow [21]; second, for several agriculturally important pathogens, both epidemiological and genetic evidence suggests very much longer range dispersal than would be expected. For example, the initial incidence of *Septoria tritici* on wheat crops does not depend in an obvious way on the location of nearby previous crops [22], and the populations are very genetically diverse [23]. This suggests that windblown ascospores, presumed to be founding the populations are well mixed on a scale of at least several hundred m. As another example, the virulence composition of *Erysiphe graminis* in the Netherlands appears to be strongly influenced by the barley varieties grown in the UK, several hundred km away [24].

The explanation of this discrepancy between physical and genetical evidence may lie in the difficulty of making measurements of rare long-distance dispersal events. In most experiments, without unique genetic markers available, it is impossible to determine spore dispersal beyond distances where 'background' disease or spore levels are reached. Even when unique markers are available, it is increasingly difficult to study long-distance dispersal simply because larger and larger areas must be searched and larger and larger samples taken. For windblown spores, available evidence is largely consistent with a two-phase dispersal distribution: a proportion of spores escape from the crop completely and are caught up in much larger scale turbulent movements, while the remainder are deposited relatively close to the source, to produce the expanding foci characteristic of mildews and rusts. The spores escaping completely from the crop (estimated as up to 95% in [25]; but 50% seems more plausible) may take part in long distance wind movement as well as being spread out randomly by atmospheric turbulence on all scales. For widely grown crops this may produce extremely long-distance pathogen movement. For example, a spores carried for 8 h in a wind of 4 ms^{-1} will have travelled 115 km. Thus, the spores arriving at a point in a field may be seen as composed of two fractions: local spores, originating within a few m, and distant spores, which may have travelled tens of thousands of times further. As wind patterns vary over a season, a site will receive spores from a very wide area, and despite strong local gradients, a population may be very well mixed.

ACKNOWLEDGEMENTS

I am very grateful to Steve Heaney of ICI at Jealotts Hill, Berkshire, for permission to use the data in Table 2.

REFERENCES

1. Van den Bosch, F., Zadoks, J.C. and Metz, J.A., Focus expansion in plant disease: 1. The constant rate of focus expansion. Phytopath., 1988, **78**, 54-58.

2. Hunter, T., Brent, K.J. and Carter, G.A., Effect of fungicide regimes on sensitivity and control of barley mildew, in Proceedings of the British Crop Protection Conference - Pests and Diseases, BCPC, Croydon, 1984 pp 471-476.

3. Bateman, G.L., Fitt, B.D.L, Creighton, N.F. and Hollomon, D.W., Changes in populations of Pseudocercosporella herpotrichoides in successive crops of winter wheat in relation to initial populations and fungicide treatments. Crop Prot., 1990, 9, 143-148.

4. Schulz, U. and Scheinpflug, M., Investigations on sensitivity and virulence dynamics of *Erysiphe graminis* f.sp. *tritici* with and without triadimenol treatment. In Proc. 1986 Brit. Crop Prot. Conf. - Pests and Diseases, BCPC, Thornton Heath, Surrey, UK, 1986, 531-538.

5. Cook, L.M., Coefficients of Natural Selection. Hutchinson, London, 1971, ch 1.

6. Shaw, M.W., A model of the evolution of polygenically controlled fungicide resistance, Plant Path., 1989, 38, 44-55.

7. May, R.M., Endler, J.A. and McMurtrie, R.E., Gene frequency clines in the presence of selection opposed by gene flow. Am. Nat., 1975, 109,659-676.

8. Malécot, G., The mathematics of heredity. Freeman, San Francisco, 1969.

9. Lande, R., Isolation by distance in a quantitative trait. Genetics, 1991, 128, 443-452.

10. Heaney, S.P., Pers. comm., 1991

11. Fisher, R.A., The wave of advance of an advantageous gene. Ann. Eugenics., 1937, 7, 355-369.

12. Slatkin, M., The rate of spread of an advantageous allele in a subdivided population. In Population Genetics and Ecology, ed. S. Karlin and E. Nevo, Academic Press, New York, 1976, 767-780.

13. Fitt, B.D.L., Gregory, P.H., Todd, A.S., McCartney, H.A. and MacDonald, O.C., Spore dispersal and plant disease gradients, a comparison between two empirical models. J. Phytopath., 1987, 118, 227-242.

14. Pasquill, F., Atmospheric Diffusion, 2nd ed. Ellis Horwood, Chichester, 1974.

15. Crute, I.R., Norwood, J.M. and Gordon, P.L., The occurrence, characteristics and distribution in the United Kingdom of resistance to phenylamide fungicides in *Bremia lactucae* (lettuce downy mildew). Plant Path., 1987, 36, 297-315.

16. Mani, G.S., Evolution of resistance with sequential application of insecticides in time and space. Proc. Roy. Soc. Lond. B, 1989, 238, 245-276.

17. Curtis, C.F., possible methods of inhibiting or reversing the evolution of insecticide resistance in mosquitoes. Pest. Sci., 1981, **12**, 557-564.

18. Via, S., Quantitative genetic models and the evolution of pesticide resistance. In Pesticide resistance: strategies and tactics for management. National Academy Press, Washington DC, 1986, 222-235.

19. Kable P.F. and Jeffrey, H. Selection for tolerance in organisms exposed to sprays of biocide mixtures: a theoretical model. Phytopath.,1980, **70**, 8-12.

20. Josepovits, G. and Dobrovolszky, A., A novel mathematical approach to the prevention of fungicide resistance. Pest. Sci., 1985, **16**, 17-22.

21. Slatkin, M., Gene flow and the geographic structure of natural populations. Science, 1987, **236**, 787-792.

22. Shaw, M.W. and Royle, D.J., Airborne inoculum as a major source of *Septoria tritici* (*Mycosphaerella graminicola*) infections in winter wheat crops in the UK. Plant Path., 1989, **38**, 35-43.

23. McDonald, B.A. and Martinez, J.P., DNA restriction fragment length polymorphisms among *Mycosphaerella graminicola* (anamorph *Septoria tritici*) isolates collected from a single wheat field. Phytopath., 1990, **80**, 1368-1373.

24. Limpert, E., Frequencies of virulence and fungicide resistance in the European barley mildew population in 1985. J. Phytopath., 1985, **119**, 298-311.

25. Sreeramulu, T. and Ramalingam, A., Experiments on the dispersal of *Lycopodium* and *Podaxis* spores in the air. Ann. Appl. Biol., 1961, **49**, 659-670.

APPENDIX

Let r_n be the rate of growth of the pathogen population in the surrounding area, and r_s and r_r the rates of the sensitive and resistant forms in the presence of fungicide. Let m be the rate of immigration as a fraction of the initial population, and x_0 the initially resistant fraction. Then let t denote time. If the external population grows exponentially, the rate of growth of the resistant population, n_r, is governed by the differential equation:

$$\frac{dn_r}{dt} = r_r \cdot n_r + m \cdot \exp(\int_0^t r_n)$$

If r_n is assumed constant, this can be solved analytically along with the analogous equation for n_s, and used to calculate the resistant fraction and the total population, as in table 1. For table 1, r_n was taken equal to r_r and r_s assumed to be 0.

MODELLING HERBICIDE RESISTANCE - A STUDY OF ECOLOGICAL FITNESS

A. M. MORTIMER, P. F. ULF-HANSEN and P. D. PUTWAIN
Department of Environmental and Evolutionary Biology
University of Liverpool
Liverpool U.K. L69 3BX

ABSTRACT

The evolution of herbicide resistance in weeds is governed by the interaction of genetic and ecological factors that determine the frequency of resistant traits in populations. To date, modelling of this process has concentrated on conceptual understanding of the relative importance of different component processes at varying levels of detail, with a view to the design of management programmes. In a case study, this paper analyses the fitness of resistant and susceptible biotypes of *Alopecurus myosuroides* under differing herbicide (chlorotoluron) selection regimes and assesses the significance of the results in the spread of resistance by simulation modelling. Fitness measured as per capita seed production of plants was shown to be both frequency and density dependent in an experimental trial comparing resistant (R) and susceptible (S) biotypes in mixture. The form of this interaction and the consequent selection against the S biotype was strongly influenced by rate of chlorotoluron application. The importance of various demographic factors was investigated using a simulation model with the assumption of dominant major gene inheritance of resistance. The model showed that, given the observed responses to selection, the interaction of density dependent and density independent regulation was critical to the spread of the resistance. Simulated management sequences rotating the intensity of selection and density independent regulation were shown to be effective in controlling the frequency of the resistance allele and prevented the emergence of resistant phenotypes.

INTRODUCTION

Interests in modelling the phenomenon of herbicide resistance in weeds stem from the need to understand the interplay of factors that influence the likelihood and rate of an evolutionary process. Such studies represent the application of population genetics and ecology in determining the frequency of a trait in weed populations that may have considerable economic implications [1]. Pre-1980 the evolution of herbicide resistance was largely confined to examples of triazine resistance [2] and interests in managing resistance in agriculture lay with the choice of alternate herbicides, albeit at higher economic costs in many instances [3,4]. More recently there has been a progressive increase both in the number of weeds displaying resistance to herbicides and in the range of chemical classes to which resistance is found, especially in relation to cross resistance. The widespread appearance of resistance in Australian phenotypes of *Lolium rigidum* [5] and the very rapid emergence of resistance to ALS inhibitor herbicides [6] has lead to an urgent need to understand the dynamics of the evolution of herbicide resistance [7,8].

There are five major evolutionary factors which interact to determine the likelihood of herbicide resistance *per se* and the speed with which it may occur. These are :

1)   the frequency of resistance traits in natural (unselected) populations ;

2)   the mode of inheritance of resistance traits ;

3)   the intensity of selection against susceptible genotypes ;

4)   the absolute fitness of resistant and susceptible genotypes ;

5)   the fitness differential between resistant and susceptible genotypes.

The aim of modelling the evolution of herbicide resistance is to attempt a synthesis of these (and other factors), to be able to - i) predict the rates of spread of resistance traits (genes) in which selection (application of herbicide) acts differentially on individual competing plants according to their genotype; and ii) to undertake sensitivity analyses to judge the relative importance of component processes in the development of rationale management programmes.

It has been elegantly argued from first principles [9], that the fundamental determinants (Table 1) of the time scale over which pesticide resistance may emerge are, the species generation time, and the logarithmic relationship between the initial frequency of the resistance allele, the choice of the threshold at which a significant level of resistance is recognised and the strength of selection (determined by dosage level and degree of dominance of the resistance allele). The overall conclusion from models such as this is that given recurrent selection, then approximate evolutionary time scales for pesticide resistance in a range of pests may be predicted.

Equation 1 was effectively extended in an independently and earlier derived model [10] for herbicide resistance in weeds to include the presence of a fraction of the population that persists over successive generations as a bank of seeds within the soil. This approach recognised the significance of the soil seed bank as a buffer to evolution, as a consequence of the recruitment of susceptible genotypes from previous generations.

More detailed models [11,12] have been used to investigate the significance of individual demographic and genetical processes in depth, in particular the role of gene flow

TABLE 1

Estimation of the approximate time (T_r) by which a specified resistance level is reached within a pest population [9]

$$T_r = T_g \ln \left(\frac{rf}{r_o} \right) / \ln \left(\frac{wRS}{wSS} \right) \qquad \text{Equation 1.}$$

$$T_r = T_g \ln \left(r_f / r_o \right) / \ln \left(1 + \left(wRS / wSS \right) / \left(fRS / fSS \right) \left(1 / T_{soil} \right) \right) \qquad \text{Equation 2.}$$

where T_g is the species generation time (years)

r_o is the initial frequency of the resistance allele.

r_f is the choice of threshold at which resistance is recognised.

(w_{RS} / w_{SS}) is the strength of selection, where w_i denotes genotypic fitness (RS=RR, resistant; SS, susceptible).

$(f_{RS}) / f_{SS})$ is the relative reproductive success of resistant and susceptible genotypes.

T_{soil} is the persistence of seeds in the seed bank (years).

[11] and relative fitness [12]. These modelling approaches differ from those based on equation 1, in that individual demographic and genetical components are explicitly defined. Thus difference equations are used to describe the dynamics of mixed populations of susceptible and resistant genotypes, which at discrete intervals (generation end) contribute genes to a gene pool. Alleles are then recombined according to specific genetic models governing the genotypic structure in successive populations. Additionally, migration of genes both in time (from persistent seed banks) and in space (from neighbouring areas) are incorporated. The absence of a complete data set to parameterise such models for most weed species has not prohibited their extensive use in simulation exercises of 'what if' in sensitivity and elasticity analysis [10, 13]. Table 2 lists some of the conclusions that have emerged from these analyses and Figure 1 illustrates some generalised relationships between four key components - selection pressure, initial frequency of resistance gene, persistence in the seed bank and species per capita rate of increase.

The emergence of herbicide resistance in the field is typically noticed as a localised occurrence within weed populations exposed to herbicide. Subsequent spread within the population may be aided by gene flow through both seed movement by agricultural implements [18,19] and by pollen flow. Initially however, resistant genotypes will be rare in the population and thus selection will occur in mixed populations of varying density. If

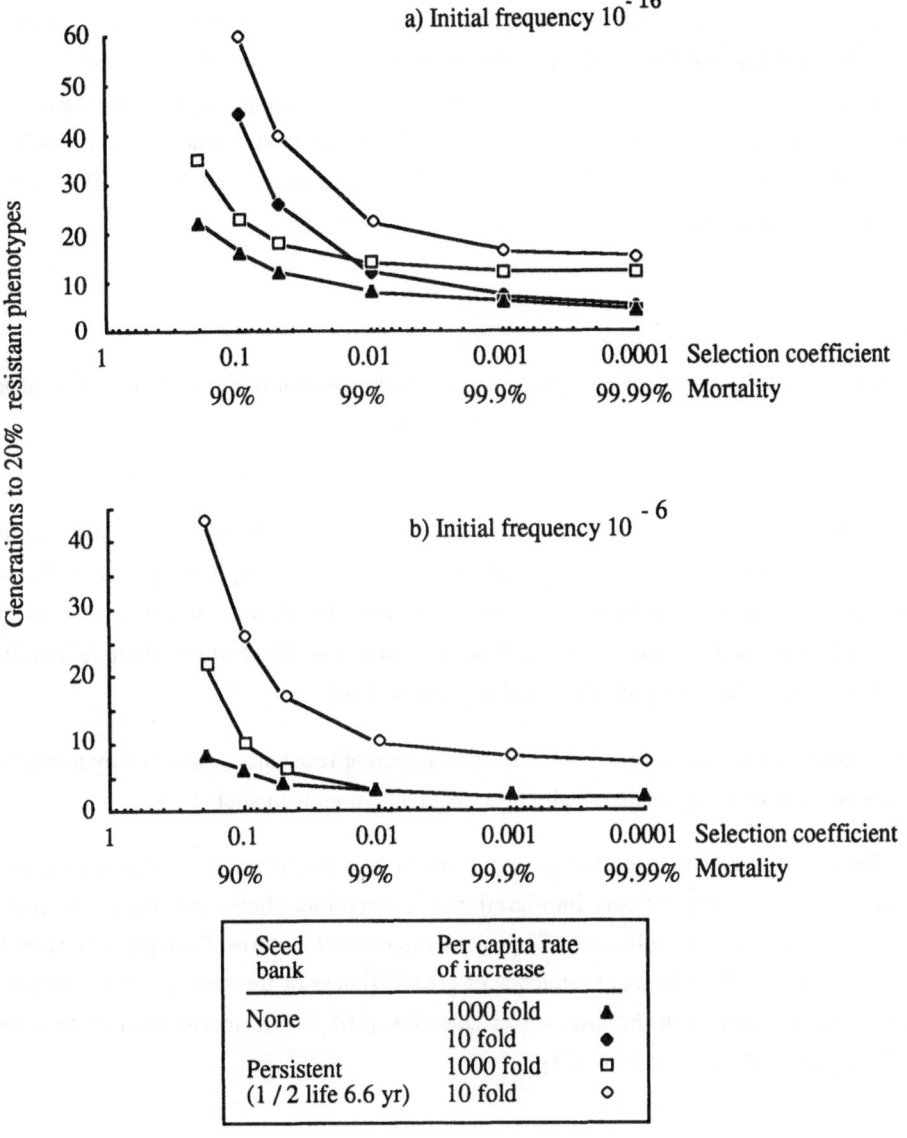

Figure 1. The relationship between the rate of evolution and selection pressure for a hypothetical weed species with differing per capita rates of increase and buried seed persistence. Data are derived by simulation with an initial resistance allele frequency of a) 1 in 10^{-16} and b) 1 in 10^{-6}, resistance being inherited as a dominant nuclear encoded gene. Selection through herbicide application is assumed to act by prohibiting the establishment of seed bearing adult plants (mortality). Note the abscissa is a logarithmic scale and that in b) ◆ and ▲ are indistinguishable.

selection against the susceptible biotype is high then the resistant genotype may come to predominate rapidly [20]. An appropriate measure of ecological fitness (as a basis for calculating selection coefficients) is the per capita rate of change of genotypes as measured over a generation of growth. This demographic statistic subsumes both mortality and reductions in seed production that may occur. Such per capita rates may be both density and frequency dependent.

TABLE 2

Some general conclusions emerging from simulation models of the evolution of herbicide resistance

1. In major gene resistance with resistance being inherited as a dominant trait, under recurrent selection, reduction of selection pressure against the susceptible phenotype will only reduce the rate of spread of resistance genes, once present in a population. If resistance is inherited as a recessive trait, evolution of resistance is unlikely to occur unless there is high initial resistance allele frequency in unselected populations [14].

2. All other factors being equal, the rate of evolution of resistance controlled by a maternally inherited trait will only be marginally faster than for a dominant allele [15] .

3. Rotational practices in which selection is intermittent may delay the evolution of resistance but it is only with significant immigration of susceptible alleles that back selection (for susceptibility) may be achieved. Such gene migration may arise through gene flow from external sources [13], as a consequence of greater fitness of the susceptible phenotype and resistant phenotypes in the absence of selection [16], or via recruitment of susceptible phenotypes from the seed bank [17].

To our knowledge there have been no detailed studies of relative fitness of susceptible and resistant genotypes, under differing herbicide dosage, in which the performance of component genotypes have been examined in mixtures over a range of densities. This is indeed a substantive task not least in the acquisition of known genotypes and in their respective bulking for population studies. This paper presents results from an experiment designed to investigate inter-biotype interactions under different selection pressures as a first step in this direction. The implications of some differing regimes for the management of resistance are then investigated by simulation modelling.

A CASE STUDY OF ECOLOGICAL FITNESS

Alopecurus myosuroides Huds., blackgrass, is a serious grass weed of winter cereals in the U.K.. The history of the occurrence and spread of resistance to the substituted urea herbicides chlorotoluron and isoproturon, together with cross resistance to other chemical groups has been well documented [21, 22, Moss, this volume]. Biotypes of resistant (R) and susceptible (S) blackgrass were grown at a range of densities and frequencies in the presence of winter wheat and were subjected to post-emergence herbicide treatment. Seed yield per plant and per capita rates of seed production were estimated. These data provided a basis for calculation of selection coefficients.

Material and Methods

Seeds (spikelets) of R and S biotypes were obtained by field collection from sites at Peldon, Essex, U.K. and Rothamsted, Hertfordshire, U.K. and bulked up as discrete populations. Confirmation of resistance and susceptibility to chlorotoluron and isoproturon in the populations has been reported elsewhere [21,23]. Seeds were sown just below soil surface into large (900 cm^{-2}) pots containing J.I. No. 1 compost at densities chosen to achieve seedling densities of 0, 11, 33, 333, and 1111 plants m^{-2} with every mixture combination being represented in addition to monoculture sowings. Seeds of the two biotypes were arranged in alternate rows at the two highest mixture densities, whilst in all other sowings they were placed at random, with biotypes identified by coloured wire rings. Wheat, cv Avalon, was sown at a constant density of 311 seeds m^{-2} at a depth of 7.5 cm. Three post-emergence herbicide treatments were applied factorially to these sowings - chlorotoluron at 1.38 kg a.i. ha^{-1} (half rate) and 2.75 kg a.i. ha^{-1} (full rate), formulated as 'Dicurane 500 FW' in 400 l ha^{-1} of water, in addition to an unsprayed control. The experiment was arranged in two randomised blocks (with additional replicates in both blocks for the mixtures 0R:1S, 1R:0S and 1R:1S) and conducted in an unheated polythene tunnel house at the University of Liverpool Botanic Gardens. Seeds were sown in January and herbicide treatments applied nine weeks later at early tillering, after which wheat was thinned to a constant density of 222 plants m^{-2}. Pots were regularly watered until seed set after which surviving blackgrass plants were harvested and the number and length of flowering spikes per plant measured. Seed number per plant was then estimated from an allometric relationship between spike length and seed number [23].

Per capita seed production was calculated as the ratio of the number of seeds produced at harvest to the number sown. Response surfaces were then fitted to observed data for each biotype - herbicide combination using the function [24]

$$f_i (x_i, x_j) = \lambda_i / \{ 1 + x_i{}^{b_{ii}} + x_j{}^{b_{ij}} \} \qquad\qquad Equation\ 3.$$

In this model, $f_i(x_i, x_j)$, which is the per capita seed production of biotype i in the presence of biotype j, is dependent on the density of both biotypes. λ_i is the per capita seed production of an uncrowded plant of biotype i and b_{ii} and b_{ij} are power parameters describing density dependent competitive ability. This model was chosen after comparative statistical analysis [25] using a range of possible models. Inspection of residuals did not justify a logarithmic transformation of these data.

Results

Per capita seed production (Figure 2) was found to decline predictably over the density range of sown seeds and in the S biotype to decline with increased rate of chlorotoluron application. In this experiment however, herbicide application at the highest rate did not however prohibit seed production at harvest and density responses were evident at all biotype-herbicide combinations. In unsprayed plants per capita seed production was both frequency and density dependent since the S biotype noticeably outcompeted the R biotype at low densities, whereas the reverse was true at high density sowings. At half and full rate applications of chlorotoluron, reductions in seed production of the S biotype due to sowing density exceeded corresponding responses in the R biotype.

The results of statistical parameter estimation are given in Table 3. For all six herbicide-biotype combinations coefficients of determination exceeded 0.66 with weakest fits being found for the S biotype sprayed with chlorotoluron. No significant differences were found amongst estimates for λ_i for the R biotype whereas significant contrasts ($p \le 0.5$) were evident for the S biotype.

Figure 3 illustrates predicted selection (S_c) against the susceptible biotype where selection is calculated as ($1 - P_S / P_R$) ; P_i (i = R, S) being the per capita seed production as calculated from parameter estimates given in Table 3 for each herbicide application. At full rate chlorotoluron, selection coefficients were in excess of 0.87 with two underlying influences - an increase towards unity with increasing sowing density of the susceptible biotype and a decrease away from unity with increasing sowing density of the resistant biotype. This flexing of the predicted response surface in two dimensions was noticeably more marked at half rate chlorotoluron application, selection always being against the susceptible biotype but lowered in magnitude. In control plots, the predicted outcome of

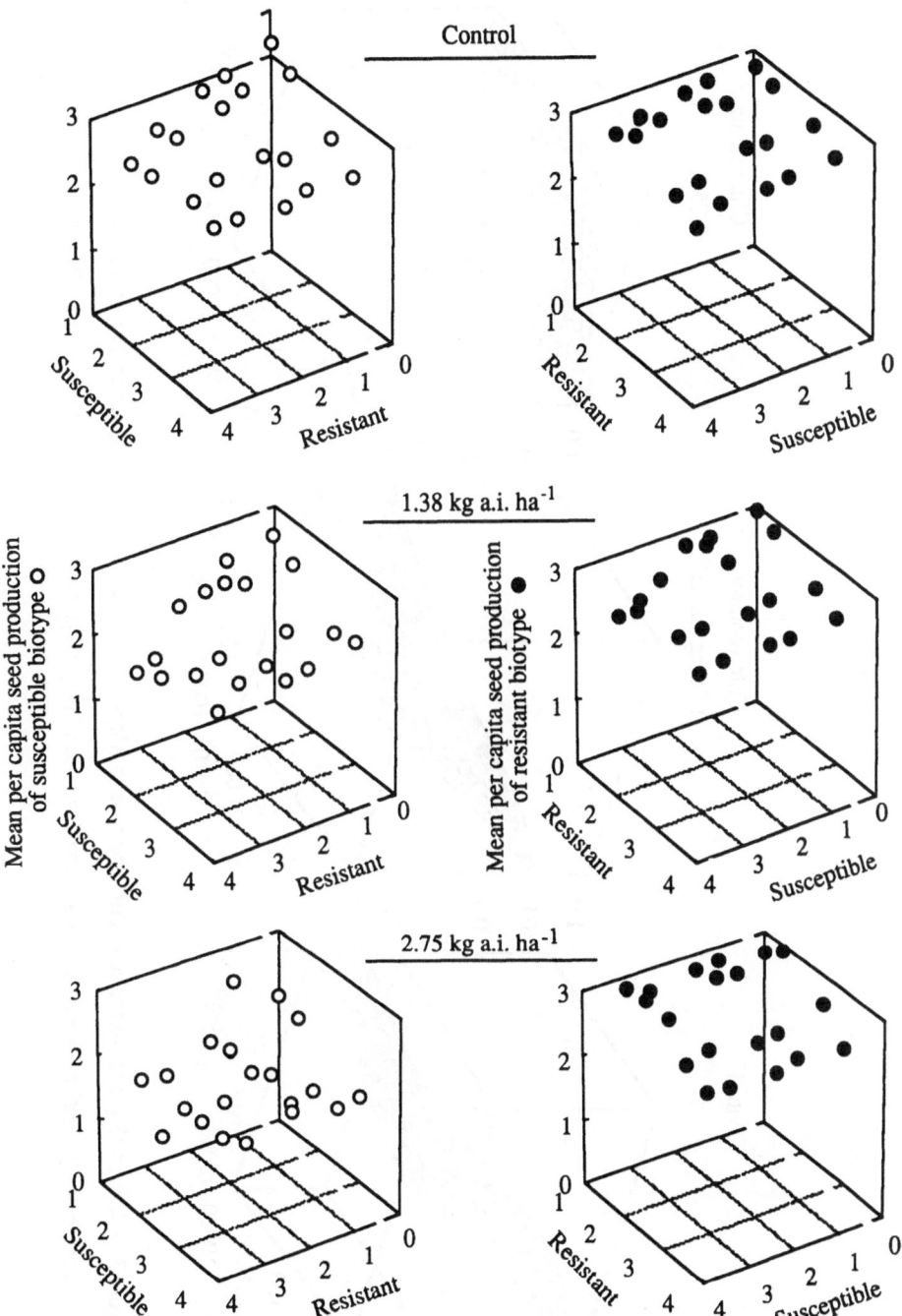

Figure 2. Ecological fitness of biotypes of *Alopecurus myosuroides* in competition under differing herbicide (chlorotoluron) selection regimes. Note abscissae are initial sowing densities and the ordinate is mean per capita seed production. All scales are logarithmic (base 10).

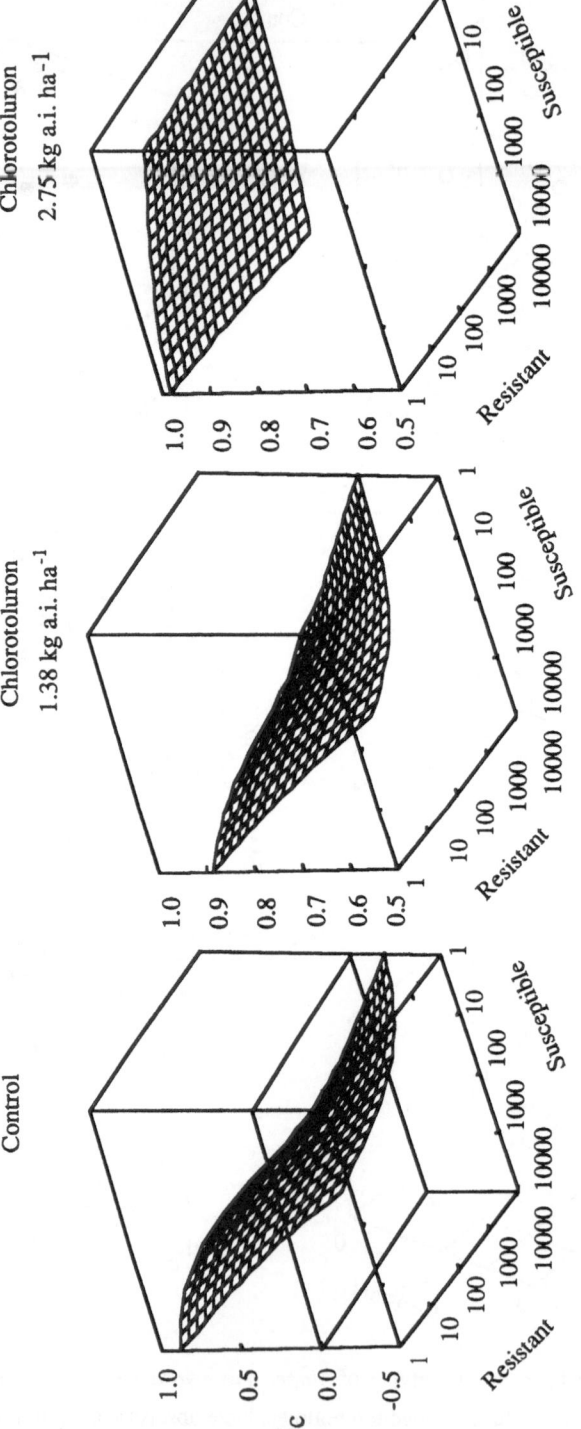

Figure 3. Selection (S_C) against the susceptible biotype in *Alopecurus myosuroides* in relation to density in mixture and herbicide application. Abscissae are sowing densities of respective biotypes on logarithmic (base 10) scales.

competitive interactions amongst biotypes was conspicuously density and frequency dependent, the susceptible biotype being selected against at high intra-biotype densities and selected for at low densities.

TABLE 3

Parameter estimates of the function $f_i(x_i, x_j) = \lambda_i / \{1 + x_i b_{ii} + x_j b_{ij}\}$ for the two biotypes in relation to herbicide application

Chlorotoluron application	BIOTYPE					
	SUSCEPTIBLE			RESISTANT		
	λ_i	b_{ii}	b_{ij}	λ_i	b_{ii}	b_{ij}
CONTROL	4182	0.427	0.334	2145	0.308	0.131
HALF RATE	1591	0.459	0.418	3534	0.382	0.300
FULL RATE	288	0.323	0.357	2552	0.345	0.013

Discussion

At present an explanation of the underlying mechanisms of density dependence determining per capita seed production is lacking. However field experimentation has shown that both mortality and seed production per surviving plant in *A. myosuroides* is negatively density dependent [23]. An explanatory hypothesis is that final per capita seed production is a consequence of differing compensatory yield responses within the population to negative density dependence. The response surfaces in Figure 3 are reflections of 'adaptive landscapes' [26] given major assumptions about the genetic architecture of the biotypes in question. Since source seed (R and S) were collected from differing populations, a common genetic background cannot be assumed. Whilst an analysis of narrow sense heritability suggested similar levels of heritable variation was present in both populations [23], the fitness differential between susceptible and resistant biotypes evident in unsprayed controls can not necessarily be seen as a fitness cost of the resistance trait. These results however do indicate that density dependent effects may be important in determining selection pressure in a weed crop mixture. The calculation of the selection coefficient surfaces (Figure 3) in relation to density hinges on statistical analysis of the response surfaces of ecological fitness. The choice of model presented here was based purely on minimisation of error variance alone in the absence of detailed knowledge on the yield-density relationships that might occur under herbicide application. It is possible that other functions [27] may be more appropriate to other species-herbicide interactions.

SYNTHESIS

The significance of the differences in response to selection as suggested by Figure 3 may be investigated by simulation modelling. We have linked population growth equations with a simple genetic model to examine the changes in gene frequency that may occur under differing management regimes. The basis of the model is i) a pair of recurrence relations of the form $y_i = x_i f_i (x_R, x_S)$ (where i = R, S) to describe the change in phenotype number in discrete generations, x_i being the density of phenotype i at some point in time, y_i the density one generation later and f_i is the growth function (*equation 3*) ; and ii) the assumption of a panmictic breeding population where resistance is inherited as a dominant nuclear encoded allele. Hence only two phenotypes (R and S) are identified and the fitness of heterozygous and homozygous resistant individuals is assumed to be identical. The genetic assumptions of this approach for *A. myosuroides* are simplistic since continuous variation has been reported amongst populations resistant to substituted-urea herbicides [23].

The following pair of equations describe the changes over a generation where the census point is taken as the point just prior to autumn germination each year.

$$y_R = \emptyset \, \lambda_R \, x'_R \, / \, \{ \, 1 + x'_R b_{RR} + x'_S b_{RS} \, \} + p \, x_R$$

$$y_S = \emptyset \, \lambda_S \, x'_S \, / \, \{ \, 1 + x'_S b_{SS} + x'_R b_{SR} \, \} + p \, x_S$$

p $(0 \le p \le 1)$ is the fraction of seed that persist from one generation to the next, in the buried seed bank and $x'_i = q \, (1 - p) \, x_i$. q $(0 \le q \le 1)$ is a parameter describing the proportion of non-dormant seed that survive to produce adult seed bearing plants. $\emptyset$ $(0 \le \emptyset \le 1)$ is a parameter describing density independent population regulation that operates after seed shedding and before seed germination at the start of the next generation. Parameter values of λ_i, b_{ii} and b_{ij} are taken from Table 3, the parameter set for the control treatment representing no herbicide selection and parameter values for full rate chlorotoluron, representing herbicide selection. At generation end, resistant and susceptible phenotypes contribute alleles to a gene pool, allelic frequencies being assumed to be in Hardy-Weinberg equilibrium. Genotypic frequencies are then computed for the next generation of growth.

Simulations were conducted for a set of parameter values (Table 4) in differing rotational sequences assuming an initial resistance allele frequency of 10^{-15}. Figure 4 A. illustrates that in the absence of herbicide selection, a mutant allele for resistance will invade the population and reach a significant frequency after 12 generations. This occurs regardless of the persistence of a seed bank. A lowering of the per capita rate of increase ($\emptyset = 0.1$) every other generation prohibited this, Figure 4 B. Where herbicide selection at full rate occurs every other generation and with no additional regulation, $\emptyset = 1$ (Figure 4 C), the frequency of the resistance allele increased exponentially towards predominance by generation 12. A rotational sequence of herbicide selection interspersed by two generations with zero selection (Figure 4 D) marginally reduced this rate as did the presence of a seed bank. Figure 4 E - H illustrates the importance of density independent regulation in determining resistance gene frequencies. Comparison of Figure 4 B and Figure 4 E, which constitute allelic counterparts, suggests that resistance alleles increase in frequency in unselected populations but do not achieve a stable mean equilibrium frequency between rotations and remain relatively scarce. Imposition of a rotation in which $\emptyset = 0.1$ every two years in three (Figure 4 F) enhanced the amplitude of oscillations on the one hand but lowered the mean resistance allele frequency. As is to be expected the possession of a transient seed bank elevated mean allele frequency. Similar underlying patterns may be discerned in the simulated rotations illustrated in Figure 4 G and H. Alternating sequences of herbicide selection followed by no selection with $\emptyset = 0.1$ for one or two generations was sufficient to halt the spread of the resistance allele in the population.

TABLE 4

Parameter values used in simulation models

Probability of seed germinating and surviving to produce an adult plant, $q = 0.4$

Proportional seed survival in the soil (p) :   persistent seed bank, $p = 0.9$

    transient seed bank, $p = 0.1$

Independent population regulation ($\emptyset$) :   none, $\emptyset = 1$.

    10 fold reduction, $\emptyset = 0.1$.

Parameters of the growth function are given in Table 3.

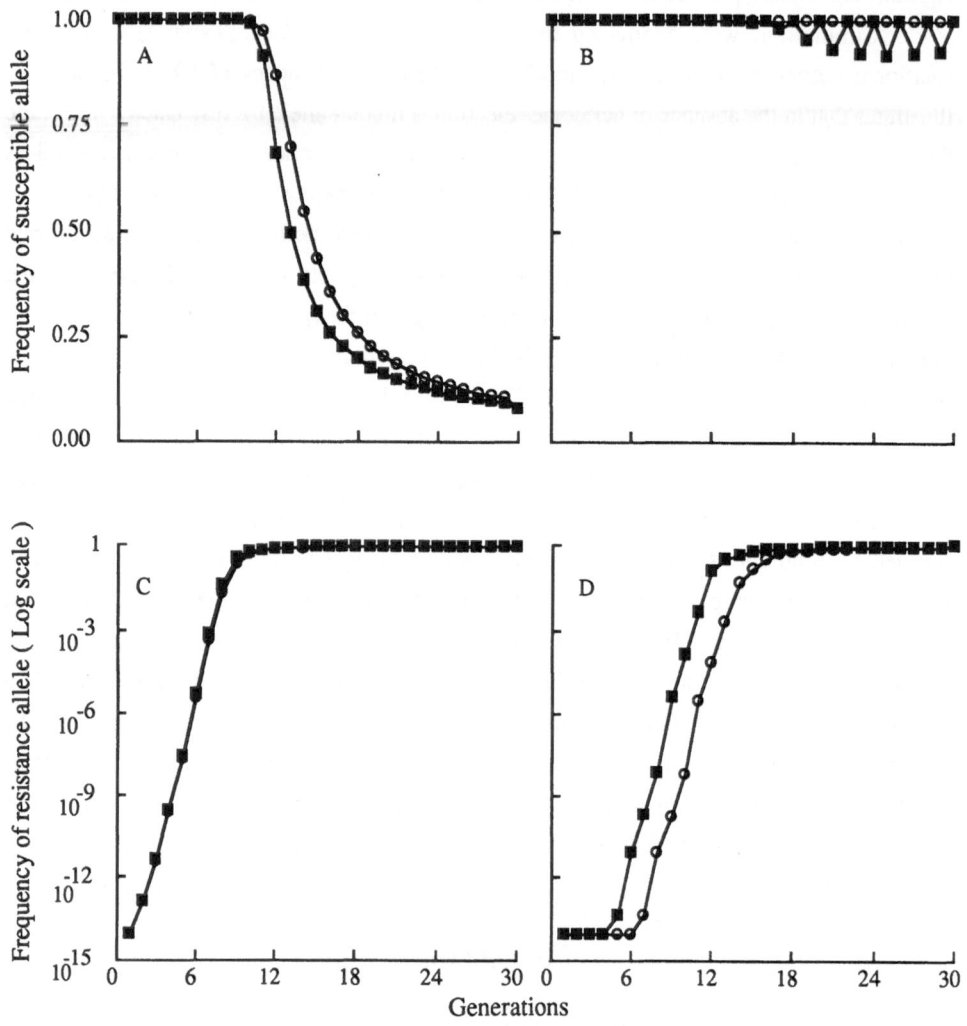

Figure 4. Changes in allele frequency under rotational sequences with time
as predicted by simulation. ■ - transient seed bank ; O - persistent seed bank .

A. No rotation   : no herbicide selection, Ø = 1 ;

B. Two year rotation   : no herbicide selection, Ø = 1 ; no herbicide selection, Ø = 0.1 ;

C. Two year rotation   : herbicide selection, Ø = 1 ; no herbicide selection, Ø = 1 ;

D. Three year rotation   : herbicide selection, Ø = 1 ; no herbicide selection, Ø = 1 ;
                           no herbicide selection, Ø = 1 ;

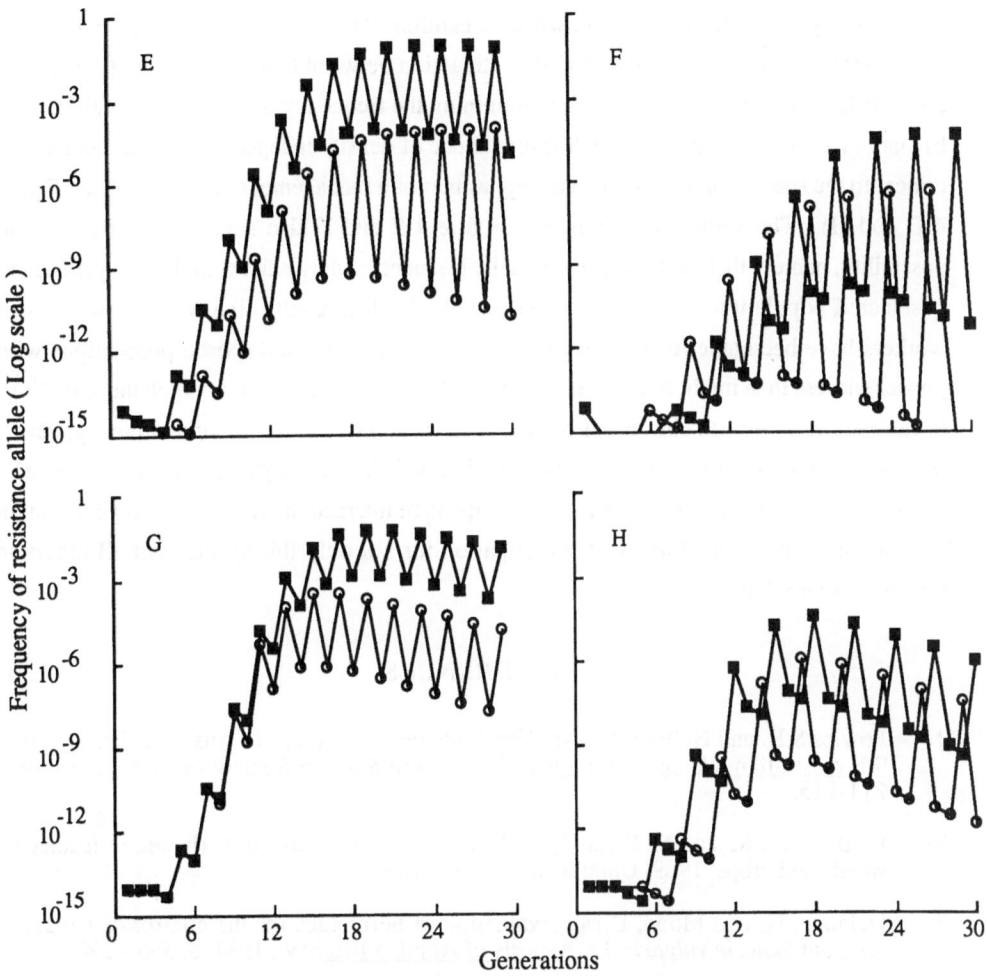

Figure 4 (Cont.). Changes in resistance gene frequency under rotational sequencies with time as predicted by simulation. ■ - transient seed bank ; O - persistent seed bank .

E. Two year rotation   : no herbicide selection, Ø = 1 ; no herbicide selection, Ø = 0.1 ;
F. Three year rotation   : no herbicide selection, Ø = 1 ; no herbicide selection, Ø = 0.1 ;
                            no herbicide selection, Ø = 0.1 ;
G. Two year rotation   : herbicide selection, Ø = 1 ; no herbicide selection, Ø = 0.1 ;
H. Three year rotation   : herbicide selection, Ø = 1 ; no herbicide selection, Ø = 0.1 ;
                            no herbicide selection, Ø = 0.1 .

Conclusions from simulation models are necessarily specific to the parameter set under investigation and the assumptions made. In the absence of appropriate analytical techniques, generalities must be drawn with caution. However the results presented here point to the possible importance of a) the interaction between density dependent regulation and density independent regulation in determining the frequency of the resistance allele; and b) that in rotational use of herbicide the magnitude of density independent regulation may be critical to the success of a rotational strategy as part of a management programme (see Figure 4 D and H). The simulations shown in Figure 4 E and F also suggest a further salient possibility, namely that the background allelic frequencies may exist at higher frequencies in unselected populations than hitherto suspected. Such a result might explain why some workers have been able to select for resistance in previously unexposed populations with greater success than might have been anticipated [28, 29]. The extent to which these findings have direct relevance to chlorotoluron resistance in *A. myosuroides* populations, however, remains open to question. The lack of detailed knowledge on the genetics of resistance in *A. myosuroides* and a critical comparison of competitive interactions amongst genotypes remain outstanding experimental areas of investigation for not only this species but all others in which resistance has been found.

REFERENCES

1. Powles S.B. and Holtum, J.A.M., Herbicide resistant weeds in Australia. Proceedings 9th Australian Weeds Conference, 1990, Crop Science Society of South Australia, 211-215.

2. Le Baron, H.M. and McFarland, J., Overview and prognosis of herbicide resistance in weeds and crops. 1989, Unpublished manuscript.

3. Delaney, H. and Moon, D.M., Evaluation of herbicides for the control of simazine resistant *Senecio vulgaris* L., Aspects of Applied Biology , 1984, 5, 385 - 390.

4. Putwain, P.D., Herbicide resistance in weeds - an inevitable consequence of herbicide use ? Proceedings 1982 British Crop Protection - Weeds, 1982, 719-728.

5. Powles S.B., Holtum J.A.M., Matthews J.M., and Liljegren, D.R, Herbicide cross-resistance in annual ryegrass (*Lolium rigidum* Gaud) : the search for a mechanism. In Managing Resistance to Agrochemicals : From Fundamental Research to Practical Strategies, eds. M.B. Green, H.M. LeBaron and W.K. Moberg, American Chemical Society, Washington DC, 1990, pp 394 - 406.

6. Thill, D.C., Mallory-Smith, C.A., Saari, L.L., Cotterman, J.C. and Primiana, M.M., Sulfonylurea herbicide resistant weeds : discovery, distribution, biology, mechanism and management. In Herbicide Resistance in Weeds and Crops eds. J.C. Caseley, G.W. Cussans and R.K. Atkin, Butterworth-Heinemann, Oxford, 1991, pp 115 - 128.

7. Gressel J., Need herbicide resistance have evolved ? - Generalisations from around the world. Proceedings 9th Australian Weeds Conference Crop Science Society of South Australia, 1990, 173 -184.

8. Gressel J., Wheat herbicides. The challenge of emerging resistance. Biotechnology Affiliates, Reading, 1988, U.K. .

9. May, R.M. and Dobson A.P., Population dynamics and the rate of evolution of pesticide resistance. In Pesticide Resistance : Strategies and Tactics for Management, National Academy Press Washington, D.C., 1986, pp 170 - 193.

10. Gressel, J. and Segel, L.A., The paucity of genetic adaptive resistance of plants to herbicides : possible biological reasons and implications. Journal of Theoretical Biology, 1978, 75, 349 - 371.

11. Maxwell, B., Roush M. and Radosevich S., Predicting the evolution and dynamics of herbicide resistance in weed populations. Weed Technology, 1990, 4, 2-13.

12. Putwain, P.D. and Mortimer, A.M., The resistance of weeds to herbicides: rational approaches for containment of a growing problem. Proceedings Brighton Crop Protection Conference - Weeds. 1989, 285-294.

13 Radesovich, S., Maxwell, B. and Roush M., Managing herbicide resistance through fitness and gene flow. In Herbicide Resistance in Weeds and Crops eds. J.C. Caseley, G.W. Cussans and R.K. Atkin, Butterworth-Heinemann, Oxford, 1991, pp 129 - 144.

14. Taylor, C.E. and Georghiou, G.P., Suppression of insecticide resistance by alteration of gene dominance and migration. Journal of Economic Entomology, 1978, 72, 105 - 109.

15. Macnair, M.R., Tolerance of higher plants to toxic materials. In Genetic consequences of man-made changes. eds J. Bishop and L. Cook, Academic Press, London, 1981, pp 177 - 207.

16. Holliday, R.J. and Putwain, P.D., Evolution of herbicide resistance in Senecio vulgaris : variation in susceptibility to simazine between and within populations. Journal of applied Ecology, 1980 , 17, 229 - 791.

17. Watson, D., Mortimer, A.M. and Putwain, P.D., The seed bank dynamics of triazine resistant and susceptible biotypes of Senecio vulgaris : implications for control strategies. Proceedings (1987) British Crop Protection Conference - Weeds. 1987 917 - 925.

18. Porterfield, J.W. Harvest equipment should address weed seed problems. Agricultural Engineering (Jan/Feb), 1989, 11.

19. Howard, C.L., Mortimer, A.M., Gould, P., Putwain, P.D., Cousens, R. and Cussans, G.W., The dispersal of weeds : seed movement in arable agriculture. Proceedings Brighton Crop Protection Conference - Weeds, 1991, 821 - 828.

20. Putwain, P.D., Scott, K.R. and Holliday, R.J., The nature of resistance to triazine herbicides : Case studies of phenology and population studies. In Herbicide Resistance in Plants. eds H.M. LeBaron and J. Gressel, Wiley, New York, 1982, pp 99 - 116.

21. Moss S.R., and Cussans, G.W., The development of herbicide resistant populations of *Alopecurus myosuroides* (Blackgrass) in England. In Herbicide Resistance in Weeds and Crops eds. J.C. Caseley, G.W. Cussans and R.K. Atkin, Butterworth-Heinemann, Oxford, 1991, pp 45 - 56.

22. Moss, S.R., Herbicide cross-resistance in slender foxtail (*Alopecurus myosuroides*). Weed Science, 1990, **38**, 492 - 496.

23. Ulf-Hansen, P.F., The dynamics of natural selection for herbicide resistance in grass weeds, Ph.D. Thesis, University of Liverpool, 1989.

24. Law R. and Watkinson, A.R., Response-surface analysis of two-species competition: an experiment on *Phleum arenarium* and *Vulpia fasciculata*. Journal of Ecology, 1987, **75**, 871 - 886.

25. SAS Institute Inc, SAS User's Guide : Statistics, Version 5. SAS Inst. Inc. Cary, N. Carolina, 1985.

26. Lewontin R.C. and White M.J.D., Interaction between inversion polymorphisms of two chromosome pairs in the grasshopper, *Moraba scurra*. Evolution, 1960, **14**, 116 - 129.

27. Mortimer, A. M., Sutton J. J., Putwain P. D. and Gould P., The dynamics of mixtures of arable weed species. Proceedings European Weed Research Society Symposium (1990), Integrated weed management in cereals, EWRS Wageningen, 1990, 19 - 26.

28. Price, S., Hill, J. and Allard, R., Genetic variability for herbicide reaction in plant populations. Weed Science, 1983, **31**, 652 - 657.

29. Price, S., Allard, R., Hill, J. and Naylor J., Associations between discrete genetic loci and genetic variability for herbicide reaction in plant populations. Weed Science, 1985, **33**, 650 - 653.

ANALYSIS OF INSECTICIDE RESISTANCE IN THE WHITEFLY, *BEMISIA TABACI*

FRANK J BYRNE, IAN DENHOLM, LINZI C BIRNIE,
ALAN L DEVONSHIRE AND MARK W ROWLAND
AFRC Institute of Arable Crops Research
Rothamsted Experimental Station
Harpenden, Herts, AL5 2JQ

ABSTRACT

The tobacco whitefly, *Bemisia tabaci*, is now a major world pest, causing extensive damage to both field and glasshouse crops. It has attained this status partly through the development of insecticide resistance resulting from over-reliance on broad-spectrum insecticides. In this paper, we examine the role of resistance in contributing to current control difficulties, focussing on recent research at Rothamsted and elsewhere into the incidence and biochemical basis of resistance mechanisms. The potential of 'field simulator' technology for studying resistance dynamics and the integration of chemical and biological control tactics under realistic exposure conditions in the laboratory is discussed.

INTRODUCTION

The tobacco whitefly, *Bemisia tabaci*, is regarded as one of the world's most destructive and intractable crop pests. It is now widespread throughout tropical and sub-tropical regions, attacking a large number of arable, horticultural and ornamental crop species. In some countries, such as Sudan [1], Israel [2], and India [3], it achieves particular significance as a pest of cotton, impairing yield through direct feeding damage, virus transmission, and the production of copious amounts of honeydew that impedes both harvesting and ginning. *B. tabaci* is also becoming increasingly important as a vector of geminiviruses on vegetable crops, especially those of the Solanaceae and Cucurbitaceae, and as such has been implicated in widespread crop failures in the Middle East, the Caribbean and southern USA [4,5].

Conventional control programmes based largely or solely on the use of chemical pesticides have often proved ineffective in suppressing whitefly infestations. Indeed, there are many examples, on cotton in particular, of pesticide usage aggravating rather than reducing whitefly problems. Insecticide resistance is by no means the only cause of these difficulties, but is unquestionably a major contributing factor. In this paper we examine the background to present problems with controlling *B. tabaci*, and review recent work at Rothamsted and elsewhere on the incidence, nature and selection of resistance in this pest.

BACKGROUND TO PROBLEMS WITH *B. TABACI*

B. tabaci provides a dramatic example of a species of relatively minor importance increasing in severity and geographical range to attain primary pest status over a short period of time. These changes have been attributed to diverse climatic, agronomic, biological and insecticide-related factors whose relative importance is still poorly understood, and likely to vary considerably between cropping regions. For example, combinations of high temperature and low rainfall, which appear to favour *B. tabaci*, were implicated as contributing to intermittent outbreaks on cotton in southern India in the 1980s [3] , but are unlikely to be a primary cause of more consistent increases in whitefly abundance. Agronomic factors refer both to cropping patterns that enable *B. tabaci* to persist on favoured hosts throughout the year [7], and to the implementation, for cotton in particular, of crop-yield maximisation strategies [8] exploiting the availability of post-war organic insecticides. Relying on these new chemicals, high-yielding varieties, usually with low resistance to pest attack, were grown closely spaced in large monocultures creating a dense canopy ideal for harbouring huge densities of well-protected immature stages. Liberal use of nitrogenous fertilizers seems to have contributed by encouraging canopy growth and generating a more favourable pH environment for whitefly development.

Problems arising from altered agricultural practice, whilst significant in their own right, may be exacerbated by more insidious changes in the pest itself. Recent identification in the USA [4,9] of an apparently novel biotype of *B. tabaci*, distinguishable by both its esterase banding pattern and ability to inflict characteristic 'silverleaf' symptoms on squash *(Curcubita pepo)* have very far-reaching consequences for the impact and containment of whitefly infestations. This biotype is not only more virulent than established ones on a wider range of hosts, but appears to be rapidly increasing its range through passive transportation on ornamental plant species [10,11].

Insecticide-Related Effects

There are three distinct ways that insecticide usage is considered to have had a marked impact on whitefly abundance. Firstly, there is a widespread belief, considered in more detail later, that broad-spectrum insecticides have contributed to outbreaks by eliminating naturally-occurring control agents such as predators and parasitoids. The second effect is that of 'acceleration' [6], whereby sublethal exposure to some chemicals may stimulate whitefly population growth by increasing fecundity or shortening development times. This effect is well-documented for mites [12], and was reported for DDT in laboratory tests with *B. tabaci* [6]. More recently, Rao *et al.* [13] evaluated several chemicals in field and laboratory experiments, and suggested that acceleratory effects are by no means confined to DDT. Sublethal exposure of whitefly adults to a range of organophosphates (OPs) and pyrethroids reportedly enhanced fecundity, reduced the incubation period of eggs and, in the case of pyrethroids at least, increased the proportion of females in resulting progeny.

The third damaging side-effect of using insecticides is resistance itself. In some cases, there is evidence of resistance appearing before the whitefly was perceived as a primary pest. For example, *B. tabaci* populations erupting to plague proportions in the Sudan Gezira at the end of the 1970s were apparently already strongly resistant to dimethoate and monocrotophos, leading to a particularly unwelcome scenario - a dramatic pest outbreak coupled with resistance sufficiently high to render valuable insecticides ineffective just when they were most needed [6]. Whether or not resistance actually preceded resurgence in other countries, there is no doubt that resistance resulting from subsequent, extensive reliance on chemicals for whitefly control now constitutes an additional threat to effective crop production in many countries.

DEVELOPMENT OF RESISTANCE

The clearest case-histories of resistance development in *B. tabaci* come from monitoring work by Dittrich and colleagues in Sudan, Central America and Turkey during the 1980s [6,14]. Their data not only demonstrate how the incidence and breadth of resistance has increased over time, but help to identify operational factors underlying these changes. In Sudan, for example, resistance to cypermethrin and deltamethrin increased dramatically to *ca.* 170- and 350-fold, respectively, between 1982 and 1985 as a consequence of heavy use of pyrethroids, but declined subsequently when fewer sprays were applied. However, resistance to monocrotophos and dimethoate remained high due to their continued usage. In California, there was a marked increase in resistance to some OPs and pyrethroids following large-scale use of these insecticides in 1983 and 1984 [15,16]. Pyrethroid resistance also rose sharply in Turkey between 1986 and 1987, when pyrethroids were reintroduced following a five-year ban imposed to combat increasing resistance in *Helicoverpa armigera* and the perceived role of pyrethroids in the resurgence of sucking pests [14]. In all these cases, monitoring proved crucial for attributing control difficulties to resistance rather than to other causes such as climatic variation or inappropriate pesticide applications.

It is notable that the high levels of pyrethroid resistance recorded in the Sudan in 1985 did not extend to bifenthrin and fenpropathrin, two newer products that remained fully effective in foliar residue bioassays [14]. In comparison, pyrethroid resistance documented in Guatemala in 1985 was more pervasive and encompassed these pyrethroids as well as the widely-used products. Without detailed studies it is impossible to assess whether these potentially exploitable differences in cross-resistance reflected geographical variation in a single mechanism, or a multifactorial basis for pyrethroid resistance in *B. tabaci*.

Due to the speed at which whitefly problems have developed and spread to new areas, current records of resistance summarised in Table 1 are undoubtedly incomplete. Even so, this list covers most areas where *B. tabaci* attains greatest importance and highlights the urgent need for research into both the nature of resistance and possible management tactics.

TABLE I

Approximate chronology of recorded resistance to insecticides in
B. tabaci

Country/ Region	Resistance first confirmed	Compounds resisted	Source
Sudan	1981	OPs, DDT	[6]
	1982	Pyrethroids	[14]
Central America	1985	OPs, Pyrethroids, DDT, Endosulfan	[14]
Turkey	1985	OPs, DDT	[14]
	1987	Pyrethroids	[14]
Israel	1985	OPs, Endosulfan	[16a]
USA	1983	OPs, Pyrethroids	[16]
Cyprus*	1991	OPs	[11]
Yemen*	1991	OPs	[11]
UK*	1991	OPs	[11]
Pakistan	1991	OPs, Pyrethroids	M.A. Shakeel pers. comm. 1991; Rothamsted, unpublished data

*Based on biochemical detection of OP-insensitive AChE

MECHANISMS OF RESISTANCE

Most attempts to resolve resistance mechanisms have relied on synergists considered to block specific metabolic detoxication pathways, coupled in some cases with esterase measurements and biochemical characterisation of acetylcholinesterase (AChE), the target enzyme of OPs and carbamates.

The first such study [6] implicated non-specific esterases and insensitive AChE in resistance of Sudanese populations to dimethoate and monocrotophos. Compared with a susceptible strain, resistant insects showed a 20-fold increase in esterase activity with 1-naphthyl butyrate as substrate, and their AChE was 10- and 2-fold less sensitive to inhibition by carbofuran and monocrotophos, respectively.

This approach was later extended to samples from Sudan, Turkey, Nicaragua and Guatemala, and disclosed some notable differences between populations [17]. Monocrotophos was synergized by both tricresylphosphate (TCP; indicative of esteratic detoxication) and piperonyl butoxide (PB; indicative of oxidative metabolism) in resistant insects from Nicaragua and Guatemala, but neither synergist was effective against populations from Sudan or Turkey. However, TCP synergised cypermethrin in all strains, its effect being greatest in the Nicaraguan strain which showed the highest esterase activity against 1-naphthyl butyrate. The AChE of Nicaraguan insects was highly insensitive to carbofuran (I_{50} ratio > 1000), but was the most sensitive of all to monocrotophos. In contrast, the AChE of Turkish insects was highly insensitive (> 2000) to monocrotophos, this being the only mechanism of resistance to monocrotophos detected. Insensitivity to carbofuran was similar in the Sudanese, Guatemalan and Turkish strains. Compared with earlier results [6], insensitivity to carbofuran and monocrotophos in Sudanese whiteflies increased 8- and 600-fold respectively between 1985 and 1988.

In an Israeli strain of unspecified resistance status, monocrotophos, profenofos and acephate synergized the toxicity of a cypermethrin formulation and prolonged efficacy under glasshouse and field conditions [18]. *In vitro* studies indicated that this synergism was due to inhibition of pyrethroid-hydrolysing esterases by the OPs. Similarly, *S,S,S*-tributylphosphorotrithioate (DEF) synergized cypermethrin and permethrin in a Californian strain, causing almost complete loss of resistance to these pyrethroids [19]. In the same strain, Prabhaker *et al.* [15] inferred the presence of several metabolic mechanisms from synergism studies. DEF not only synergized permethrin, but also malathion, methyl parathion and sulprofos. There was also synergism of malathion by triphenyl phosphate (TPP), taken to indicate the presence of malathion carboxylesterase(s). The three OPs and permethrin were further synergized by PB, indicating oxidative metabolism, but the combined synergistic effects of DEF, TPP and PB did not eliminate resistance entirely and it was suggested, without experimental support, that glutathione S-transferases and/or insensitive AChE might account for the remainder of the resistance.

Although the presence of additional mechanisms cannot be discounted, biochemical work at Rothamsted on several *B. tabaci* strains has supported earlier findings of Dittrich and co-workers indicating a general involvement of non-specific esterases and insensitive AChE in resistance to pyrethroids and OPs respectively [11,20]. The first conclusion is supported by both a consistent relationship between total esterase activity (measured using 1-naphthyl butyrate as substrate) and pyrethroid resistance, and by limited *in vitro* metabolism studies with radio-labelled pyrethroids [F.J. Byrne, unpublished data]. In most strains, however, it is still impossible to equate resistance with specific esterase bands disclosed by PAGE; this must be borne in mind when attempting to exploit electrophoresis for resistance monitoring purposes (see later).

AChE variation within and between populations has been characterised by a kinetic microplate assay measuring insensitivity profiles of AChE in single male whiteflies [11]. The three AChE variants detected so far are best distinguished by correlating their responses to discriminating concentrations of the OP inhibitors paraoxon and azamethiphos (Figure 1). Two of these variants, SUD-S (fully sensitive) and NIC-R (doubly insensitive), appear to be rare in contemporary field populations. Indeed, virtually all recent collections have contained only the third, paraoxon-insensitive variant (e.g. SUD-R) originally characterised in strains from Guatemala and the Sudan. Work to isolate strains homozygous for different esterase and AChE phenotypes, and to correlate these with cross-resistance patterns disclosed by bioassays, is currently underway.

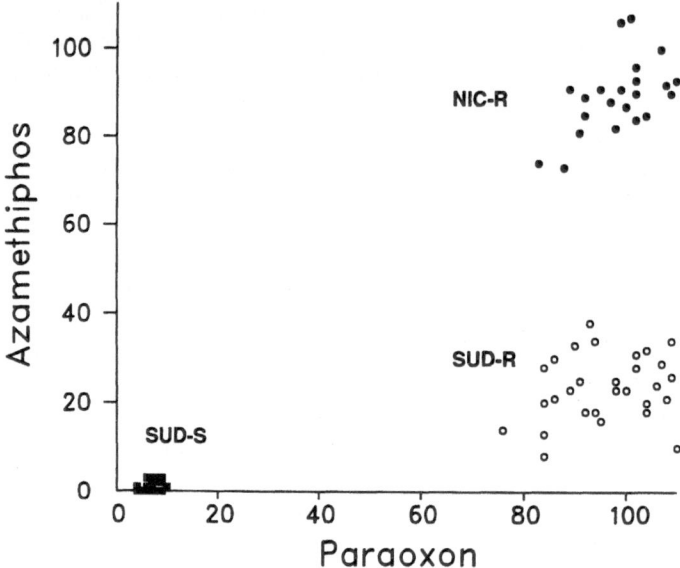

Figure 1. Bivariate plot of mean % activity remaining during inhibition of AChE by 30 μM paraoxon and 100 μM azamethiphos for three AChE variants identified in populations of adult male whiteflies.

IMPLICATIONS FOR BIOCHEMICAL MONITORING OF RESISTANCE

As for other insects, an improved understanding of resistance mechanisms in *B. tabaci* introduces exciting opportunities for developing biochemical assays for diagnosing resistance and monitoring its development under selection with insecticides. In recent years, much effort has been expended on the development of sensitive microplate technology to enable large-scale and rapid throughput of test insects. However, the limitations of this approach compared to coarse-grained bioassay procedures must also be appreciated [21, 22]. In particular, it is vital that these techniques, which usually depend on the use of model substrates for enzymes in order to achieve the desired sensitivity, are fully validated to establish the exact contribution to resistance of the enzyme being assayed [21]. Work at Rothamsted illustrates some advantages and potential pitfalls of such assays.

Assays requiring least validation, and which can be transferred with greatest confidence from one species to another, are those based on measurement of the inhibition kinetics of a known target enzyme such as AChE. Hence the microplate assay being used to characterise AChE insensitivity in single insects [11] provides a robust and reliable means of diagnosing resistant heterozygotes and homozygotes, establishing their frequency, and even distinguishing between enzyme variants coded by different alleles at the AChE locus [see also 23]. The major drawback is that absence of an insensitive enzyme need not signify the absence of resistance to a particular OP or carbamate, which could be conferred by metabolic enzymes not disclosed by this assay.

Exploiting total esterase activity as a diagnostic for resistance requires much greater care, even if a correlation between these characters has been established. In resistant whiteflies from the Sudan and Nicaragua, specific esterases have been identified by PAGE which contribute most to the high esterase activities associated with these populations [11]. In the Sudanese resistant strain, esterase measurements are normally 20-fold higher than in the susceptible strain, but susceptible and resistant insects surviving on profenofos-treated cotton leaves cannot be distinguished because the esterases in the resistant strain are inhibited and give esterase activities equal to, and below, those in unexposed susceptible insects [24]. Only by transferring insects to clean leaves, to allow the enzyme activity to recover, can the two forms be distinguished. As long as insects remain in contact with residues on treated leaves, the esterase activity will not recover sufficiently to be a useful diagnostic marker. This effect may explain the finding of Wool and Greenberg [25] that resistant *B. tabaci* individuals collected from insecticide-treated cotton fields showed, contrary to expectations, lower levels of esterase activity than those from untreated control populations.

APPLICATION OF 'FIELD SIMULATOR' TECHNOLOGY TO *B. TABACI*

A second component of whitefly research at Rothamsted has entailed the development and use of new technology for simulating the field selection of resistance, and for evaluating control tactics against resistant populations. With this system, a known number of adults of prescribed genotypic composition are released onto cotton plants held in large 'field simulator' chambers [26]. Insects breed continuously and distribute themselves on the plants as under field conditions, and can be treated as and when desired by applying formulated insecticides from a sprayer running along rails in the roof of each cage. Sophisticated techniques, including biochemical assays, are employed to monitor the effects of these treatments on both frequencies of resistance and population size.

By reproducing, within limits, ways that populations develop and are controlled in the field, this system has several important applications for studying the selection and management of resistance in *B. tabaci*. Principal applications have been (i) investigating how resistance is expressed and selected under different control regimes; (ii) evaluating tactics such as insecticide alternations and mixtures for combating resistance; (iii) establishing the efficacy of new products against contemporary populations, including the potential for cross-resistance;

(iv) examining the effects of sub-lethal exposure to residues on whitefly demography and behaviour, thereby providing insight into the mechanisms of whitefly resurgence; and (v) most recently, analysing interactions between *B. tabaci* and its primary parasitoids in the presence and absence of pesticides. The first and last of these are illustrated below.

Expression and Selection of Insecticide Resistance

An important requirement for interpreting and managing resistance is to establish the magnitude of a selective advantage conferred by resistance genes under field exposure conditions, and to identify the stage(s) in a treatment regime at which it is expressed [27,28]. Conventional laboratory bioassays, in which insects are dosed topically or confined onto treated surfaces for prescribed periods, may offer very limited assistance in this respect. Certainly, there is increasing evidence that resistance ratios can be highly dependent on the bioassay method employed [eg. 29], and that standard tests can greatly misjudge the degree of protection conferred by different genotypes under conditions prevailing in the field [30,31,32].

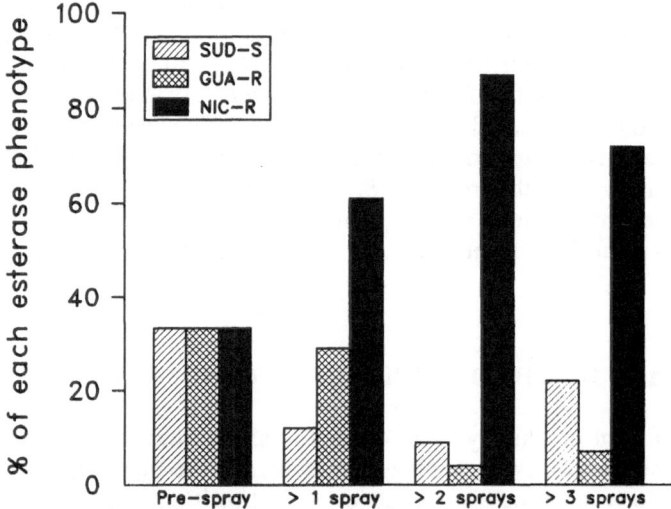

Figure 2. Changes in frequency of three whitefly strains selected with cypermethrin in field-simulators. Strains were identified using esterase activity as a diagnostic biochemical marker (see text for details).

Discrepancies that can arise between exposure regimes were well illustrated by work comparing the responses of *Bemisia* adults of a susceptible (SUD-S) and resistant (SUD-R) strain, both originating in the Sudan, to cypermethrin in a conventional leaf-dip bioassay, with their survival in simulators sprayed with the same chemical at different application rates [33]. Resistance was very clearly expressed in the standard bioassay, yielding a resistance ratio of 83-fold at LC_{50}. In the simulators, however, the two strains exhibited almost identical survivorship at each cypermethrin concentration. Reason(s) for this similarity in response are still unclear, though a behavioural explanation was suggested based on sub-lethal exposure to cypermethrin disrupting flight activity or causing insects to avoid treated foliage [33].

Longer-term experiments exposing mixed populations to repeated insecticide sprays have disclosed similar anomalies in the relationship between selection rates and bioassay data. In the example shown (Figure 2), a population initially containing equal frequencies of three strains - one susceptible (SUD-S) and two resistant (GUA-R and NIC-R, of Guatemalan and Nicaraguan origin respectively) - was sprayed in each generation with cypermethrin ('Polytrin'; Ciba-Geigy AG) at the manufacturer's recommended rate (50 g/ha). Changes in their frequency as a consequence of treatment were monitored biochemically, exploiting different levels of total esterase activity as diagnostic markers for the three strains. As expected, NIC-R insects possessing the highest esterase activity, and strongest resistance to pyrethroids in leaf-dip bioassays, rapidly came to predominate in the population. However, GUA-R insects, which were biochemically and toxicologically indistinguishable from those of the SUD-R strain referred to previously, showed no overall advantage over SUD-S, despite their *ca.* 80-fold resistance in the foliar bioassay.

Clearly, the increased esterase activity and intrinsic resistance to pyrethroids of GUA-R and SUD-R insects must be of some practical significance to have been selected in the first place. Such findings nonetheless demonstrate that factors determining the selective superiority of resistance genes may be far more subtle than initially expected, and highlight challenges that the simulators are well-equipped to address in more detail.

Integration of Control Tactics

Field simulators also provide a means of evaluating broader aspects of pest management, in particular the prospects for integrating chemical and biological control tactics. There is, as noted earlier, a strong suspicion that overuse of broad-spectrum insecticides has contributed to problems in controlling *B. tabaci* by eliminating natural enemies. Certainly, such enemies do exist, the most widely-reported and specific being Aphelinid parasitoids within the genera *Eretmocerus* and *Encarsia*. This parasitoid complex achieves its greatest diversity in the Middle East and the Indian Subcontinent, where *B. tabaci* is likely to have originated, but it extends throughout the geographical range of the whitefly and includes apparently endemic and host-specific taxa in Japan and the south-western USA [reviewed in [34]. Despite difficulties with the taxonomy of parasitoids, two species, *Eretmocerus mundus* and *Encarsia lutea*, have clearly predominated in records from much of the Old World distribution of *B. tabaci*. *E. mundus* in particular appears to be a relatively constant associate of the whitefly, having been reported, often at appreciable densities, from the Sudan, Egypt, Israel, Jordan, East Africa, and India. Although information on its effectiveness in controlling the whitefly is sparse and often anecdotal, reports of 40-90% parasitism in Jordan and Egypt indicate that *E. mundus* can exert significant mortality on field populations [35,36]. Conversely, the scarcity of parasitism in populations treated frequently with pesticides supports the view that some compounds at least are incompatible with IPM programmes. Unfortunately, data on the impact of different chemicals on *E. mundus* is again fragmentary; the only comprehensive data of direct relevance being that collected under very artificial laboratory conditions for *Encarsia formosa*, the standard biocontrol agent for glasshouse whitefly, *Trialeurodes vaporariorum* [eg. 37].

To address this shortcoming, simulator experiments have been extended to study in more detail interactions between *B. tabaci* and *E. mundus*, with the aim of evaluating some widely-used chemicals for their compatibility with IPM schemes. Figure 3 shows one such experiment investigating the impact of *E. mundus* (cultured from insects collected in Israeli cotton fields) on whitefly population growth in the absence of pesticide treatments. Female parasitoids were added 10 days after the initial introduction of whiteflies (100 females/cage), to coincide with the first cohort of immature whiteflies reaching 2nd instar, the preferred stage for parasitism. Whitefly numbers were then monitored at regular intervals until day 54, just prior to the expected emergence of the third generation of whitefly adults.

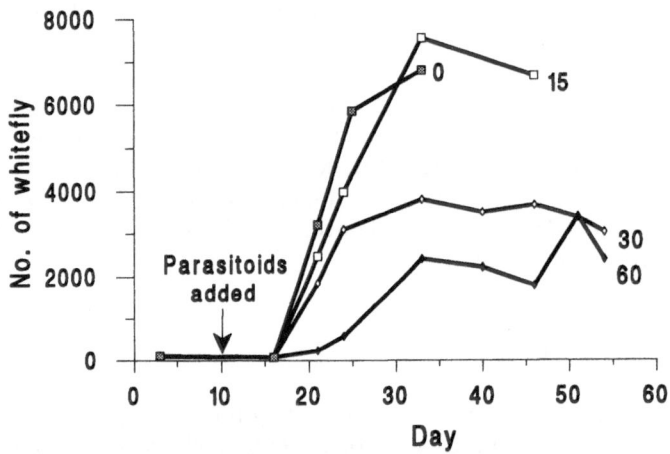

Figure 3. Impact of different starting densities of *E. mundus* (0-60 females per cage) on whitefly numbers in field simulators over two successive generations.

In all cages, there was an initial slow decline in the size of the founding population, followed by a sharp increase in numbers from day 17 as adult progeny emerged. Addition of 30 or 60 parasitoids/cage not only substantially reduced the size of this first generation, but completely suppressed any further increase due to the emergence of second generation adults between days 36 and 40. By this stage, numbers in the control cage had reached a level (several tens of thousands) at which monitoring was no longer feasible. Confirmation that *E. mundus* can be a powerful moderator of whitefly population growth in the simulator has led to series of experiments, currently underway, in which both established insecticides and newer products are being evaluated for their disruptive effects on both host and parasitoid. Such data should in turn assist in improving programmes aimed at conserving parasitoids as an additional, naturally-occurring mortality factor.

CONCLUSIONS

Work on the nature and selection of resistance in *B. tabaci* can undoubtedly help define specific tactics for preserving the effectiveness of established and new products. However, integrating these into coherent strategies that contend also with resistance in coexisting pests such as bollworms is likely to prove a complex task, and depend closely on progress with IPM-orientated approaches of improving scouting methods, setting realistic economic thresholds, and developing cultural control practices that reduce whitefly abundance and overall reliance on pesticide applications [2,14,38].

With respect to insecticide usage, however, some general recommendations are possible. Firstly, new or substitute insecticides must be evaluated realistically for their ability to control contemporary populations, and must be used in strict moderation to minimize additional resistance risks. Secondly, compounds that accelerate whitefly reproduction should be identified and deployed with great care in areas where *B. tabaci* is a current or perceived threat. Thirdly, use of more selective chemicals that preserve whitefly natural enemies should be exploited as widely as possible, and, finally, tactics such as insecticide alternations and use of synergists must be optimised on the basis of cross-resistance and biochemical studies. Research on these topics is now of utmost priority for meeting the challenge of whitefly resistance on cotton and other crops.

REFERENCES

1. Eveleens, K.G. Cotton-insect control in the Sudan Gezira: Analysis of a crisis *Crop Prot.*, 1983, **2**, 273-287.

2. Forer, G. Whitefly management in Israel to prevent honeydew contamination. *Proceedings of 49th Plenary Meeting of the Internation Cotton Advisory Committee*, Montpellier, France, 1990, 33-37.

3. Basu, A.K. Resurgence of whitefly in cotton and strategies for its management *Proceedings of the National Symposium on Resurgence of Sucking Pests*, (T.N.A.U. Coimbatore), 1986, 129-133.

4. Cohen, S., Duffus, J.E. and Liu, H.Y. A new *Bemisia tabaci* biotype in the southwestern United States and its role in silverleaf of squash and transmission of lettuce infectious yellow virus. *Phytopath.*, 1992, **82**, 86-90.

5. Perring, T.M., Cooper, A., Kazmer, D.J., Shields, C. and Shields, J. New strain of sweetpotato whitefly invades California vegetables. *Calif Agric.*, 1991, **45**(6), 10-12.

6. Dittrich, V., Hassan, S.C. and Ernst, G.E. Sudanese cotton and the whitefly: A case study of the emergence of a new primary pest, *Crop Prot.*, 1985, **4**, 161-176.

7. Byrne, D.N., Bellows, T.S. and Parrella, M.P. Whiteflies in agricultural systems. In *Whiteflies: Their Bionomics, Pest Status and Management,* ed D. Gerling, Intercept, Andover, England, 1990, pp. 227-261..

8. Bottrell, D.G. and Adkisson, P.L. Cotton insect pest management. *Ann. Rev. Entomol.* 1977, **22**, 451-481.

9. Costa, H.S. and Brown, J.K. Variation in biological characteristics and esterase patterns among populations of *Bemisia tabaci*, and the association of one population with silverleaf symptom induction. *Entomol. exp. appl.,* 1991, **61**, 211-219.

10. Cohen, S. Induction of silverleaf of squash by *Bemisia tabaci* whitefly from California desert whitefly populations. *Plant Disease,* 1991, **75**(8), 862.

11. Byrne, F.J. and Devonshire, A.L. Insensitive AChE and esterase polymorphism in susceptible and resistant populations of the tobacco whitefly *Bemisia tabaci* (Genn.), Submitted to *Pestic. Biochem. Physiol.*

12. Dittrich, V., Streibert, P. and Bathe, P.A. An old case reopened: mite stimulation by insecticide residues. *Environ. Ent.,* 1974, **3**, 534-540.

13. Rao, N.V., Reddy, A.S. and Reddy, D.D.R. Impact of some insecticides on *Bemisia tabaci* on cotton. *Jour. Pl. Trop.,* 1990, **7**(2), 77-86.

14. Dittrich, V., Uk, S., Ernst, G.E. Chemical control and insecticide resistance in whiteflies In *Whiteflies: Their Bionomics, Pest Status and Management,* ed D. Gerling, Intercept, Andover, England, 1990, pp. 263-285.

15. Prabhaker, N., Coudriet, D.L. and Toscano, N.C. Effect of synergists on organophosphate and permethrin resistance in sweetpotato whitefly (Homoptera:Aleyrodidae). *J. Econ. Entomol.,* 1988, **81**, 34-39.

16. Prabhaker, N., Coudriet, D.L. and Meyerdirk, Insecticide resistance in the sweetpotato whitefly *Bemisia tabaci* (Homoptera: Aleyrodidae). *J. Econ. Entomol.,* 1985, **78**, 748-752.

16a. Perry, A.S. The relative susceptibility to several insecticides of adult whiteflies *(Bemisia tabaci),* from various cotton-growing areas in Israel. *Phytoparasitica,* 1985, **13**, 77-78.

17. Dittrich, V., Ernst, G.E., Ruesch, O. and Uk, S. Resistance mechanisms in sweetpotato whitefly (Homoptera:Aleyrodidae) populations from Sudan, Turkey, Guatemala and Nicaragua. *J. Econ. Entomol.,* 1991, **83**, 1665-1670.

18. Ishaaya, I., Mendelson, Z., Ascher, K.R.S. and Casida, J.E. Cypermethrin synergism by pyrethroid esterase inhibitors in adults of the whitefly *Bemisia tabaci. Pestic. Biochem. Physiol.,* 1987, **28**, 155-162.

19. Horowitz, A.R., Toscano, N.C., Youngman, R.R. and Georghiou, G.P. Synergism of insecticides with DEF in sweetpotato whitefly (Homoptera:Aleyrodidae). *J. Econ. Entomol.,* 1988, **81**, 110-114.

20. Byrne, F.J., Devonshire, A.L., Rowland, M.R. and Sawicki, R.M. Biochemical study of insecticide resistance in the cotton whitefly, *Bemisia tabaci* (Genn.). *Proc. 7th Int. Congr. Pestic. Chem., Hamburg,* 1990, **2**, 85.

21. Devonshire, A.L. Biochemical and DNA probing techniques for monitoring insecticide resistance genes. In *Pesticides and Alternatives: Innovative Chemical and Biological Approaches to Pest Control,* ed. J. Casida, Elsevier, Amsterdam, 1990, pp. 421-431.

22. Denholm, I. Monitoring and interpreting changes in insecticide resistance. *Funct. Ecol.,* 1990, **4**, 601-608.

23. Moores, G.D., Devonshire, A.L. and Denholm, I. A microtitre plate assay for characterising insensitive acetylcholinesterase genotypes of insecticide-resistant insects. *Bull. Entomol. Res.,* 1988, **78**, 537-544.

24. Byrne, F.J. and Devonshire, A.L. *In vitro* inhibition of esterase and acetylcholinesterase activities by profenofos treatments in the tobacco whitefly *Bemisia tabaci* (Genn.): implications for routine biochemical monitoring of these enzymes. *Pestic. Biochem. Physiol.,* 1991, **40**, 198-204.

25. Wool, D. and Greenberg, S. Esterase activity in whiteflies *(Bemisia tabaci)* in Israel in relation to insecticide resistance. *Entomol. exp. appl.,* 1990, **57**, 251-258.

26. Rowland, M.W., Pye, B., Stribley, M., Hackett, B., Denholm, I. and Sawicki, R.M. Laboratory apparatus and techniques for the rearing and insecticidal treatment of whitefly, *Bemisia tabaci*, under simulated field conditions. *Bull. Entomol. Res.,* 1990, **80**, 209-216.

27. Roush, R.T. and McKenzie, J.A. Ecological genetics of insecticide and acaricide resistance. *Ann. Rev. Entomol.,* 1987, **32**, 361-380.

28. Denholm, I., Sawicki, R.M. and Farnham, A.W. Laboratory simulation of selection for resistance. In *Combating Resistance to Xenobiotics*, eds. M.G. Ford, D.W. Hollomon, B.P.S. Khambay and R.M. Sawicki, Ellis Horwood, Chichester, England, 1987, pp. 138-149.

29. Dennehy, T.J., Granett, J. and Leigh, T.F. Relevance of slide-dip and residual bioassay comparisons to detection of resistance in spider mites. *J. Econ. Entomol.,* 1983, **76**, 1225-1230.

30. Rawlings, P., Davidson, G., Sakai, R.K., Rathar, H.R., Aslamkan, K.M. and Curtis, C.G. Field measurement of the effective dominance of an insecticide resistance in anopheline mosquitoes. *Bull. WHO,* 1981, **59**, 631-640.

31. McKenzie, J.A. and Whitten, M.J. Estimation of the relative viabilities of insecticide resistance genotypes of the Australian sheep blowfly, *Lucilia cuprina*. *Aust. J. Biol. Sci.*, 1984, **37**, 45-52.

32. Daly, J.C., Fisk, J.H. and Forrester, N.W. Selective mortality in field trials between strains of *Heliothis armigera* (Lepidoptera:Noctuidae) resistant and susceptible to pyrethroids: functional dominance of resistance and age-class. *J. Econ. Entomol.*, 1988, **81**, 1000-1007.

33. Rowland, M.W., Hackett, B. and Stribley, M. Evaluation of insecticides in field control simulators and standard laboratory bioassays against resistant and susceptible *Bemisia tabaci* (Homoptera:Aleyrodidae) from Sudan. *Bull. Entomol. Res.*, 1991, **81**, 189-199.

34. Gerling, D. Natural enemies of whiteflies: predators and parasitoids. In *Whiteflies: Their Bionomics, Pest Status and Management,* ed D. Gerling, Intercept, Andover, England, 1990, pp. 147-185.

35. Hafez, M., Awadallah, K.T., Tawfik, M.F.S. and Sarhan, A.A. Impact of the parasite *Eretmocerus mundus* on populations of the cotton whitefly *Bemisia tabaci* in Egypt. *Bull. Soc. Ent. Egypt*, 1978, **62**, 23-32.

36. Sharaf, N.S. Parasitisation of the tobacco whitefly *Bemisia tabaci* on *Lantana camera* in the Jordan Valley. *Z. Ang. Ent.*, 1982, **94**, 263-271.

37. Oomen, P. Guideline for the evaluation of side-effects of pesticides: *Encarsia formosa. Bull. OEPP/EPPO*, 1985, **15**, 257-265.

38. Denholm, I. and Birnie, L. Prospects for managing resistance to insecticides in the whitefly. *Proceedings 49th Plenary Meeting of the International Cotton Advisory Committee,* Montpellier, 1990, 37-40.

NEGATIVE CROSS-RESISTANCE IN FUNGICIDES :
FROM THE LABORATORY TO THE FIELD

PIERRE LEROUX
INRA, Station de Phytopharmacie
Etoile de Choisy, 78000 Versailles, France

ABSTRACT

When emergence of resistance to one fungicide coincides with increased sensitivity to another chemical, it may be a case of negative cross-resistance. Examples are given for carboxamides, sterol C-14 demethylation inhibitors and benzimidazoles. The mechanisms of such negative relationships and their practical implications are discussed.

INTRODUCTION

Resistance to fungicides in plant pathogens is one of the main problems encountered in chemical crop protection. This phenomenon involves the most effective fungicides available for fungal disease control (eg. benzimidazoles, dicarboximides, phenylamides, sterol C-14 demethylation inhibitors). The most common strategy for preventing the onset of fungicide resistance in practice restricts use of "at-risk" product, and combines and alternates it with other types of fungicides (1). Of course, the mixture partners must not present positive cross-resistance. On the other hand the use of negative cross-resistance companions may be a practical strategy.

Negative cross-resistance arises when the same genetic factor confers simultaneously resistance to a toxicant and increased sensitivity to another (2). This phenomenon, which can occur in laboratory mutants or field strains of various fungi, concerns antifungal compounds : (a) having the same target and related chemical structures, (b) inhibiting the same metabolic process but at different sites or (c) having different modes of action (3).

This subject has been reviewed in 1984 by De Waard (3) and this paper will deal with recent findings, concerning the negative cross-resistance in fungal strains resistant to : carboxamides, sterol C-14 demethylation inhibitors or benzimidazoles.

CARBOXAMIDES

Carboxin, and more generally the carboxamide fungicides having the basic cis-crotonanilide structure, are systemic fungicides used to control Basidiomycetes. Their fungitoxic activity is due to inhibition of succinate oxidation in the mitochondria chain. This effect on complex II (succinate ubiquinone reductase) or SDC (succinate dehydrogenase complex) results probably from the binding of these carboxamides to peptides C II-3 and C II-4 in a phospolipid environment (4).

In Ustilago maydis, carboxin-resistant strains selected in the laboratory exhibit various resistance levels. Amongst them, two allelic mutations of the gene oxr 1 result in moderate or high resistance to carboxin. The highly carboxin-resistant strains carrying the oxr 1B mutations are resistant to many carboxamides whereas those carrying the oxr 1A mutations exhibit greater sensitivity to some of them. This negative cross-resistance is observed with carboxamides having substituent groups at the 4'-position on the phenyl ring of the anilide (eg. 4'- phenylcarboxin). On the contrary, the 3'-substituted carboxamides (eg. mepronil) present a positive cross-resistance towards carboxin (5-7 ; Table 1). With few exceptions the inhibition by carboxamides of cell growth of wild-type and carboxin-resistant strains of U. maydis closely parallels the inhibition of their respective SDCs (5-7). It has been suggested that the mechanism of carboxin resistance is some modification of the site of carboxin action in mitochrondrial complex II. Alterations which decrease the affinity of carboxin to its binding site, may either decrease or increase binding of other carboxamides leading to positive or negative cross-resistance.

Table 1

Cross-resistance between carboxamides in laboratory selected mutants of U.maydis and field strains of U. nuda

Strains	Fungicide concentrations in μM[a]			
	carboxin	fenfuram	mepronil	4'-phenyl-carboxin
Ustilago maydis				
wild-type	0.36	4.2	1.8	4.8
oxr-1A	5	27[b]	21	0.22
oxr-1B	37	-	-	75
Ustilago nuda				
wild-type	0.85	1.7	29	>35
UR1	3.8	25	120	>35
UR2	8.5	> 50	1.0	>35

(a) Concentration inhibiting by 50 % the SDC activity in mito-chondrial preparations of U. maydis [after ref. 5-7] or the growth of U. nuda cultivated on PDA [after ref. 9]
(b) - not tested

Recently, carboxin-resistant isolates of U. nuda have been detected on winter barley crops in France and in several other European countries (8, 9). The most common strains (UR2-type) are resistant to carboxin, fenfuram, methfuroxam, pyracarbolid but exhibit high sensitivity towards mepronil and flutolanil which both possess an iso- propyloxy group at the 3'-position of the anilide ring. In the other phenotype (UR1) there is a positive cross-resistance between all these carboxamides. The 4'-phenylcarboxin has no effect on the various isolates of U. nuda (Table 1). In preliminary experiments, Hollomon and Carter (10) have shown that the SDC from a wild-type isolate was less sensitive to carboxin than that from a resistant one (UR2-type) ; the reverse occurred with flutolanil. However, differences in sensitivity to these carboxamides observed in bioassays were greater than differences in degree of SDC inhibition indicating that other mechanisms might be involved. In field trials, carboxin (100 g a.i./100 kg seeds) controls totally loose smut when infection is only due to sensitive strains ; efficacy is about 30 % when resistant strains (UR2-type) are present. Mepronil (100 g a.i./100 kg seeds) is not effective against the wild-type strains, whereas it gives an efficacy of about 65 % against carboxin-resistant isolates (UR2- type) (8, 9). To my knowledge no field tests have been conducted with mixtures of carboxamides exhibiting negative cross-resistance, but in practice, carboxin is replaced by sterol C-14 demethylation inhibitors. In France, four triazole derivatives are available : diniconazole (30 g a.i./100 kg seeds), flutriafol (7.5 g a.i./100 kg seeds), myclobutanil (12.5 g a.i./100 kg seeds), triadimenol (10 g a.i./100 kg seeds).

STEROL C-14 DEMETHYLMATION INHIBITORS

Amongst the agricultural fungicides which inhibit the biosynthesis of sterols (SBIs), the most numerous ones are sterol C-14 demethylation inhibitors (DMIs). These DMIs have various chemical structures but all contain an heterocycle which is an imidazole (eg. prochloraz, triflumizole) a pyrimidine (eg. fenarimol) a pyridine (eg. pyrifenox) or a triazole (eg. cyproconazole, propiconazole, triadimenol, tebuconazole). The first step of sterol C-14 demethylation is a hydroxylation of the C-14 methyl group which is mediated by a cytochrome P-450 monooxygenase. The DMIs bind to a site of cytochrome P-450 which is normaly occupied by the natural substrate (lanosterol or eburicol). One of the nitrogen atoms in the heterocycle of these DMIs coordonates with the protohaem iron atom of the cytochrome P-450, while the lipophilic substituents interact with nearby regions of the enzyme to increase binding affinity. The size of these substituents determines the mobility of the DMIs within the substrate binding site and thus their inhibitory effects. (11, 12).

DMI-resistant strains collected in the field, or selected in the laboratory, are normally resistant to all DMIs. However the resistance factors can vary over a wide range. This situation occurs in mutants of U. avenae, U. maydis, Aspergillus nidulans, Cladosporium cucumerinunum, Monilinia fructicola, Penicillium italicum, Pseudocercosporella herpotrichoides, and in field strains of Erysiphe graminis, Sphaerotheca fuliginea and Rhynchosporium secalis (1, 3, 13-16).

In a few cases, absence of cross-resistance or even negative cross-resistance has been observed with some DMIs. Negative cross-resistance occurs in a mutant of Nectria haematoccoca var pisi selected on tebuconazole (17) and N. haematoccoca var cucurbitae selected on triadimenol (18), strains of P. herpotrichoides (Rye-type) isolated from wheat fields treated with prochloraz (19), a strain of Uncinula necator collected in a vineyard intensively treated with triadimenol (20) and a field isolate of Pyrenophora teres highly resistant to triadimenol, fenarimol and nuarimol (13). From the data in Table 2, it appears that DMIs

which exhibit a negative cross-resistance are different in the various
fungi.

Table 2
Resistance factors[a] to DMIs in laboratory selected
mutants of N. haematoccoca var pisi (A) and var
cucurbitae (B) and in field strains of P. teres (C),
P. herpotrichoides (D) and U. necator (E)
[after ref. 13, 17-20]

fungicides	Fungal strains				
	(A) teb-4	(B) teb-1	(C) 37/841	(D) IIp	(E) MAO2
prochloraz	1.4	-[b]	1.0	30	-
imazalil	0.3	1.0	0.3	7.5	-
triflumizol	-	0.3	-	0.05	-
fenarimol	0.1	-	200	0.35	0.4
cyproconazole	0.08	-	-	0.3	1.7
diclobutrazol	-	-	0.2	1.0	-
flusilazole	1.0	-	0.3	5.0	6.2
penconazole	0.07	-	3.0	2.4	0.4
propiconazole	-	3.5	0.5	3.2	-
tebuconazole	7	-	-	2.6	0.9
triadimenol	-	50	12	0.3	16

(a) The resistance factors are the ratios of the ED50 values for
the resistant strains to that for the wild-type strains.
Resistance factors below 1 and above 1 indicate negative and
positive cross-resistance respectively.
(b) - not tested

Many DMIs have one or two asymmetric carbons. The various isomers
can exhibit similar antifungal activities (eg. cyproconazole, propiconazol)
or not (ex. tebuconazole, triadimenol). In C. cucumerinum, P.
herpotrichoides and U. avenae positive cross-resistance occurs between all
isomers of various DMIs (14, 16, 21). With tebuconazole, however, the
situation is different in Saccharomycopsis lipolytica and Pyricularia
oryzae, because against wild-type strains the most active form is the (-)
isomer whereas in triadimenol - resistant mutants, it is the (+) isomer.
This result indicates that the racemate tebuconazole can be considered as a
mixture of compounds exhibiting negative cross-resistance (22).

In weakly DMI-resistant mutants of A. nidulas, P. italicum and N.
haematoccoca the mechanism of resistance is based on a constitutive energy-
dependent efflux (3, 11, 21). This process does not seem to operate in DMI-
resistant field isolates or in strains in which negative cross-resistance
between DMIs exists. In such cases modifications of the target C-14
demethylase could increase binding affinity with some DMIs, and a reduced
binding affinity with some others. According to Hollomon (23) changes may
involve the substrate binding site rather than the heme-binding region.

In practice when resistance to a DMI occurs, efficacy of all the other DMIs can be reduced (eg. E. graminis, S. fuliginea) or not (eg. R. secalis, P. teres, P. herpotrichoides, U. necator). However in the latter case, the situation can evolve if new phenotypes highly resistant to most DMIs appear as seems to be the case in U. necator (Steva, personal communication).

In A. nidulans, M. fructicola, Penicillium casecolum and U. maydis, some DMI-resistant mutants showed positive cross-resistance to fenpropimorph and/or tridemorph (3, 15, 24, 25). These morpholine derivatives, as well as the piperidine, fenpropidin, are inhibitors of sterol $\Delta14$ reductase and/or $\Delta8 \rightarrow \Delta7$ isomerase (11, 12). In P. italicum and U. maydis several strains selected in the laboratory for resistance to DMIs, show a negative cross-resistance to fenpropimorph and tridemorph (3, 24). In many other laboratory mutants, and in all DMI-resistant strains so far isolated in the field there is no cross-resistance with the inhibitors of sterol $\Delta14$ reductase or $\Delta8 \rightarrow \Delta7$ isomerase. The latter situation is of great importance for development of antiresistance strategies, especially in the case of cereal powdery mildews (26).

DMI-resistant isolates of various fungi sometimes show negatively cross-resistance to chemicals without any effect on sterol biosynthesis. This phenomenon has been found in laboratory selected mutants for acri-flavin, cycloheximide, neomycin, carboxin, doguadine and guazatine (3). However in field strains of Venturia inaequalis or P. herpotrichoides there is no cross-resistance between DMIs and doguadine or guazatine (Leroux, unpublished data). So, the practical interest of DMI-doguadine or DMI-guazatine combinations in antiresistance strategies seems to be restricted.

An interesting example of negative cross-resistance in DMI-resistant strains, concerns ethirimol which inhibits specifically the enzyme adenosine-deaminase in powdery mildews (27). According to Butters et al (28), in E. graminis, isolates highly resistant to DMIs are sensitive to ethirimol, whereas those highly resistant to ethirimol are sensitive to DMIs. However, isolates sensitive or highly resistant to both types of fungicides can been found. Absence of the double highly resistant isolates is probably correlated with reduced fitness. As the genetic factors which determine resistance to DMIs and ethirimol seem to be different (29, 30), the negative relathionship between these two types of fungicides can not be considered as a case of negative cross-resistance. However, previous observations suggest that use of DMI-ethirimol mixtures may delay development of resistance to both compounds, and some limited data from field experiments with barley powdery mildew support this view (26). It seems that this type of mixture can select strains moderatly resistant to both compounds, and lead to the elimination of the most sensitive and resistant strains, without any adverse effects on the efficacy (Heanay, personal communication).

BENZIMIDAZOLES

Benzimidazoles (including thiophanates) were the first family of systemic fungicides active against a wide range of pathogenic fungi. At low concentrations inhibition of mycelial growth is accompanied by typical distortions of hypha. Such effects which probably arise through perturbations in the functioning of cellular microtubules, as a result of binding these fungicides to tubulin (1, 31).

After only a few years of use, field resistance to benzimidazoles was reported in a number of pathogens. For example, in France, serious problems were encountered in <u>Botrytis cinerea</u>, <u>P. herpotrichoides</u> and <u>V. inaequalis</u> (1). Positive cross-resistance occurs between benzimidazoles in field strains and most laboratory mutants. However, some thiabendazole-resistant strains selected in the laboratory, showed negative cross-resistance to benomyl and carbendazim (3).

Table 3

Spectrum of negative cross-resistance to diphenylamine and N-phenylcarbamates in benzimidazoles-resistant, strains of <u>P. herpotrichoides</u> [after ref. 32 ; Leroux, unpublished data]

compounds[b]	EC 50 values[a] of phenotypes :				
	PS	PR1	PR2	PR3	PR4
carbendazim	0.03	25	> 50	27	4
diphenylamine	9	1.5	4	1.5	9
MMPC	> 30	12	> 30	> 30	> 30
MCPC	30	7.5	10	3.0	27
MDPC	10	0.3	0.4	1.0	10
chlorpropham	20	3.0	6.0	2.0	20
diethofencarb	> 50	0.03	0.2	> 50	> 50

(a) EC 50 values (mg/l) for inhibition of germ tube elongation.
(b) MMPC : methyl N-(3-methoxyphenyl) carbamate ; MCPC : methyl N-(3-chlorophenyl) carbamate ; MDPC : methyl N-(3,5-dichloro-phenyl) carbamate.

In several fungal species, some benzimidazole-resistant phenotypes exhibit negative cross-resistance to N-phenylcarbamates, such as the herbicides barban and chlorpropham, or the fungicides MDPC and diethofencarb, and to the scald inhibitor, diphenylamine (Table 3 ; 32). Generally this phenomenon concerns the higly benzimidazole-resistant isolates, whereas the moderately and weakly benzimidazole-resistant strains remain insensitive to N-phenylcarbamates. The exeperimental antifungal N-phenylformamidoximes DCPF and CDPF act similarly to that of diethofencarb (33). Morever, various other pesticides including triazine herbicides (eg. terbutylazine, terbutryn, terbumeton), diphenylether herbicides (eg. binefox, nitrofen), organophosphorous insecticides (eg. phentoate, phoxime) and carbamate insecticides (eg. fenoxycarb, promecarb) can also exhibit negative cross-resistance to benzimidazoles. Within a chemical family the spectra of negative cross-resistance can differ from one coumpound to another. This is the case with N-phenylcarbamates or triazines in <u>P. herpotrichoides</u> (table 3 ; 32).

When tests are conducted on germinating spores, of benzimidazole-resistant strains, N-phenylcarbamates produced changes similar to those produced in wild-type strains by carbendazim. These observations suggest that the N-phenylcarbamates may also affect the functioning of microtubules in fungal cells (32, 33, 36).

It is generally assumed that resistance to benzimidazoles in fungi, is associated with changes in β-tubulin which confer a decreased binding affinity to these fungicides (31). Such modifications can lead to an increased binding affinity of tubulin to the chemicals involved in negative cross-resistance in benzimidazole-resistant strains. This phenomenon was demonstrated with the N-phenylformamidoxime DCPF (34) but not with ethyl N-(3,5-dichlorophenyl) carbamate, a structural analogue of MDPC (35). According to Davidse (31) this phenomenon of negative cross-resistance might be attributed to altered microtubule stability rather than affinity changes of tubulin to these various toxicants.

In A. nidulans and N. crassa, mutations to benzimidazole resistance are found almost exclusively in the β-tubulin gene. Of interest is the recent finding of Fujimura et al. (37) showing that a substitution in the β-tubulin sequence of glutamic acid by glycine at position 198 confers resistance to carbendazim and increased sensitivity to diethofencarb in N. crassa. In another carbendazim-resistant allele confering insensitivity to diethofencarb, phenylalanine at position 167 in wild-type is changed to tyrosine (38). This information will probably contribute to understanding the molecular basis of resistance to benzimidazoles, and increased sensitivity to N-phenylcarbamates and perhaps other toxicants having similar effects.

As mentioned previously, in most fungi, benzimidazole resistant phenotypes can be classified in two categories, according to whether they are sensitive (RS-type) or resistant (RR-type) to diethofencarb. The respective proportions of these phenotypes in natural populations depend upon the fungus. For instance in French vineyards, the RR-type was not detected in B. cinerea before the introduction of diethofencarb, whereas all the benzimidazole-resistant strains of Fusarium sambucinum collected in France from potato tubers have the RR-type. In P. herpotrichoides and Fusarium nivale 90 % of the isolates collected in France on winter wheat were of RS-type (Leroux, unpublished data). Sometimes, for a particular pathogen the relative proportion of both phenotypes can differ from country to country. For instance in Cercospora beticola most benzimidazole-resistant strains collected in France have the RS-type whereas in Greece the most common ones were of RR-type (39). In practice a carbendazim-diethofencarb mixture will be effective on benzimidazole-sensitive and RS-type isolates but not on RR-type ones. It is one reason why such a combination is actually used only against B. cinerea (40).

A carbendazim-diethofencarb mixture has been introduced in France in 1987, to control grey mould in vineyards ; the recommended rate of application was 500 g a.i. of both compounds per hectare. In 1987, at the harvest no strains resistant to both fungicides (RR-type) were detected in French vinyards. The following season (1988), with a programme involving two applications of the previous mixture, RR-type strains were detected in Burgundy, Loire Valley, Alsace and Champagne ; in the last region their average frequency was 2.4 %. Consequently it was recommended to use the carbendazim-diethofencarb combination only once per year. Although this strategy was followed, in Champagne, the frequencies of RR-type strains increased ; in 1989 and 1990 the respective average values were 22.3 and 43.7 % (this increase was less important in the other French vineyards). However, in long-term trials it appears that if the carbendazim-diethofencarb mixture is stopped, the frequency of RR-type strains decreases rapidly (Leroux, unpublished data) suggesting that the RR-type isolates are less fit than the RS-type ones and the wild-type benzimidazole-sensitive isolates.

According to Rosenberger and Meyer (41), the first commercial application of negative cross-resistance in fungi concerns combinations of diphenylamine (an antiscald agent) and benomyl, thiabendazole or thiophanate-methyl. Such mixtures provide better control of benzimidazole-resistant P. expansum in inoculated apples stored btween 2 and 5°C than either diphenylamine or the fungicides used alone. This result is due to the fact that most benzimidazole-resistant strains are negative cross-resistance to diphenylamine. However the effectiveness of diphenylamine-benzimidazole mixtures can be nullified in the presence of double resistant isolates.

CONCLUSION

The various examples detailed in this paper, as well as those described by De Waard (3), indicate that for most fungicides, compounds showing negative cross-resistance has been discovered. Consequently, development of fungicide resistance in the field might (in principle) be counteracted by application of mixtures of two compounds with negative cross-resistance. This will be only possible in practice if all fungal strains respond in this way to a combination of such chemicals. Unfortunately, in all cases studied so far, double resistant strains have been observed in nature (before or after the introduction of these combinations). Even though practical applications of negative cross-resistance phenomena might limited, knowledge of the mechanism involved in such phenomena might provide informations on the molecular interaction between fungicides and their target in fungal cells.

REFERENCES

1. Leroux, P. La résistance des champignons aux fongicides, 1987, Phytoma, 385, 6-14 and 386, 31-35.

2. Georgopoulos, S.G. Development of fungal resistance to fungicides. In Antifungal compounds, eds. M.R., Siegel and H.D., Sisler. Marcel Dekker, New-York, 1977, 2, 439-495.

3. De Waard, M.A. Negatively correlated cross-resistance and synergism as strategies in coping with fungicide resistance. Br. Crop Prot. Conf. Pests and Dis., 1984, 89-95.

4. Kuhn, P.J. Mode of action of carboxamides. In Mode of action of anti-fungal agents, eds. A.PJ., Trinci and J.F. Riley. Cambridge University Press, Cambridge, 1984, pp. 155-183.

5. White, G.A., Thorn, G.D. and Georgopoulos, S.G. Oxathiin carboxamides highly active against carboxin resistant succinate dehydrogenase complexes from carboxin-selected mutants of Ustilago maydis and Aspergillus nidulans. Pestic. Biochem. Physiol, 1978, 9, 176-182.

6. White, G.A. Substituted benzanilides : structural variations and inhibition of complex II activity in mitochondria from a wild-type strain and a carboxin-selected mutant strain of Ustilago maydis. Pestic. Biochem. Physiol., 1987, 27, 249-260

7. White, G.A. Furan carboxamide fungicides : structure-activity relationships with the succinate dehydrogenase complex in mitochondria from a wild-type strain and a carboxin-resistant mutant strain of Ustilago maydis. Pestic. Biochem. Physiol., 1988, 31, 129-145.

8. Leroux, P. and Berthier, G. Resistance to carboxin and fenfuram in Ustilago nuda, the causal agent of barley loose smut. Crop Protec., 1988, 7, 16-19.

9. Leroux, P. and Berthier, G. Phenomènes de résistance du charbon nu de l'orge (Ustilago nuda) à la carboxine et au fenfurame. 2nd International Conference on Plant Disease, Annale ANPP, Paris, 1988, 1283-1291.

10. Hollomon, D.W. and Carter, G.A. Resistance within U.K. isolates of Ustilago nuda to carboxanilide fungicides. ISPP, Chemical control Newsletter, 1989, 12, 37-39.

11. Koller, W. Sterol demethylation inhibitors : mechanism of action and resistance. In Fungicide resistance in North America, ed. C.J. Delp, APS Press, St. Paul., 1988, pp. 79-88.

12. Leroux, P. and Benveniste, P. Mode d'action des fongicides inhibiteurs des stérols. In Proc. Bordeaux Mixture Centenary Meeting, BCPC monograph n° 31, 1985, pp. 67-78.

13. Kendall, S.G. Cross-resistance of triadimenol-resistance fungal isolates to other C-14 demethylation inhibitors fungicides. Br. Crop Prot. Conf. Pests and Dis., 1986, 539-546.

14. Leroux, P. Gredt, M. and Boéda, P. Resistance to inhibitors of sterol biosynthesis in field isolates or laboratory strains of the eyespot pathogen Pseudocercosporella herpotrichoides, Pestic. Sci., 1988, 23, 119-129.

15. Nuninger-Ney, C., Schwinn, F.J. and Staub, T. In vitro selection of sterol-biosynthesis inhibitor (SBI) resistant mutants in Monilinia fructicola. Neth. J. Pl. Path., 1989, 137-150.

16. Krämer W., Berg, D. and Köller W. Chemical synthesis and fungicide resistance. In Combating Resistance to xenobiotics, ed. M.G., Ford, D.W. Hollomon, B.P.S. Khambay and R.M. Sawicki. VCH, Horwood, Chichester, 1987, pp. 291-305.

17. Akallal, R. Résistance de Fusarium solani f. sp. pisi, à des fongicides inhibant la C-14 déméthylation des stérols. Thèse, Université de Paris Sud, France, 1989, 136 p.

18. Kalamarakis, A.E., Demopoulos, V.P., Ziogas B.N. and Georgopoulos, S.G. A highly mutable major gene for triadimenol resistance in Nectria haematoccoca var. cucurbitae. Neth J. Pl. Path. 1989, 95, 109-120.

19. Leroux, P. et Marchegay P. Caractérisation des souches de Pseudocercosporella herpotrichoides, agent du piétin-verse des céréales, résistantes au prochloraze, isolées en France sur blé tendre d'hiver. Agronomie, 1991 (submitted).

20. Stéva, H. et Clerjeau, M. Cross-resistance to sterol biosynthesis inhibitor fungicides in strains of Uncinula necator isolated in France and Portugal. Med. Fac. Landbouww Rijksuniv. Gent, 1990, 55, 983-988.

21. Fuchs, A. Implications of stereoisomerism in agricultural fungicides. In Stereoselectivity of pesticides, ed. E.J. Ariëns, J.J.S. Van Rensen and W. Welling. Elsevier, Amsterdam, 1988, pp. 203-262.

22. Berg, D., Born, L., Büchel, K.H., Holmwood, G. and Kaulen J. HWG 1608-Chesmistry and Biochemistry of a new azole fungicide. Pflanzen schutz. Nachrichten Bayer, 1987, 40, 111-132.

23. Hollomon, D.W. Molecular approaches to understanding the mechanisms of fungicide resistance. Brighton Crop Prot. Conf. Pests and Dis., 1990, 881-888.

24. Barug, D. and Kerkenaar, A. Resistance in mutagen-induced mutants of Ustilago maydis to fungicides which inhibit ergosterol biosynthesis. Pestic. Sci., 1984, 15, 78-84.

25. De Falandre, A., Bouvier, I. Seng, J.M. and Leroux, P. Induction and characterization of Penicillium caseicolum mutants resistant to ergosterol biosynthesis inhibitors. Appl. Environ. Microbiol., 1987, 53, 1500-1503.

26. Brent, K.J., Carter, G.A., Hollomon, D.W., Hunter, T., Locke, T. and Proven, M. Factors affecting build-up of fungicide resistance in powdery mildew in spring barley. Neth. J. Pl. Path., 1989, 95, 31-41.

27. Hollomon, D.W. Antifungal activity of substituted 2-aminopyrimidines. In Mode of action of antifungal agents. ed. A.P.J. Trinci and J.F. Riley Cambridge University Press, Cambridge, 1984, pp. 185-205.

28. Butters, J.A., Clark, J. and Hollomon, D.W. Resistance to inhibitors of sterol biosynthesis in barley powdery mildew. Med. Fac. Landbouww. Rijksuniv. Gent, 1984, 49, 143-151.

29. Hollomon, D.W. Genetic control of ethirimol resistance in a natural population of Erysiphe graminis f. sp. hordei. Phytopathology, 1981, 71, 536-540.

30. Hollomon, D.W., Butters, J.A. and Clark, J. Genetic control of triadimenol resistance in barley powdery mildew. Br. Crop. Prot. Conf. Pests. and Dis., 1984, 477-482.

31. Davidse, L.C. Benzimidazole fungicides mechanism of action and biological impact. Ann. Rev. Phytopathol., 1986, 24, 43-65.

32. Leroux, P. and Gredt, M. Negative cross-resistance of benzimidazole-resistance strains of Botrytis cinerea, Fusarium nivale and Pseudocercosporella herpotrichoides to various pesticides. Neth. J. Pl. Path., 1989, 95, 121-127.

33. Nakata, A., Sano, S., Hashimoto, S., Hayakawa, K., Nishikawa, H. and Yasuda, Y. Negatively correlated cross-resistance to N-phenylformamidoximes in benzimidazole-resistant phytopathogenic fungi. Ann. Phytopath. Soc., Japan, 1987, 53, 659-662.

34. Ishii, H. and Takeda, H. Differential binding of a N-phenylformamido-xime compound in cell-free extracts of benzimidazole-resistant and-sensitive isolates of Venturia nashicola, Botrytis cinerea and Fusarium fujikuroi. Neth. J. Pl. Path., 1989, 95, 99-108.

35. Groves, J.D., Fox, R.T.V. and Baldwin, B.C. Mode of action of carbenda-zim and ethyl N-(3,5-dichlorophenyl) carbamate on field isolates of Botrytis cinerea. Brighton Crop. Prot. Conf. Pests and Dis., 1988, 397-402.

36. Suzuki, K., Kato, T., Takahashi, J. and Kamoshita, K. Mode of action of methyl N-(3,5-dichlorophenyl) carbamate in the benzimidazole-resis-tant isolate of Botrytis cinerea, J. Pesticide Sci., 1984, 9, 497-501.

37. Fujimura, M. Oeda, K., Inoue, H. and Kato, T. Mechanism of action of N-phenylcarbamates in benzimidazole-resistant Neurospora strains. In Managing Resistance to Agrochemicals : from fundamental research to practical strategies. ed. M.B., Green, H.M., Lebaron and W.K., Moberg. ACS Symposium Series 421, Washington, 1990, 224-236.

38. Orbach, M.J., Porro, E.B. and Yanofsky, C. Cloning and characterization of the gene for -tubulin from a benomyl-resistant mutant of Neurospara crassa and its use as a dominant selctable marker. Moll. Cell Biol., 1986, 6, 2452-2461.

39. Demakopoulou, M.G. and Georgopoulos, S.G. Sensitivity to N-phenylcarba-mates as related to benzimidazole resistance in Cercospora beticola. 6th IUPAC Congress, Ottawa, Canada, 1986, 3E-05 (Abs).

40. Leroux, P., Gredt, M., Massenot, F. and Kato, T. Activité du phényl-carbamate S 32165 sur Botrytis cinerea, agent de la pourriture grise de la vigne. In Proc. Bordeaux Mixture Centenary Meeting, BCPC monograph n° 31, 1985, pp. 443-446.

41- Rosenberger, D.A. and Meyer, F.W. Negatively correlated cross-resis-tance to diphenylamine in benomyl-resistant Penicillium expansum. Phytopathology, 1985, 75, 74-79.

THE CONTRIBUTION OF GENETIC STUDIES TO UNDERSTANDING FUNGICIDE RESISTANCE

IAN R. CRUTE
Horticulture Research International,
East Malling, West Malling, Kent ME19 6BJ, UK

INTRODUCTION

The genetics of fungicide resistance is a topic almost as broad as the subject of fungal genetics itself and information accumulated over the last decade has been synthesised within recent reviews by Georgopoulos (13) and Grindle (15). This paper does not attempt to provide a similarly exhaustive treatment. Rather, by restricting the presentation of findings to certain fungicides and fungi, this paper will seek to illustrate the central role that genetic studies can have in providing an understanding of the biochemical mechanisms of resistance, while also impacting on the conception and implementation of management strategies designed to minimise practical problems with disease control. Successful resistance management strategies need to be based on a continuum of information; studies on fungal populations (genetics and dynamics) on the one hand, and on the molecular mechanism of action or resistance on the other, benefit from information on the inheritance of variation for response to fungicide.

Fungi

From a genetic standpoint, fungi are a diverse group of organisms. Following the summary presented by Burnett (2) plant pathogenic fungi can be grouped on the basis of four quite distinct life cycles. Genetic studies with fungi have been concentrated on those having a predominantly haploid life cycle. In particular, two non-pathogenic filamentous ascomycetes: *Aspergillus nidulans* and *Neurospora crassa* have been the laboratory systems of choice. Research on resistance to fungicides in these saprotrophic species has had great relevance to related plant pathogenic species with similar life cycles. For this reason, the focus in this paper will be on the highly analogous *Venturia inaequalis*, the cause of apple scab.

Fungicides

There is no attempt in this paper to deal exhaustively with the genetics of fungal resistance to all chemical groups of fungicides where the phenomenon has been studied. For the purpose of drawing out some general principles, therefore, emphasis will be on contrasting experiences with the non-aromatic cationic fungicide dodine (dodecylguanidinium acetate), the benzimidazole group (MBC) (including negatively correlated cross-resistance to N-phenylcarbamates, NPC) and the sterol 14∝demethylation inhibiting group (DMI).

Genetics

There are five components of the genetics of variation for response to fungicides that are of major concern and where need for information focuses research (it is worth noting at this juncture that, for the fungi under consideration, dominance is not on this list): a, the number of loci involved; b, the extent of allelic variation at each locus involved; c, the occurrence and importance of additivity; d, the occurrence and importance of epistasis and e, the occurrence and importance of pleiotropic effects on fitness and negative or positive cross-resistance. Changes in response to fungicide use that might occur at the levels of pathogen population genetics and dynamics are critically dependent on each of these genetic components. Furthermore, during investigations of the molecular and biochemical bases for pathogen variation, hypothesis construction and testing becomes sharply focused when detailed genetic information is available.

DODINE

Occurrence of resistance

Resistance to dodine was first reported to be a problem in control of *Venturia inaequalis* on apple in north-eastern USA (53) and subsequently elsewhere in North America (23, 36, 43, 48) and in South Africa (45). The loss of effective disease control only became evident after about a decade of sustained use, reflecting an estimated 80 or so applications of the chemical and at least 9 sexual and more than 100 asexual generations of the fungus (16). Populations of ascospores and conidia of *V. inaequalis* exhibited a high degree of variability in response to dodine; even fungal populations considered to be sensitive (from sites never exposed to dodine or orchards where control remains effective) contained individuals with up to 100 fold differences in the concentration required to inhibit germination (0.05 - 5.0μg/ml) (37). There was a considerable overlap between the responses of populations considered to be sensitive and those considered resistant (ie when field control was no longer effective). Gilpatrick and Blowers (16) reported mean LD_{50} values ranging from 0.30 - 1.25 μg/ml for ascospore populations. McKay & MacNeill (36) working with conidial populations reported mean ED_{50} values in the range 0.12 - 1.85 μg/ml. In both studies, a mean value of approximately 0.7μg/ml coincided with inadequate control. Ross & Newbery (44) compared the sensitivities of two sets of isolates, one obtained prior to the introduction of dodine in 1957 and the other in 1974 from orchards which included ones where the fungicide was no longer effective. A few isolates were markedly inhibited in the presence of 0.25μg/ml while others were capable of growth at >2.0μg/ml. The mean sensitivity of isolates obtained in 1957 was greater than those obtained in 1974 but marked variation in response to dodine was evident in both samples.

Mechanism of action

One component of the mode of action of dodine may relate to its cationic surface-active properties, but it is a good deal more active than other n-alkylamines and this suggests that the guanidine nucleus plays a role. Membrane permeability is known to be affected by dodine and effects on various enzyme systems have also been reported. The precise mechanism of action is uncertain but the concensus is that this fungicide acts on more than one target site (3, 12).

Mechanisms of resistance

Nothing is known about mechanisms underlying the observed variation in response to dodine.

Genetics of resistance

The first study of the genetic control of variation in *V. inaequalis* for response to dodine was conducted by Polach (42). Five crosses were made between isolates expressing clear differences in response as determined by the occurrence or absence of growth on media containing 0.25 and 0.5μg/ml dodine. The responses of 223 F_1 progeny were determined. Ability or inability to grow at 0.25μg/ml segregated 1:1 among progeny indicating the action of a single locus. Ability or inability to grow at 0.5μg/ml segregated 1:3 among progeny from four out of five crosses and in the fifth cross the data were not significantly different from 1:7. The simplest interpretation of these data is that at least 3 loci control response to dodine and that variation is additive. Alleles for decreased sensitivity are required at at least two and sometimes three loci for growth at 0.5μg/ml while a single such allele at one locus permits growth at 0.25μg/ml. Yoder & Klos (61) used tetrad analysis to examinate the F_1 generation from crosses between several isolates highly sensitive to dodine and an isolate expressing a 2-3 fold level of reduced sensitivity. Variation for response among parents and progeny was determined by measuring the inhibition zone round discs impregnated with 50 and 300μg/ml dodine, or by percentage inhibition of germination at 1μg/ml. Two pairs of each of four classes of ascospore progeny were evident in 12 tetratype asci of the 20 asci analyzed. Two classes were the high and low sensitivity parental types and the other two were distinguishable intermediates. The data were consistent with the occurrence of additive alleles regulating sensitivity to dodine at two independent loci.

Prior to this work with *V. inaequalis*, reduced sensitivity to dodine in UV-induced mutants of *Nectria haematococca* var. *cucurbitae* had similarly been shown by Kappas & Georgopoulos (26) to be controlled by alleles at a few (four) loci which acted additively.

BENZIMIDAZOLES AND N-PHENYLCARBAMATES

Occurrence of resistance

Resistance of *V. inaequalis* to benzimidazole fungicides emerged independently at several locations a few years after MBCs became widely used. The first report of failure to control *V. inaequalis* due to benzimidazole resistance was from Australia (59) followed soon by similar experiences in the USA (23, 52), New Zealand (54), Germany (29, 30), Poland (40), South Africa (45) and Canada (39, 44, 48). It is clear that benzimidazole resistance will persist in populations in the absence of selection (60, 35). However, there is evidence from field and experimental studies that some resistant variants may carry a fitness deficit causing them to be less competitive in the absence of benzimidazoles than sensitive components of the population (34).

Mode of action

Benzimidazole fungicides act by binding to the ß monomer of tubulin, the heterodimeric subunit of microtubules that constitute a major component of the cytoskeleton. This binding results in inhibition of microtubule assembly and consequently of nuclear division, intercellular transport function and maintenance of cell shape (6).

Mechanisms of resistance

Resistance to benzimidazole fungicides can result from at least two different phenonema. Resistant isolates of fungi produce variant forms of ß-tubulin which have either a lower binding affinity for benzimidazoles or form hyperstable microtubules that are resistant to disassembly and therefore remain functional (6).

Genetics of resistance

Depending on the organism, one or several functional genetic loci may code for α and ß tubulin polypeptides (5). For example, *Aspergillus nidulans* has two functional loci coding for α-tubulin (tubA and tubB) and two for ß-tubulin (benA and tubC) (38). Allelic variation at loci coding for α and ß tubulin can result in structural changes in polypeptides and altered sensitivity to benzimidazoles and N-phenylcarbamates (see below) without destroying the function of tubulin. For example, mutants benA15, benA4, benA31 and benA33 all confer resistance to carbendazim and thiabendazole. In the case of benA15, this is due to a lower binding affinity than the wild-type whereas the phenotype of benA33 has been shown to be due to hyperstability of microtubules (6). The benA gene from *A. nidulans* has now been cloned (25) and the single amino acid changes responsible for the altered phenotypes of several mutants have been determined. Thus, benA4 results from a change in codon 6 (histidine to tyrosine); benA31 in codon 50 (tyrosine to serine) and benA38 in codon 134 (glutamic acid to lysine). The mutant allele benA16 interestingly results in resistance to thiabendazole but increased sensitivity to carbendazim demonstrating that cross-resistance to all benzimidazoles, although usually experienced in practice, is not invariable (57). This phenotype has been shown to result from a single amino acid change in codon 165 (alanine to valine). Interestingly, resistance to benzimidazole in *Neurospora crassa* can also result from an amino acid alteration in the nearby codon 167 (phenylalanine to tyrosine) (41).

Benzimidazole resistant isolates of several fungi exhibit increased sensitivity to NPC (28, 9) (negatively correlated cross-resistance). This has recently been the subject of a detailed classical and molecular genetic study in *N. crassa* (11). Mutants with benzimidazole resistance and high sensitivity to NPC were produced (one of these was referred to as F914) and crossed with a phenotypically wild-type isolate (benzimidazole sensitive and NPC resistant) and also the Bm1511 mutant already mapped on the ß-tubulin gene (see above) (41); Bm1511 was resistant to both benzimidazole and NPC. F_1 progeny from crosses between isolates with these distinct phenotypes all segregated 1:1 for parental phenotype. This indicated that the three phenotypes result from allelic forms of the ß-tubulin gene. A fourth mutant phenotype was selected from F914. These mutants retained benzimidazole resistance and were also markedly less sensitive to NPC (but not to the same extent as Bm1511). Isolates with this new mutant phenotype were crossed with a wild-type isolate and F914 from which they were derived. The F_1 progeny from these crosses also segregated 1:1 for parental phenotypes. This indicated that a further allelic form of the ß-tubulin gene carrying two mutations was responsible for the expression of benzimidazole resistance and moderate resistance to NPC. Using the cloned Bm1511 ß-tubulin gene located on pSV50 as a probe, the ß-tubulin gene of F914 was cloned into pEF50. Chimeric ß-tubulin genes were then constructed and the phenotypes of transformants determined. The mutation conferring the F914 phenotype was localised on the same BamHI - EcoRI fragment as the previously mapped Bm1511 mutation (in codon 167). Sequencing of this fragment detected a change in the second base of codon 198 resulting in a change from glutamic acid to glycine. The proximity of these mutations suggests that this region could be the binding domain for both benzimidazoles and NPC.

Isolates of *V. inaequalis* from the USA, Germany, Israel, Australia, Chile, Italy and New Zealand have been used in a series of studies on the inheritance of resistance to benzimidazoles and negatively correlated cross-resistance to NPC. Kiebacher & Hoffman (31) carried out a tetrad analysis on seven asci in addition to studying random F_1 ascospore progeny from crosses among two benzimidazole sensitive and three resistant isolates. The earlier report of Jones & Ehret (22) that two alleles at a single locus were responsible for the two distinct phenotypes was confirmed. Subsequently, Shabi *et al.* (46) recognised four phenotypic responses to benzimidazoles: sensitive (S), moderate resistance (MR), high resistance (HR) and very high resistance (VHR). In crosses between isolates with the same phenotype, no segregation was observed among F_1 progeny and only parental types segregated, in a 1:1 ratio, among F_1 progeny from crosses between phenotypically different isolates. The same range of phenotypic variants (except MR) was located from orchards in New York State together with isolates expressing a further phenotype: low resistance (LR) (27). Crosses between phenotypically identical isolates from Israel and USA, resulted in no segregation among F_1 progeny and segregation for parental phenotypes only, in a 1:1 ratio, was again observed among progeny from crosses between isolates of different phenotype. A set of crosses with an extensive set of isolates expressing similar phenotypic diversity but from six different countries yielded the same results (50). All isolates expressing the VHR phenotype, regardless of their diverse origins, proved to be highly sensitive to NPC (24), the same was true of isolates from Israel with the MR phenotype (47) although not of MR isolates from elsewhere. Isolates expressing all other types of phenotypic variation for benzimidazole sensitivity were insensitive to NPC. Studies on the inheritance of sensitivity to NPC in a range of VHR isolates revealed that the two characters were controlled by the same allele at a single locus or by alleles at closely linked loci. It seems reasonable to conclude that these studies with *V. inaequalis* have revealed six allelic forms of a gene at a single locus controlling response to benzimidazoles and NPC. Although speculative, it seems likely this gene codes for ß-tubulin.

DEMETHYLATION INHIBITING (DMI) FUNGICIDES

Occurrence of resistance

Practical control problems caused by resistance to DMI fungicides have occurred for a few diseases (20) and most notably powdery mildews. Over the last six years, there have also been several reports that indicate the occurrence of isolates of *V. inaequalis* with reduced sensitivity to DMI fungicides; in one case at least, a shift in the mean sensitivity of the population has resulted in inadequate control (18). Warnings that this might occur had previously been given by Jones (21).

The first isolates of *V. inaequalis* found with reduced sensitivity to DMI fungicides originated in a German orchard (51) where these fungicides had been used experimentally. In comparison with seven sensitive isolates from Michigan and Germany, nine isolates from the treated orchard were shown to be less sensitive to a range of several DMI fungicides by a factor of about five. Isolates with reduced sensitivity showed no evidence of impaired fitness and despite suggestions that they may exhibit negatively correlated cross resistance (8) their response to dodine was not significantly different. Isolates of *V. inaequalis* with reduced sensitivity to DMI fungicides have also been obtained from orchards in France (55, 56) Italy (10), Belgium (4), and Austria (17). Recently, Smith *et al.* (49) conducted an extensive survey of *V. inaequalis* isolates from three orchards, only one of which had received DMI fungicides. They confirmed the existence of considerable variation among isolates for sensitivity although the mean sensitivity of the populations was unrelated to fungicide use.

A study conducted in an experimental orchard in Canada over 9 years (18) has confirmed beyond doubt that efficacy of DMI fungicides for control of *V. inaequalis* can be eroded as components of the population with decreased sensitivity are selected and become more frequent. Isolates with only slightly reduced sensitivity to bitertanol (x2) were detected in 1983 but DMI fungicides were still as effective as they had been for the five previous years of use. In the succeeding years (1984-1987), control became less effective and failed completely in 1987. This was accompanied by the more frequent occurrence of isolates expressing higher levels of insensitivity with factors ranging from two to more than twenty.

Mode of action

DMI fungicides act on a specific step in the synthetic pathway for ergosterol, the major sterol in most fungi. This results in ergosterol depletion of membranes and interference with their normal function; altered or abnormal sterol patterns also result which could affect sterol-mediated regulation of essential cell processes. The biosynthetic process affected is the hydroxylation step in the removal of the 14α-methyl group from lanosterol (32). This is mediated by a cytochrome P-450 monooxygenase (lanosterol 14α demethylase) and there is good evidence that DMI fungicides bind to both the haem system of the enzyme and the site normally occupied by the lanosterol substrate.

Mechanisms of resistance

Several different mechanisms have been proposed to account for reduced sensitivity to DMI fungicides and this is a subject still under active investigation (20). It has been commonly assummed that alteration in the binding affinity of the target enzyme for DMI fungicides would destroy, or markedly affect, enzyme activity and thereby be lethal or markedly reduce fitness (7, 33). However, this contention awaits confirmation currently being sought by molecular genetic analysis (20).

Genetics of resistance

So far, few studies of the inheritance of variation for response to DMI fungicides have been conducted. However, in a now classical study, Van Tuyl (58) investigated laboratory induced mutants of *Aspergillus nidulans* resistant to imazalil. From a genetic analysis involving twenty-one mutant isolates, at least eight loci in six linkage groups were found to control response to imazalil. At two loci, the existence of a range of alleles was demonstrated by their different phenotypic responses to fungicide. At any single locus, the maximum decrease in sensitivity contributed by one allele was relatively small and never more than a factor of ten. However, additivity was clearly demonstrated in recombinant strains selected to carry a mutant allele at more than one locus. In addition, pleiotropic effects on cross-resistance and/or increased sensitivity to other toxophores (including fenarimol) differed between loci. It has yet to be clearly established how comparable this study with laboratory mutants of a non-pathogen will prove to be with natural field variants of plant pathogens.

The individual effects of single loci could not be discerned in crosses between strains of *Erysiphe graminis* f.sp. *hordei* differing in sensitivity to triadimenol (19) and evidence of transgressive segregation was indicative of additive genetic variation. The implication is that a number of loci of individually small effect are involved in the expression of observed high levels of insensitivity to DMI fungicides in cereal powdery mildews.

The only genetic analysis of variation for response to a DMI fungicide yet reported for *V. inaequalis* was conducted by Stanis & Jones (51) utilising nine isolates (derived from the same orchard in Germany; see above) that expressed a reduced sensitivity of approximately 5-fold to several DMI fungicides. No segregation was observed for response to fenarimol among the 361 F_1 progeny from sensitive x sensitive crosses nor among the 1394 F_1 progeny from reduced sensitive x reduced sensitive crosses. Segregation into two classes was observed among the 1867 F_1 progeny from sensitive x reduced sensitive crosses and the authors interpreted their data as conforming to the 1:1 ratio expected if a single locus regulated difference in response to fungicide. However, the chi-squared values for half of the ten crosses indicated that the observed values were significantly different ($P = 0.05$) from those expected for a 1:1 segregation. There was also highly significant (Chi-squared = 44.7, 9 degrees of freedom; $P = 0.001$) heterogeneity between the crosses which invalidates summation of the total segregation. Confirmation that variation for response to DMI fungicides is under the control of individually identifiable loci awaits further investigation.

GENERAL CONCLUSIONS

It should be evident from the case studies dealt with above that genetic studies can play a major part in providing a fuller understanding of both the molecular mechanisms of resistance to fungicides and also explaining, or even predicting field phenomena. The vital importance of investigations with laboratory mutants of saprophytic fungi must not be underestimated even though not all findings have proved to be of relevance to practical problems with fungicide resistant variants of plant pathogens. These studies can rapidly provide research leads and guidance for the efficient conduct of studies with less amenable, but important plant pathogenic species. Concepts associated with determining the risks of practical problems with fungicide resistance are still developing (14, 1) but it is clear that genetic control of observed variation is a critical parameter. It remains to be seen whether research will progress sufficiently quickly to enable a prediction, for a newly introduced toxophore, of the occurrence of resistance associated for example with alteration of the binding properties of a target peptide. In retrospect, this would have been difficult to achieve in the case of benzimidazoles given the speed with which resistance emerged. Georgopoulos (13) has, however, suggested that where resistance is likely to result by slow drift as a consequence of the additive action of alleles at several loci, there should be ample warning for changes in disease control strategies to be made before complete control failure ensues. It is pertinent to enquire whether sufficient effort is being expended on genetic studies or the critical interpretation of monitoring exercises to transform this optimistic statement into a practical reality.

REFERENCES

1. Brent, K.J., Hollomon, D.W. and Shaw, M.W., Predicting the evolution of fungicide resistance. In *Managing Resistance to Agrochemicals: from Fundamental Research to Practical Strategies,* ed. M.B. Green, H.M. LeBaron and W.K. Moberg, American Chemical Society, Washington DC, USA, 1990, Chapter 21, pp. 303-319.

2. Burnett, J.H., *Mycogenetics.* John Wiley & Sons, London, 1975.

3. Byrde, R.J.W., Nonaromatic organics. In *Fungicides - Volume 2,* ed. D.C. Torgeson, Academic Press, New York & London, 1969, Chapter 12, pp. 531-578.

4. Creemers, P., Vandergeten, J. and Vanmechelen, A., Variability in sensitivity of field isolates of *Venturia* spp. to demethylation inhibitors. *Mededelingen van de Faculteit Landbouwwetenschappen Rijksuniversiteit Gent,* 1988, **53**, 577-587.

5. Cleveland, D.W. and Sullivan, K.F., Molecular biology and genetics of tubulin. *Annual Review of Biochemistry,* 1985, **54**, 331-365.

6. Davidse, L.C., Benzimidazole fungicides: mechanism of action and biological impact. *Annual Review of Phytopathology,* 1986, **24**, 43-65.

7.   Davidse, L.C., Mechanisms of resistance in fungi to benzimidazoles, inhibitors of sterol biosynthesis and phenylamides. In: *Combating Resistance to Xenobiotics: Biological and Chemical Approaches*, ed. M.G. Ford, D.W. Hollomon, B.P.S. Khambay and R.M. Sawicki, Ellis Horwood, Chichester, UK, 1987, Chapter 18, pp. 216-227.

8.   De Waard, M.A. and Van Nistelrooy, J.G.M., Negatively correlated cross-resistance to dodine and fenarimol-resistant isolates of various fungi. *Netherlands Journal of Plant Pathology*, 1983, **89**, 67-73.

9.   De Waard, M.A., Negatively correlated cross-resistance and synergism as strategies in coping with fungicide resistance. *Proceedings of the 1984 British Crop Protection Conference - Pests and Diseases*, 1984, pp. 573-584.

10.   Fiaccadori, R., Gielink, A.J. and Dekker, J., Sensitivity to inhibitors of sterol biosynthesis in isolates of *Venturia inaequalis* from Italian and Dutch orchards. *Netherlands Journal of Plant Pathology*, 1987, **93**, 285-287.

11.   Fujimura, M., Oeda, K., Inoue, H. and Kato, T., Mechanism of action of N-phenylcarbamates in benzimidazole-resistant *Neurospora* strains. In *Managing Resistance to Agrochemicals: from Fundamental Research to Practical Strategies*, ed. M.B. Green, H.M. LeBaron and W.K. Moberg, American Chemical Society, Washington DC, USA, 1990, Chapter 15, pp. 224-236.

12.   Gasztonyi, M. and Lyr, H., Other fungicides. In *Modern selective fungicides - properties, applications, mechanisms of action*, ed. H. Lyr, Longman Group, UK, 1987, Chapter 21, pp. 309-336.

13.   Georgopoulos, S.G., The genetics of fungicide resistance. In *Modern selective fungicides - properties, applications, mechanisms of action*, ed. H. Lyr, Longman Group, UK, 1987, Chapter 4, pp. 53-61.

14.   Gisi, U. and Staehle-Cseh, U., Relative risk evaluation of new candidates for disease control. In *Fungicide Resistance in North America*, ed. C.J. Delp, APS Press, St Paul, USA, 1988, Chapter 32, pp. 101-106.

15.   Grindle, M., Genetic basis of fungicide resistance. In *Combating Resistance to Xenobiotics*, ed. M.G. Ford, D.W. Hollomon, B.P.S. Khambay and R.M. Sawicki, Ellis Horwood, Chichester, UK, 1987, Chapter 7, pp. 74-93.

16.   Gilpatrick J.D. and Blowers, D.R., Ascospore tolerance to dodine in relation to orchard control of apple scab. *Phytopathology*, 1974, **64**, 649-652.

17.   Hermann, M., Szith, R. and Zinkernagel, V., Verringerte sensitivitat einiger isolate von *Venturia inaequalis* aus der steiermark fur EBI-fungizide. *Gartenbauwissenschaft*, 1989, **54**, 160-165.

199

18. Hildebrand, P.D., Lockhart, C.L., Newberry, R.J. and Ross, R.G., Resistance of *Venturia inaequalis* to bitertanol and other demethylation-inhibiting fungicides. *Canadian Journal of Plant Pathology*, 1988, **10**, 311-316.

19. Hollomon, D.W., Butters, J.A. and Clark, J., Genetic control of triadimenol resistance in barley powdery mildew. *Proceedings of the 1984 British Crop Protection Conference - Pests and Diseases*, 1984, pp. 477-482.

20. Hollomon, D.W., Butters, J.A. and Hargreaves, J.A., Resistance to sterol biosynthesis-inhibiting fungicides: current status and biochemical basis. In *Managing Resistance to Agrochemicals: From Fundamental Research to Practical Strategies*, ed. M.B. Green, H.M. LeBaron and W.K. Moberg, American Chemical Society, Washington DC, USA, 1990, Chapter 13, pp. 199-214.

21. Jones, A.L., Fungicide resistance: past experience with benomyl and dodine and future concerns with sterol inhibitors. *Plant Disease*, 1981, **65**, 990-992.

22. Jones, A.L. and Ehret, G.R., Tolerance to fungicides in *Venturia* and *Monilinia* of fruit trees. *Proceedings of the American Phytopathological Society*, 1976, **3**, 84-89.

23. Jones, A.L. and Walker, R.J., Tolerance of *Venturia inaequalis* to dodine and benzimidazole fungicides in Michigan. *Plant Disease Reporter*, 1976, **60**, 40-44.

24. Jones, A.L., Shabi, E. and Ehret, G.R., Genetics of negatively correlated cross-resistance to a N-phenylcarbamate in benomyl-resistant *Venturia inaequalis*. *Canadian Journal of Plant Pathology*, 1987, **9**, 195-290.

25. Jung, M.K., Dunne, P.W., Suen, I.H. and Oakley, B.R., Sequence alterations in ß-tubulin mutations of *Aspergillus nidulans*. *Journal of Cell Biology*, 1987, **105**, Abstract 1560.

26. Kappas, A. and Georgopoulos, S.G., Genetic analysis of dodine resistance in *Nectria haematococca* (Syn. *Hypomyces solani*). *Genetics*, 1970, **66**, 617-622.

27. Katan, T., Shabi, E. and Gilpatrick, J.D., Genetics of resistance to benomyl in *Venturia inaequalis* isolates from Israel and New York. *Phytopathology*, 1983, **73**, 600-603.

28. Kato, T., Suzuki, K., Takahashi, J. and Kamashita, K., Negatively correlated cross-resistance between benzimidazole fungicides and methyl N-(3,5-dichlorophenyl) carbamate. *Journal of Pesticide Science*, 1984, **9**, 489-495.

29. Kiebacher, H. and Hoffmann, G.M., Benzimidazol-resistenz bei *Venturia inaequalis*. *Zeitschrift fur Pflanzenkrankheiten und Pflanzenschutz*, 1976, **83**, 352-358.

30.   Kiebacher, H. and Hoffmann, G.M., Qualitative und quantitative untersuchungen zur resistenz von *Venturia inaequalis* gegen benzimidazole-fungizide. *Zeitschrift für Pflanzenkrankheiten und Pflanzenschutz*, 1980, **87**, 705-716.

31.   Kiebacher, H. and Hoffmann, G.M., Zur genetik der benzimidazol-resistenz bei *Venturia inaequalis*. *Zeitschrift für Pflanzenkrankheiten und Pflanzenschutz*, 1981, **88**, 189-205.

32.   Köller, W., Sterol demethylation inhibitors: mechanism of action and resistance. In *Fungicide Resistance in North America*, ed. C.J. Delp, APS Press, St. Paul, USA, 1988, Chapter 27, pp. 79-88.

33.   Köller, W. and Scheinpflug, H., Fungal resistance to sterol biosynthesis inhibitors: a new challenge. *Plant Disease*, 1987, **71**, 1066-1074.

34.   Lalancette, N., Hickey, K.D. and Cole, H., Parasitic fitness and intrastrain diversity of benomyl-resistant subpopulations of *Venturia inaequalis*. *Phytopathology*, 1987, **77**, 1600-1606.

35.   McGee, D.C. and Zuck, M.G., Competition between benomyl-resistant and sensitive strains of *Venturia inaequalis* on apple seedlings. *Phytopathology*, 1981, **71**, 529-532.

36.   McKay, M.C.R. and MacNeill, B.H., Tolerance of *Venturia inaequalis* to the fungicide dodine in Ontario orchards. *Proceedings of the American Phytopathological Society*, 1977, **4**, 193.

37.   McKay, M.C.R. and MacNeill, B.H., Spectrum of sensitivity to dodine in field populations of *Venturia inaequalis*. *Canadian Journal of Plant Pathology*, 1979, **1**, 76-78.

38.   Morris, N.R., The molecular genetics of microtubule proteins in fungi. *Experimental Mycology*, 1986, **10**, 77-82.

39.   Northover, J., Characterization and detection of benomyl resistant *Venturia inaequalis* in Ontario apple orchards. *Canadian Journal of Plant Pathology*, 1986, **8**, 117-122.

40.   Novacka, H., Karolczak, W. and Millikan, D.F., Tolerance of the apple scab fungus to the benzimidazole fungicides in Poland. *Plant Disease Reporter*, 1977, **61**, 346-350.

41.   Orbach, M.J., Porro, E.B. and Yanofsky, C., Cloning and characterization of the gene for ß-tubulin from a benomyl-resistant mutant of *Neurospora crassa* and its use as a dominant selectable marker. *Molecular and Cell Biology*, 1986, **6**, 2452-2461.

42. Polach, F.J., Genetic control of dodine tolerance in *Venturia inaequalis*. *Phytopathology*, 1973, **63**, 1189-1190.

43. Ross, R.G. & Newberry, R.J., Tolerance of *Venturia inaequalis* to dodine in Nova Scotia. *Canadian Plant Disease Survey*, 1977, **57**, 57-60.

44. Ross, R.G. and Newberry, R.J., Tolerance to benomyl of *Venturia inaequalis* in Nova Scotia. *Canadian Journal of Plant Pathology*, 1985, **7**, 435-437.

45. Schwabe, W.S., Tolerance of *Venturia inaequalis* to benzimidazole fungicides and dodine in South Africa. *Phytophylactica*, 1977, **9**, 47-54.

46. Shabi, E., Katan, T. and Marton, K., Inheritance of resistance to benomyl in isolates of *Venturia inaequalis* from Israel. *Plant Pathology*, 1983, **32**, 207-211.

47. Shabi, E., Koenraadt, H. and Dekker, J., Negatively correlated cross-resistance to phenylcarbamate fungicides in benomyl-resistant *Venturia inaequalis* and *V. pirina*. *Netherlands Journal of Plant Pathology*, 1987, **93**, 33-41.

48. Sholberg, P.L., Yorston, J.M. and Warnock, D., Resistance of *Venturia inaequalis* to benomyl and dodine in British Columbia, Canada. *Plant Disease*, 1989, **73**, 667-669.

49. Smith, F.D., Parker, D.M. and Köller, W., Sensitivity distribution of *Venturia inaequalis* to the sterol demethylation inhibitor flusilazole: baseline sensitivity and implications for resistance monitoring. *Phytopathology*, 1991, **81**, 392-396.

50. Stanis, V.F. and Jones, A.L., Genetics on benomyl resistance in *Venturia inaequalis* from North and South America, Europe, and New Zealand. *Canadian Journal of Plant Pathology*, 1984, **6**, 283-290.

51. Stanis, V.F. and Jones, A.L., Reduced sensitivity to sterol-inhibiting fungicides in field isolates of *Venturia inaequalis*. *Phytopathology*, 1985, **75**, 1098-1101.

52. Sutton, T.B., Failure of combinations of benomyl with reduced rates of non-benzimidazole fungicides to control *Venturia inaequalis* resistant to benomyl and the spread of resistant strains in North Carolina. *Plant Disease Reporter*, 1978, **62**, 830-834.

53. Szkolnik, M. and Gilpatrick, J.D., Apparent resistance of *Venturia inaequalis* to dodine in New York apple orchards. *Plant Disease Reporter*, 1969, **53**, 861-864.

54. Tate, K.G. and Samuels, G.J., Benzimidazole tolerance in *Venturia inaequalis* in New Zealand. *Plant Disease Reporter*, 1976, **60**, 706-710.

55.   Thind, T.S., Clerjeau, M. and Olivier, J.M., First observations on resistance in *Venturia inaequalis* and *Guignardia bidwellii* to ergosterol-biosynthesis inhibitors in France. *Proceedings of the 1986 British Crop Protection Conference - Pests and Diseases*, 1986, pp. 491-498.

56.   Thind, T.S., Clerjeau, M. and Olivier, J.M., Travelure du pommier: mise en evidence d'une resitance aux fongicides inhibiteurs de la biosynthese de l'ergosterol. *Phytoma Defense des Cultures*, 1986, **381**, 13-16.

57.   Van Tuyl, J.M., Davidse, L.C. and Dekker, J., Lack of cross-resistance to benomyl and thiabendazole in some strains of *Aspergillus nidulans*. *Netherlands Journal of Plant Pathology*, 1974, **80**, 165-168.

58.   Van Tuyl, J.M., Genetics of fungal resistance to systemic fungicides. *Mededelingen Landbouwhogeschool Wageningen*, 1977, 77-2, pp. 1-136.

59.   Wicks, T., Tolerance of the apple scab fungus to benzimidazole fungicides. Plant *Disease Reporter*, 1974, **58**, 886-889.

60.   Wicks, T., Persistence of benomyl tolerance in *Venturia inaequalis*. *Plant Disease Reporter*, 1976, **60**, 818-819.

61.   Yoder, K.S. and Klos, E.J., Tolerance to dodine in *Venturia inaequalis*. *Phytopathology*, 1976, **66**, 918-923.

MECHANISMS OF RESISTANCE TO HERBICIDES

ALAN D. DODGE
School of Biological Sciences,
University of Bath, Bath BA2 7AY, UK

ABSTRACT

Evolution of resistance to herbicides by weeds in field crops
has become an increasing worldwide problem during the last 20
years. Although over 100 biotypes have been reported, our
understanding of the mechanisms that enable these weeds to
survive normally toxic concentrations of herbicide is far
from complete. In this paper resistance mechanisms will be
discussed that involve (i) reduced herbicide movement (ii)
modifica- tion of membrane function and structure (iii) a
change in the sensitivity of the key target enzyme (iv)
enchanced production of a herbicide target (v) enhanced
metabolic breakdown and conjugation (vi) enhanced generation
of herbicide generated toxic products.

INTRODUCTION

Growers have become accustomed to the remarkable selectivity
many herbicides show when applied to crops and weeds, leaving
the former untouched but killing the weeds. Selectivity of
action may arise through differential uptake, movement or
metabolism, and enhanced in some instances by the addition of
synergists or safeners. In the last 20 years it has become
increasingly apparent that some weeds are not killed by
normally toxic herbicide concentrations. Over 100 weed
biotypes have been identified that might now be described as
resistant or tolerant to a particular herbicide or a range of
herbicides. Recent research has attempted to identify the
structural, physiological or biochemical reasons that enable
these biotypes to tolerate the otherwise lethal concentration
of herbicide. This paper summarises some of these
mechanisms.

REDUCED HERBICIDE MOVEMENT

Reduced uptake and movement appear to be involved in the
resistance mechanism of some paraquat tolerant biotypes.
During the last 10 years a number of such biotypes have been
identified, ranging from Conyza bonariensis from Egypt,
Conyza canadensis from Hungary, Erigeron philadelphicus from
Japan, Hordeum glaucum and H. leporinum from Australia, and
Epilobium ciliatum from England [1]. In all instances
resistant biotypes have been identified after many years of
continuous field application of paraquat.

Bishop, Powles and Cornic [2] showed that there was no
difference in cuticle penetration of ^{14}C- paraquat in
sensitive and resistant H. glaucum, but uptake of paraquat
into cut leaves was severely restricted to the vascular
tissue in resistant plants. Similar restricted movement was
shown by the autoradiographic studies of Tanaka, Chisaka and
Saka [3] with E.philadelphicus, (Figure 1) and Fuerst,
Nakatani, Dodge, Penner and Arntzen [4] with C. bonariensis.
In all of these experiments, apoplastic movement of the
herbicide appeared to be unaffected, but symplastic movement
was limited, probably because of some form of sequestering or
immobilisation before movement into, or through, the
mesophyll.

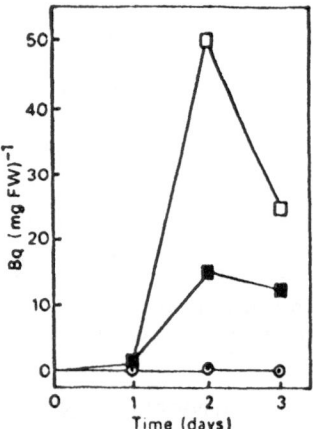

Figure 1   Time course movement [^{14}C]-paraquat into
excised leaves of Erigeron philadelphicus.
Susceptible biotype in the light □ ; or
darkness ■ . Tolerant biotype incubated in
light ○ ; or darkness ● . Redrawn from Tanaka
et al. (3)

Paraquat acts by diverting electron flow from chloroplast
photosytem I, which stops NADP⁺ reduction and CO_2 fixation.
Failure of paraquat to reach the chloroplast in resistant
plants was indicated by a failure to inhibit CO_2
incorporation, [2] or to quench chlorophyll fluorescence [4].
(Figure 2)

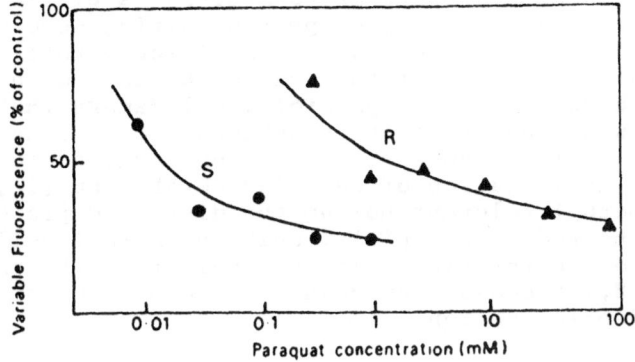

Figure 2   Variable fluorescence of leaves
            susceptible ● and   resistant ▲ biotypes of
            Conyza bonariensis after incubation with
            variation concentrations of paraquat.
            Redrawn from Fuerst et al. (4).

Thus far no specific mechanism has been identified to account
for the restricted movement or sequestration. It is possible
that it could be associated with the cationic properties of
paraquat which, like calcium and polyamines, could bind to
cell wall components such as pectins [5]. This would limit
movement into the symplast. This is further supported by the
fact that there appeared to be no difference in the
penetration of paraquat into isolated protoplasts of tolerant
and susceptible H. glaucum plants [6].

MODIFICATION OF MEMBRANE FUNCTION AND STRUCTURE

A diclofop-methyl resistant biotype of Lolium rigidum has
been identified in Australian wheat fields [7]. Subsequent
work has shown that this biotype is also tolerant to a range
of other herbicides of unrelated chemistry and action [8].
Although the primary action of diclofop-methyl is apparently
related to an inhibition of acetyl-CoA carboxylase [9], a key
enzyme in fatty acid biosynthesis, there was no difference in
sensitivity of this enzyme from tolerant or susceptible
Lolium plants [10].

Antagonistic action between diclofop-methyl and certain
phenoxy acetic acid herbicides such as 2,4-D is well known
[11].   Recent work by Shimabukuro and Hoffer [12] has shown
that 2,4-D partially reverses growth inhibition of diclofop-
methyl sensitive plants (Figure 3).   IAA and synthetic
hormones such as 2,4-D appear to activate a plasmalemma based
ATPase that leads to enhanced proton pumping across the
membrane and into the cell wall.   Diclofop-methyl, however,
leads to a dissipation of the transmembrane proton gradient
[13].   Although the exact physiological damage that follows
dissipation of the gradient is unknown, it is likely to
affect the active transport of solutes and lead to a
disruptive modification of cellular metabolism [11].   In the
diclofop-methyl tolerant Lolium the disturbed proton gradient
may be corrected more rapidly, enabling normal cellular
function to recover [12].   It is likely that the antagonistic
effect of 2,4-D could slow down the diclofop-methyl enhanced
membrane depolorization.

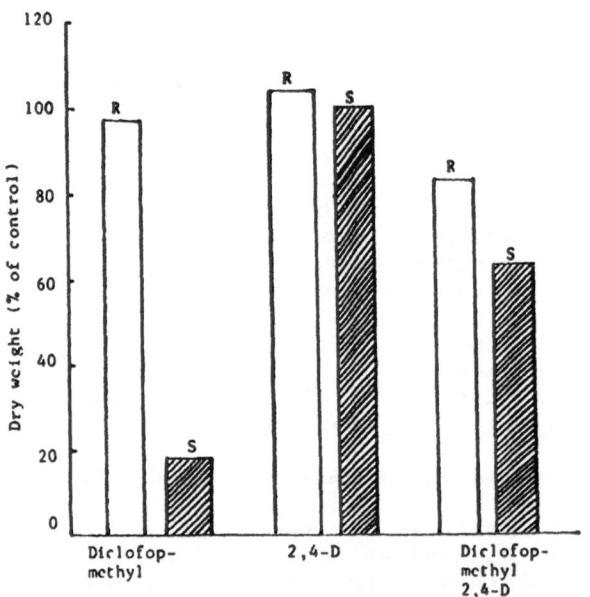

Figure 3   Antagonism of diclofop-methyl (DM) by 2,4-D
           in Lolium rigidum. DM was applied at
           0.84kg/ha and 2,4-D at 0.56 kg/ha.
           R, resistant biotype; S, susceptible
           biotype. Redrawn from Shimabukuro and
           Hoffer; (12).

A change in auxin sensitivity of the plasmalemma could be involved in the resistance mechanism of mecoprop tolerant Stellaria media. Such plants have been identified in fields near Bath U.K. [14]. Barnwell and Cobb [15] showed that shoot tip elongation of mecoprop tolerant plants was less susceptible to lower concentrations of mecoprop. Subsequent work [16] with etiolated hypocotyls from susceptible and resistant plants showed that mecoprop induced H^+ efflux was less sensitive to the herbicide in the "resistant" hypocotyls. This would restrict ATPase activation, H^+ pumping, and cell wall acidification, resulting in more limited growth aberrations.

Historically and numerically, triazine resistant plants are of considerable importance. Atrazine resistant biotypes of Senecio vulgaris were identified in Washington State USA in 1968, and at least 55 resistant biotypes have now been identified throughout the world [17]. Although the triazine herbicides have been used extensively as selective weedkillers in maize for many years, the mechanism of resistance of the photosynthetic inhibitors in the crop plant is predominantly due to glutathione conjugation via glutathione-S-transferase. This mechanism does not appear to be involved in the triazine-resistant weed biotypes. A detailed account of the mechanism of resistance is reported elsewhere in this volume [18], and only brief mention will be made here.

Triazine herbicides, like phenylureas, uracils and hydroxybenzonitriles function by displacing the quinone QB that normally acts as an intermediate electron carrier between QA and plastoquinione, and which is bound to a specific protein (D_1) of the reaction centre of photosystem 2. Binding to this protein is by a high affinity non-covalent bond which totally disrupts electron transport. The D_1 protein consists of 353 amino acid residues and the specific site of QB interaction is between the IVth and V[th] helices [19]. (Figure 4). Resistance to a range of photosynthetic inhibitor herbicides is due to mutations of the psbA gene that codes for the protein, and are predominantly between valine 217 and leucine 275. The first mutant identified was a change at 264 from serine to glycine in atrazine resistant Amaranthus hybridus [20]. Modifications to the D_1 protein, in addition to causing a change in the potential for herbicide binding, have in some instances been shown to reduce photosynthetic efficiency by up to 21% [21].

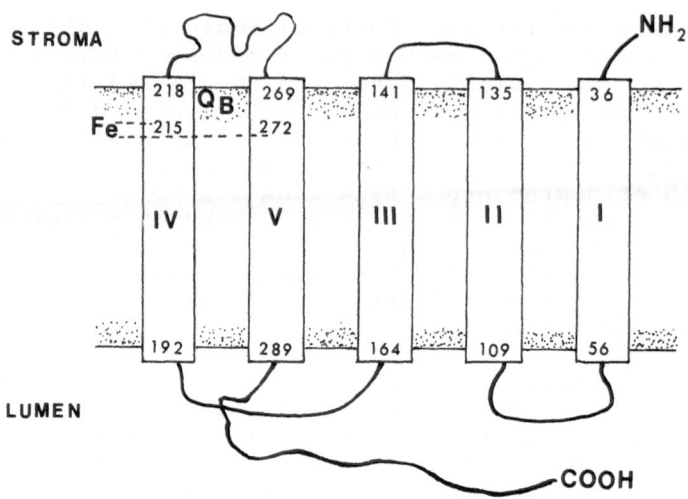

Figure 4   Diagrammatic scheme showing the possible
            folding of the D₁ polypeptide of photo-
            system 2 in five transmembrane segments.
            Numbers refer to the amino acid residues.

CHANGE IN SENSITIVITY OF KEY TARGET ENZYME

Very recently a number of biotypes of S.media, Kochia
scoparia, Salsola iberica and Lactuca serriola have been
identified in North America that are resistant to
chlorsulfuron [22].   This herbicide, like other
sulfonylureas, imidazoliniones and triazolo- pyrimidines,
inhibits acetolactate synthase, a key enzyme in the synthesis
of the branched chain amino acids leucine, isoleucine and
valine.   Using acetolactate synthase isolated from resistant
and susceptible biotypes of S. media, it was shown that the
enzyme from the resistant biotype was less sensitive to the
herbicidal inhibitor. (Figure 5).   The enzyme acetolactate
synthase is chloroplast located, and in addition to thiamine
pyrophosphate, has an absolute requirement for FAD, although
unusually not as a redox cofactor [23].   The thiamine
pyrophosphate combines with one of the pyruvate molecules to
produce hydroxyethyl thiamine pyrophosphate.   Schloss,
Ciskanik and Van Dyk [24] have proposed that the herbicides
operate in a non-competitive manner at an enzyme site that is
an evolutionary vestige of a non-functional quinone binding
site.   The pattern of cross resistance between sulfonylureas,
imidazolinines and triazolopyrimidines has led to the
suggestion of a hypothetical scheme for herbicide binding and
resistance [22], in which mutations in acetolactate

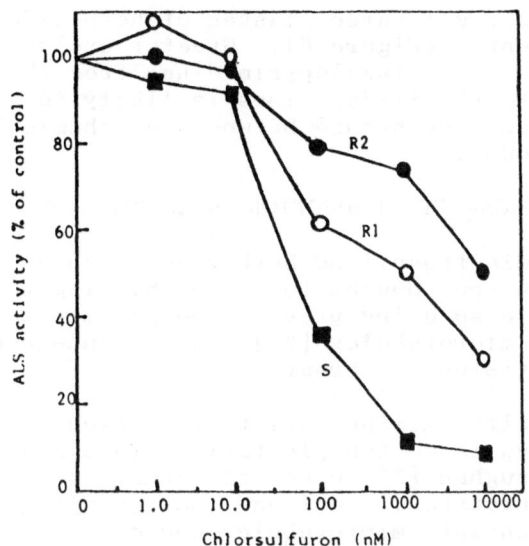

Figure 5   Inhibition of acetolactate synthase (ALS) in a susceptible (S) and two resistant biotypes (R_1) and (R_2) of _Stellaria media_ by chlorsulfuron.   Redrawn from Devine _et al._

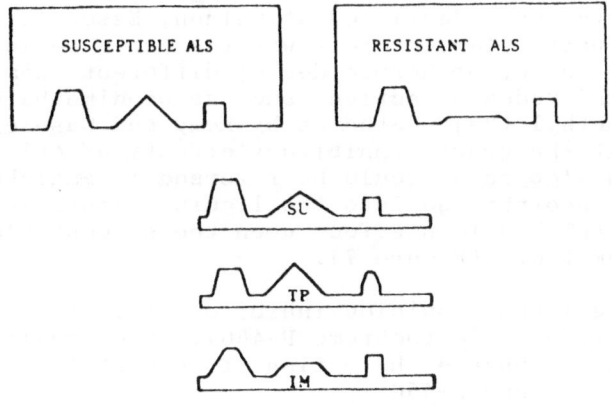

Figure 6   Hypothetical binding model of sulfonylurea (SU), triazolopyrimidine (TP) and imidazolinone (IM) herbicides in resistant and susceptible acetolactate synthase (ALS). Redrawn from Devine _et al._ (22).

synthase affect all three classes of herbicide, but not to the same extent. (Figure 6). Greater exclusion of sulfonylureas and triazolopyrimidines from the binding site in contrast to the imidazolines is likely to affect the degree of cross resistance between the chemically unrelated herbicide groups.

ENHANCED PRODUCTION OF HERBICIDE TARGET

A number of dinitroaniline herbicides such as trifluralin and pendimethalin are thought to act by binding to α and β tublin, and in so doing prevent the polymerization of the tubulin into microtubules [25]. As a consequence mitosis and thus cell division is arrested.

In 1984 a dinitroaniline resistant biotype of Eleusine indica was identified in cotton plantations in S. Carolina USA [26]. Vaughn and Vaughan [27] examined tubulin from resistant and susceptible biotypes, and found that in vitro tubulin polymerization into microtubules would proceed if GTP and dimethyl sulphoxide was added. When the dinitroaniline herbicide oryzalin was added, microtubule formation was severely disrupted in susceptible Eleusine biotypes. The resistant biotype was shown to possess a novel set of β - tubulin genes and thus a novel form of β -tubulin. This was postulated to be responsible for the herbicidal resistance [27].

ENHANCED METABOLIC BREAKDOWN AND CONJUGATION

A biotype of blackgrass Alopecurus myosuroides resistant to chlorotoluron was identified at Peldon, Essex, U.K. in 1984 [28]. Of particular interest was the cross resistance shown to a number of other herbicides of different chemical families and modes of action, such as pendimethalen and diclofop-methyl [29]. Studies by Kemp and Caseley [30] showed that the growth inhibitory effects of chlorotoluron on the Peldon Alopecurus could be reversed if amniobenzotriazole (ABT) was incorporated into the liquid culture medium. By contrast, ABT had less effect upon the susceptible Alopecurus from Rothamsted. (Figure 7).

ABT is a well-known suicide inhibitor of microsomal mixed function oxidases (cytochrome P-450). Previously, ABT has been shown to enhance the action of chlorotoluron [31], and another cytochrome P-450

inhibitor, tetcylasis, to inhibit the metabolism of chlorotoluron in maize and cotton suspension cell cultures [32].

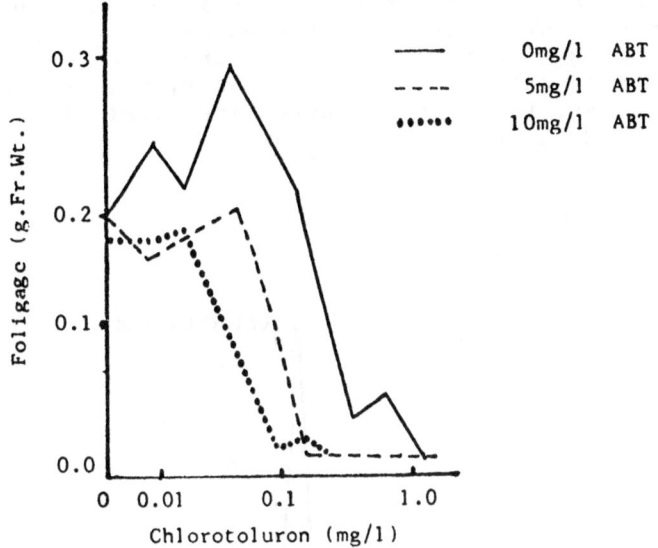

Figure 7   The synergistic effect of aminobenzo-
            triazole (ABT) on phytoxicity of chloro-
            toluron in "Peldon" <u>Alopecurus</u>
            <u>myosuroides.</u>   Redrawn from Kemp <u>et al.</u>
            29).

Evidence suggests that the Peldon strain of <u>Alopecurus</u> has
increased cytochrome P-450 activity, and greater herbicide
detoxification.   The metabolic events associated with
chlorotoluron detoxification are predominantly N-
monodemethylation, N-didemethylation and ring
methyloxidation.   This is probably followed by glucose
conjugation.   The enhanced production of chlorotoluron
metabolites was demonstrated convincingly by Kemp, Moss and
Thomas [29] after treatment of resistant and susceptible
blackgrass for 24 hours with labelled herbicide, followed by
extraction and radio-TLC analysis [Figure 8].

Recent work with microsomes extracted from resistant and
susceptible cell cultures of <u>A. myosuroides</u>, showed a much
greater carbon monoxide induced difference spectra at 450 n m
in the "resistant" microsomes.   This is good evidence for
enhanced cytochrome P-450 activity [33].

The sequence of detoxification reactions involving (phase I)
demethylation or hydroxylation catalyzed by cytochrome P-450,
followed by conjugation (phase II), and the subsequent
deposition of conjugates within the vacuole, may well apply
in whole or in part to the other herbicides that are much
less effective on the Peldon <u>A. myosuroides.</u>

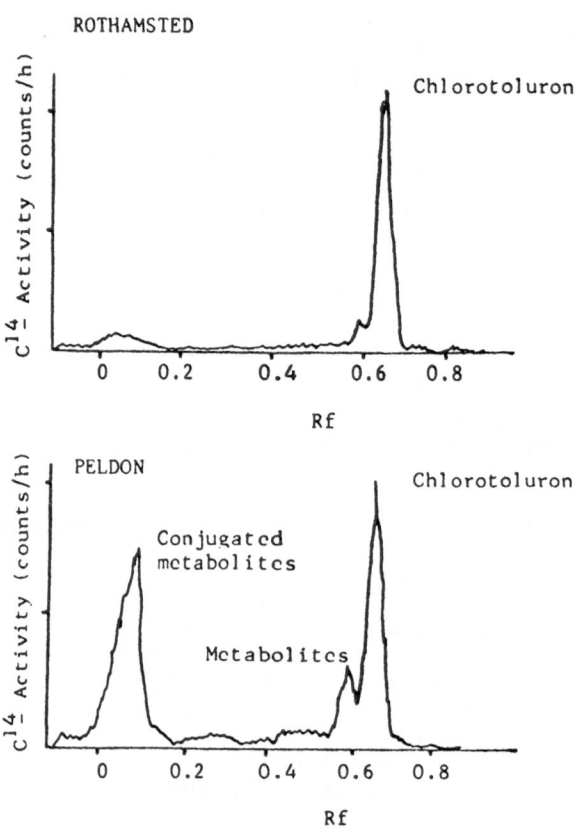

Figure 8   Radio-TLC analysis of components extracted
            by 20% aqueous MeOH from "Peldon" and
            "Rothamsted" <u>Alopecurus</u> <u>myosuroides</u> after
            24 hour treatment with ^{14}C-chlorotoluron.
            Redrawn from Kemp <u>et al.</u> (29).

Reference has already been made to mecoprop resistant
S. media from Bath, and a possible resistance mechanism
involving a change in auxin sensitivity [15]. An alternative
or related mechanism could involve the enhanced conjugation
of mecoprop in the resistant plants, with the possibility of
enhanced vacuolar compartmention. Coupland, Lutman and Heath
[34] showed that S. media plants treated with ^{14}C-mecoprop
metabolised the herbicide more readily in resistant biotypes.
This also corresponded to an increase in the production of
TLC immobile polar metabolites, such as sugar or amino acid
conjugates. This enhanced metabolism could therefore play an
important role in herbicide resistance mechanisms.

ENHANCED BREAKDOWN OF HERBICIDE GENERATED TOXIC PRODUCTS

The action of bipyridinium herbicides, paraquat and diquat is
dependent upon penetration into the chloroplast, followed by
reduction and reoxidation to produce superoxide [35]. It is
envisaged that excess superoxide will be generated, and the
very toxic hydroxyl radical produced, damage lipids, proteins
and nucleic acids.

Although superoxide is probably formed to a limited extent
during "normal" photosynthesis as a result of an endogenous
Mehler-Reaction, its toxicity is controlled by a range of
protective enzymes and scavenging molecules. These include
superoxide dimutase, ascorbate peroxidase, dehydroascorabate
reductase, and glutathione reductase, together with ascorbate
and glutathione [36]. Experiments with some paraquat
tolerant biotypes have indicated that enhanced levels of some
of these enzymes and scavengers could contribute, at least in
part, to tolerance mechanisms in Conyza [37] and Erigeron
[38] biotypes. Recent work with genetically engineered
petunia plants with 50 times higher than normal level of
superoxide dismutase, were however not more resistant to
paraquat treatment.

CONCLUSION

The increasing appearance of herbicide tolerant weeds during
the last 20 years, has turned an interesting scientific
curiosity, into a serious agronomic problem.
This paper has identified some of the biochemical/
physiological mechanisms that have evolved in natural
populations of weeds after a number of years of herbicide
selection pressure. Older simplistic ideas to overcome
resistance by using "another herbicide" have been overtaken
dramatically by the appearance of cross resistance. This has
become evident as resistance to one class of herbicide is
also effective against other classes of herbicide that
operate at a similar site of action, or are detoxified by a

similar metabolic mechanism. Potentially more serious is the appearance of multiple resistance where one plant has more than one resistance mechanism for one herbicide, or more than one mechanism for two or more herbicides or herbicide classes. The consequences of these developments are that in addition to the environmental advantage of using lower herbicide doses, the use of alternative herbicides or herbicide mixtures must be considered with much more care.

REFERENCES

1. Dodge, A.D. Mechanisms of paraquat tolerance. In Herbicide Resistance in Weeds and Crops, ed. Caseley, J.C., Cussans, G.W. and Atkin, R.K., Butterworth-Heinemann, Oxford, 1991, pp. 165-175.

2. Bishop, T., Powles, S.B. and Cornic, G. Mechanism of paraquat resistance in Hordeum glaucum:2. Studies in paraquat uptake translocation. Aust. J. Plant Physiol., 1987, 14, 539-547.

3. Tanaka, Y., Chisaka, H. and Saka, M. Movement of paraquat in resistant and susceptible biotypes of Erigeron philadelphicus and E. canadensis. Physiol. Plant., 1986, 66, 605-608.

4. Fuerst, E.P., Nakatani, H.Y., Dodge, A.D., Penner, D. and Arntzen, C.J. Paraquat resistance in Conyza. Plant Physiol., 1985, 77, 984-989.

5. Fuerst, E.P. and Vaughn, K.C. Mechanisms of paraquat resistance. Weed Tech., 1990, 4, 150-156.

6. Powles, S.B. and Cornic, G. Mechanisms of paraquat resistance in Hordeum glaucum I. Studies with isolated organelles and enzymes. Aust. J. Plant Physiol. 1986, 14, 81-89.

7. Heap, I.M. and Knight, R. The occurrence of herbicide cross-resistance in a population of annual ryegrass Lolium rigidum, resistant to diclofop-methyl. Aust. J. Agric. Res., 1986, 37, 149-156.

8. Heap, I.M. and Knight, R. Variation in herbicide cross-resistance among populations of annual ryegrass (Lolium rigidum) resistant to diclofop-methyl. Aust. J. Agric. Res., 1990, 41, 121-128.

9. Secor, J. and Cseke, C. Inhibition of acetyl-CoA carboxylase activity by haloxyfop and tralkoxydim. Plant Physiol., 1988, 86, 10-12.

10. Powles, S.B., Holtum, J.A.M., Matthews, J.M. and
    Liljegren, D.R. Herbicide cross-resistance in
    annual ryegrass (Lolium rigidum Gaud). In
    Managing Resistance to Agrochemicals, ed.
    M.B. Green, H.M. Le Baron and W.K. Moberg,
    ACS, Washington, 1990, pp. 394-406.

11. Shimabukuro, R.H., Walsh, W.C. and Wright, J.P.
    Effect of diclofop-methyl and 2,4-D on trans-
    membrane proton gradient: A mechanism for their
    antagonistic interaction. Physiol. Plant., 1989,
    77, 107-114.

12. Shimabukuro, R.H. and Hoffer, B.L. Metabolism of
    diclofop-methyl in susceptible and resistant bio-
    types of Lolium rigidum. Pestic. Biochem.
    Physiol., 1991, 39, 251-260.

13. Shimabukuro, R.H. and Hoffer, B.L. Diclofop
    action: Physiological evidence for reversible bind-
    ing to a receptor-like molecule. Weed Sci. Soc.
    Am., Las Vegas, 1988, Abstract 182.

14. Lutman, P.J.W. and Lovegrove, A.W. Variations in
    the tolerance of Galium aparine (cleavers) and
    Stellaria media (chickweed) to mecoprop. Proc.
    Brit. Crop Prot. Conf. - Weeds, 1985, 2, 411-418.

15. Barnwell, P. and Cobb, A.H. Physiological studies
    of mecoprop-resistance in chickweed (Stellaria
    media L.). Weed Res., 1989, 29, 135-140.

16. Barnwell, P., Early, C. and Cobb, A.H. Differ-
    ential auxin-sensitivity and mecoprop-resistance
    in chickweed (Stellaria media). Proc. Brighton
    Crop Prot. Conf. - Weeds, 1989, 2, 427-432.

17. Le Baron, H.M. and McFarland, J. Herbicide
    resistance in weeds and crops. In Managing
    Resistance to Agrochemicals, ed. M.B. Green,
    H.M. Le Baron and W.K. Moberg. ACS. Washington
    1990, pp. 336-352.

18. Van Rensen, J.J.S. Biochemical mechanisms of
    resistance to photosystem II herbicides. (These
    proceedings).

19. Trebst, A. The topology of the plastoquinone and
    herbicide binding peptides of photosystem II in
    the thylakoid membrane. Z. Naturforsch, 1987,
    42c, 742-750.

20. Hirschberg, J. and McIntosh, L. Molecular basis
    of herbicide resistance in <u>Amaranthus</u> <u>hybridus</u>.
    <u>Science</u>, 1983, **222**, 1346-1348.

21. Van Oorschot, J.L.P. and Van Leeuwen, P.H. Photo-
    synthetic capacity of intact leaves of resistant
    and susceptible cultivars of <u>Brassica</u> <u>napus</u> L.
    to atrazine. <u>Weed</u> <u>Res.</u>, 1989, **29**, 29-32.

22. Devine, M.D., Marles, M.A.S. and Hall, L.M.
    Inhibition of acetolactate synthase in susceptible
    and resistant biotypes of <u>Stellaria</u> <u>media</u>.
    <u>Pestic</u> <u>Sci.</u>, 1991, **31**, 273-280.

23. Hawkes, T.R., Howard, J.L. and Pontin, S.E. Herbi-
    cides that inhibit the biosynthesis of branched
    chain amino acids. In <u>Herbicides and Plant</u>
    <u>Metabolism</u>, ed. A.D. Dodge. Cambridge University
    Press, Cambridge, 1989. pp. 113-136.

24. Schloss, J.V., Ciskanik, L.M. and Van Dyk, D.E.
    Origin of the herbicide binding site of aceto-
    lactate synthase. <u>Nature</u>, 1988, **331**, 360-362.

25. Strachan, S.D. and Hess, F.D. The biochemical
    mechanism of action of the dinitroaniline herbicide
    oryzalin. <u>Pestic. Biochem. Physiol.</u>, 1983, **20**,
    141-150.

26. Mudge, L.C., Gossett, B.J. and Murphy, T.R.
    Resistance of goosegrass <u>(Eleusine indica)</u> to
    dinitroaniline herbicides. <u>Weed</u> <u>Sci.</u>, 1984,
    32, 591-594.

27. Vaughn, K.C. and Vaughan, M.A. Structural and
    biochemical characterization of dinitroaniline-
    resistant <u>Eleusine.</u> In <u>Managing</u> <u>Resistance</u> <u>in</u>
    <u>Agrochemicals</u>, ed. M.B. Green, H.M. Le Baron
    and W.K. Moberg, ACS, Washington, 1990, pp. 364-
    375.

28. Moss, S.R. Herbicide resistance in black-grass
    <u>(Alopecurus</u> <u>myosuroides</u>). <u>Proc.</u> <u>Brit.</u> <u>Crop</u> <u>Prot.</u>
    <u>Weeds</u>, 1987, **3**, 879-886.

29. Kemp, M.S., Moss, S.R. and Thomas, T.H. Herbicide
    resistance in <u>Alopecurus</u> <u>myosuroides</u>. In <u>Managing</u>
    <u>Resistance to Agrochemicals,</u> ed. M.B. Green,
    H.M. Le Baron and W.K. Moberg, ACS, Washington,
    1990, pp. 376-393.

30. Kemp, M.S. and Caseley, J.C.   Synergistic effects
    of 1-aminobenzotriazole on the phytotoxicity of
    chlorotoluron and isoproturon in a resistant
    population of black-grass (Alopecurus myosuroides).
    Proc. Brit. Crop Prot. Conf. - Weeds, 1987, 3,
    895-899.

31. Cabanne, F., Gaillardon, P., Scalla, R. and
    Durst, F. Aminobenzotriazole as a synergist of
    urea herbicides.   Proc. Brit. Crop Prot. Conf. -
    Weeds,., 1985, 3, 1163-1170.

32. Cole, D.J. and Owen, W.J.   Influence of mono-
    oxygenase inhibitors on the metabolism of the
    herbicides chlorotoluron and metolachlor in cell
    suspension cultures. Plant Sci., 1987, 50, 13-20.

33. Caseley, J.C., Kueh, J., Jones, O.T.G., Hedden, P.
    and Cross, A.R.   Mechanism of chlorotoluron
    resistance in Alopecurus myosuroides.   In Proc.
    7th Int. Cong. Pestic. Chem., Hamburg, ed.
    E. Kessler-Schnitz and S. Conway, 1990, pp. 417.

34. Coupland, D., Lutman, P.J.W. and Heath, C.
    Uptake, translocation, and metabolism of meco-
    prop in a sensitive and resitant biotype of
    Stellaria media. Pestic. Biochem. Physiol.,
    1990, 36, 61-67.

35. Dodge, A.D.   Herbicides interacting with photo-
    system I.   In Herbicides and Plant Metabolism.
    ed. A.D. Dodge, 1989, pp. 37-50.

36. Kunert, K.J. and Dodge, A.D. Herbicide-induced
    radical damage and antioxidative systems.   In
    Target Sites of Herbicide Action.   ed. P. Boger
    and G. Sandmann, CRC Press, Boca Raton, 1989,
    pp. 45-63.

37. Shaaltiel, Y. and Gressel, J.   Multienzyme oxygen
    radical detoxifying system correlated with paraquat
    resistance in Conyza bonariensis.   Pestic.
    Biochem. Physiol., 1986, 26, 22-28.

38. Matsunaka, S. and Ito, K.   Paraquat resistance in
    Japan.   In Herbicide Resistance in Weeds and Crops,
    ed. J.C. Caseley, G.W. Cussans and R.K. Atkin,
    Butterworth-Heinemann, 1991, pp. 77-86.

39. Tepperman, J.M. and Dunsmuir, P.   Transformed
    plants with elevated levels of chloroplast super-
    oxide dismutase are not more resistant to super-
    oxide toxicity.   Plant Mol. Biol., 1990, 14, 501-
    511.

MECHANISMS OF RESISTANCE TO FUNGICIDES

M. J. Henry
DowElanco Research Laboratories,
2001 W. Main St., Greenfield, Indiana USA 46140.

ABSTRACT

Fungicide resistance is a problem which followed the development of antifungals with specific biochemical modes of action. It is not surprising that the primary mechanism of resistance to these new specific fungicides is target site based. The most characterized example of fungicide resistance based on a modified target is resistance to the benzimidazole class of fungicides such as carbendazim. Several laboratory studies have clearly shown by genetic and biochemical analysis that modified tubulin is the mechanism of benzimidazole resistance. Resistance mechanisms to compounds such as the phenylamides, carboxamides, polyoxins, and sterol biosynthesis inhibitors have been investigated and will be discussed to exemplify the diversity of mechanisms used by fungi to circumvent the toxicity of modern disease control agents.

INTRODUCTION

Prior to the mid 1960's nearly all fungicides used in agriculture acted at multiple subcellular targets and possessed limited biochemical specificity which restricted their application to external plant therapy. These fungicides with multiple sites of action fall into two distinct classes. The first includes the inorganic Bordeaux mixture and the sulfur and copper compounds. The second class includes the organic compounds such as maneb, dichlone, captan, and chlorothalonil. The discovery of fungicides with specific modes of action brought about the benefit of systemic protection of the plant as well as curative and eradicant disease control, but also the problem of fungicide resistance.

The mechanism of fungicide resistance is often related to the mechanism of action. In the case of fungicides, it is not a coincidence that the invention of fungicides with specific biochemical mechanisms, led to fungicide resistance and resistance mechanisms based on target site changes in the fungus. The diversity of fungal populations and their high fecundity provide ample opportunity for the selection of naturally resistant populations and spontaneously arising mutants that are tolerant to the fungicide. Resistance to fungicides can be attributed to one or more of the following basic mechanisms: 1) reduced affinity of target site; (2) reduced uptake or increased efflux of fungicide; (3) detoxification; (4) lack of conversion to active compound; (5) compensation, such as increased production of target enzyme; and (6) circumvention of block through an alternate pathway. A detailed understanding of the mechanism of fungicidal action and of resistance can help in devising strategies for countering a fungicide resistance problem. Knowledge on the mechanism of action and resistance not only provides a basis for the practical plant pathologist to recommend various mixtures and alternate fungicide application schedules, but gives the fungicide discovery scientist a starting point to initiate a research program to find a new alternative fungicide. This paper, while not attempting to be an exhaustive review, will briefly discuss some of the mechanisms of fungicide resistance to the organic fungicides with specific modes of action and describe some of the known mechanisms of resistance.

DISCUSSION

Benzimidazoles

The benzimidazole class of fungicides, while not the first class of systemic fungicides, were the first broad spectrum systemic compounds that ushered in a new age of plant disease control (1). These compounds control a variety of plant pathogenic fungi belonging to the Ascomycete and Basidiomycete classes of fungi (2). Benomyl and carbendazim (MBC) are two representatives of this class. The target site of carbendazim and other benzimidazole analogs is fungal tubulin. Microtubules are composed of alternating helixes of α and β tubulin. Mechanism studies that led to this finding first discovered that carbendazim inhibited DNA synthesis in *Neurospora, Ustilago* (3) and *Apergillus* (4). The inhibition of mitosis was observed in synchronous cultures of yeast when DNA synthesis was allowed to continue for one cell cycle after treatment with MBC, but was interrupted as the cells were unable to separate thus indicating that mitosis not DNA synthesis *per se* was blocked by carbendazim (5). The inhibition of mitosis and other microtubular dependent cellular functions was confirmed to be due to MBC binding to tubulin with direct evidence of the binding of carbendazim to tubulin in sensitive and resistant strains of *Aspergillus nidulans* (4). The affinity of the tubulin containing preparations for ^{14}C-carbendazim was proportional to the inhibition of growth in these genetically defined strains. The specific location of the mutation has been identified in the β-tubulin gene and identified by DNA sequence analysis in *Aspergillus* (6), *Neurospora* (7), and *Saccharomyces* (8). A single amino acid modification in the β-tubulin protein from phenylalanine to tyrosine resulting from a mutation in codon 167 provides resistance to carbendazim in *Neurospora*, while in *Saccharomyces* a change in codon 241 causes an amino acid change from arginine to histidine. Such variation indicates that this protein which is highly conserved across species is still mutable to result in significant fungicide resistance. In addition to resistance, it has been found that additional mutations in the β-tubulin gene can impart increased sensitivity to the benzimidazole fungicides (9).

Mutations that result in resistance to one benzimidazole fungicide usually show cross-resistance to other benzimidazoles, but in some cases the mutation may lead to increased sensitivity to certain benzimidazoles and related compounds (negative cross-resistance). An example of negative cross-resistance was first found with certain phenylcarbamate herbicides that inhibit nuclear division in higher plants by affecting microtubular functioning (10). These are much more toxic to benzimidazole-resistant mutants of *Botrytis cinerea* than to the benimidazole-sensitive wild-type strains. This phenomenon led to the development of non-phytotoxic compounds such as N-(3,5-dichlorophenylcarbamate (MDPC) which control benzimidazole-resistant *B. cinerea* and several other fungi (11). Negative cross-resistance has been reported for species such as *Cercospora beticola*, *Fusarium nivale*, *Mycosphaerella melaonis*, *Pseudocercosporella herpotrichoides*, and *Venturia nashicola* (11,12,13).

Phenylamides

The phenylamide fungicides includes metalaxyl, furalaxyl, benalaxyl, ofurace, cyprofuram, and oxadixyl (14). Metalaxyl is the best known and most studied member of the phenylamide fungicides, but all of the aforementioned compounds appear to act by the same mechanism. They selectively inhibit ribosomal RNA synthesis in Peronosporales fungi. Initial investigations found RNA biosynthesis to be affected in sensitive *Pythium* and *Phytophthora* species (15, 16). The specific target for inhibition of RNA synthesis is the RNA polymerase I complex and modification of this target site is the basis of resistance in *Phytophthora*. Resistance to the inhibition of RNA-polymerase by phenylamides has been shown to occur in isolated RNA-polymerase preparations from both laboratory-induced and field resistant isolates of *Phytophthora* (17). Similarly, these same resistant isolates show essentially no binding of radio-labeled metalaxyl to cell free preparations containing RNA-polymerase I, confirming the target site nature of the resistance (17).

Carboximides

These fungicides, used primarily to control plant-pathogenic Basdiomycetes, were the first to show systemic disease control activity (18). Substantial evidence indicates that the fungitoxic activity of carboxin and related analogs is due to inhibition of electron flow through succinate-ubiquinone oxidoreductase (complex II) in the mitochondrial electron transport pathway. This complex consists of 4 polypeptides of 70K, 30K, 15K and 7.5K daltons in molecular weight with both an FAD ligand and a 2Fe-2S cluster. Specifically, the receptor for carboxin within complex II appears to consist of the iron-sulfur cluster S3 complexed with a small ubiquinone binding polypeptide(s) in a membrane environment (19, 20).

Carboximide resistant strains of *Ustilago maydis* selected in the laboratory are considered to be tolerant due to target site mutations. Two allelic mutations in carboxin-resistant *Ustilago* result in moderate (oxr-1A) and high (oxr-1B) resistance to the inhibition of succinate-dichlorophenolindophenol reductase activity by carboxin and moderate resistance with both alleles to 3'-octyloxycarboxin. In contrast, 4'-phenylcarboxin and carboxin showed negative-cross resistance with mutants containing the oxr-1A allele and moderate cross resistance in mutants containing the oxr-1B allele. In comparison to *Ustilago,* allelic mutations in *Aspergillus nidulans* providing moderate (cbx A) and high (cbx B) degrees of resistance to carboxin also show moderate and high degrees of resistance to 3'-octyloxycarboxin and 4'-phenylcarboxin (19). These genes apparently code for components of complex II which when modified by mutation have decreased or increased affinities for the inhibitors. The variability between species indicate that this site, as with tubulin, contains multiple mutable locations which can be altered to change the affinity of related inhibitors.

An oxycarboxin-resistant strain of *Puccinia horiana* has been reported on chrysanthemums at several locations (21), and there appear to be carboxin-resistant strains of *Ustilago nuda* infecting barley in the United Kingdom and France (22). Although not confirmed, it is possible that resistance in these reported cases may be due to target site modifications.

Polyoxins

Polyoxins are specific inhibitors of chitin synthase (23). The mechanism of inhibition of chitin synthase is by competitive inhibition with the structurally similar substrate UDP-N-acetyl-glucosamine (Glu-N-Ac) (24, 25). Field isolates of *Alternaria kikuchiana* were found that varied in sensitivity to polyoxin B up to 15 fold. This differential sensitivity correlated with the incorporation of ^{14}C-Glu-N-Ac into chitin of intact mycelial cultures but not with the isolated chitin synthase indicating a mechanism independent of the fungicidal target site(24, 25). The variations in sensitivity of *Alternaria* do correlate with the uptake of polyoxin A and the dipeptide glycyl-glycine (25). This change in permeability responsible for resistance suggests that the uptake of polyoxin is via an active dipeptide system which if altered to no longer transport dipeptides, provides resistance to the polyoxin fungicide.

Kitazin P

The organophosphorus compound S-benzyl-O-O-diisopropylphoshorothiolate known as IBP or Kitazin P is thought to act by inhibition of phospholipid biosynthesis. IBP has been shown to inhibit the methyltransferase step in the synthesis of phosphotidylcholine from phosphotidylethanolamine (26). The mechanism of resistance to this fungicide in some field isolates of *Pyricularia oryzae* is not considered to be target site related. Both sensitive and resistant strains are able to metabolize IBP but the moderately resistant strains metabolize IBP faster and via a different pathway than the sensitive strains (27). Although insecticide resistance is frequently based on differential metabolism rates in sensitive and tolerant insects, metabolism based fungicide resistance is very rarely observed.

Pyrazophos

In contrast to metabolic degradation as a mechanism of resistance, the mechanism of resistance to pyrazophos in a laboratory isolate of *Pyricularia oryzae* is associated with the isolates inability to metabolize the parent molecule to an active metabolite. The cleavage of the phosphothioester to yield the active 2-hydroxypyrazolopyrimidine does not occur in the resistant isolate (28). This mechanism is identical with the mechanisms of resistance of naturally resistant fungi *Pythium debaryanum* and *S. cerevisiae* (29).

Sterol Biosynthesis Inhibitors (SBI's)

Due to their widespread use in agriculture and medicine, resistance to sterol biosynthesis inhibiting fungicides (SBI's) is of continuing concern (30). Fungicides of this general group, inhibit sterol biosynthesis at one or more specific points in the biosynthetic pathway of the typical fungal sterol ergosterol (31, 32).

The first class of SBI's are the 14α-demethylation inhibitors (DMI's) which include various substituted nitrogen containing heterocycles (33). Although during the past two decades several reports of laboratory induced resistance surfaced, the actual field resistance to these fungicides has been relatively slow and gradual (34). Furthermore, the mechanisms of resistance in field populations of fungi resistant to DMI's have been difficult to ascertain especially in the obligate powdery mildew parasites of cereals and vines.

Currently there are no satisfactory explanations of the mechanisms of resistance to SBI fungicides in tolerant isolates from the field. We must therefore rely on possible explanations based on laboratory models. One mechanism of resistance to DMI's has been well characterized in laboratory induced resistant mutants of *Aspergillus* and *Penicillium*. This mechanism involves the energy-dependant efflux of the toxicant from the fungal cells. (35, 36). This mechanisms may involve a change in the regulation of this inducible efflux mechanism in resistant strains since it is present in both sensitive and resistant strains, but constitutively operational in the resistant mutants.

A second mechanism of resistance to DMI fungicides which operates at least under laboratory conditions is the over production of the P-450 14α-demethylase enzyme. In *Saccharomyces* the inclusion of a high copy number plasmid containing the P-450 14α-demethylase gene is adequate to provide resistance to the DMI ketoconazole (37). Such mechanisms could possibly function under field conditions where reduced efficacy of the DMI fungicides is observed.

Another possible mechanism of resistance to DMI fungicides is the alteration of fungal membrane structure (38). In laboratory produced mutants of *Cercospora beticola*, it was found that 24 of 25 mutants selected for resistance to flusilazole contained an altered sterol composition. The modification of sterol composition may in fact alter the over all membrane such that the deleterious effects of 14α-methyl sterols is attenuated.

The second group of SBI fungicides includes various morpholine and piperidine derivatives such as tridemorph, fenpropimorph, fenpropidin and piperalin which inhibit primarily sterol Δ8-Δ7 isomerization, Δ14 reduction, or both, depending on the organism and inhibitor involved (39). Due to he nature of these inhibitors acting as carbocationic mimics, other steps in the sterol biosynthetic pathway that involve carbocationic intermediates such as the cyclization of 2,3-oxidosqualene and the transmethylation of lanosterol at the 24 position may also be affected at higher rates of these compounds (40). The morpholines have a somewhat narrower antifungal spectrum than the DMIs but they are highly effective for control of powdery mildews and a number of other fungal pathogens.

Little is known about the mechanisms of resistance to morpholine fungicides. Currently there is no significant resistance problems to the morpholine SBI fungicides. This may be related to the ability of these compounds to inhibit multiple steps in the biosynthesis of sterols in fungi (39, 40). Since these compounds may inhibit up to 4 different enzymes in the sterol biosynthetic pathway, they can act as multi-site fungicides rather than single-site fungicides.

An additional group of SBIs are effective pharmaceutical antifungals, but as of yet no compounds of this type have not been developed as agricultural fungicides. This group includes the allylamines and the naphthyl thiocarbamates. These compounds inhibit the epoxidation of squalene in *Candida* (41, 42) and *Ustilago* (43).

Resistance to the squalene epoxidase inhibitors in practice has not become a serious problem even though tolnaftate has been in use for over 20 years. A recent laboratory study has shown in *Ustilago maydis* that resistant mutants may be obtained with two possible mechanisms of resistance (43). The first is the alteration of the target enzyme squalene epoxidase which showed resistance to terbinafine inhibition *in vivo* and *in vitro*. This mutant showed the highest degree of resistance to growth inhibition and to the inhibition of squalene epoxidase in cell free preparations of the mutant enzyme. The second mechanism of resistance is facilitated by the increased enzyme activity in the mutant. The higher enzyme activity is probably due to the over production of the target enzyme which contributed to a moderate level of resistance. This mutant contains nearly twice the specific enzyme activity as the parent strain.

CONCLUSION

Knowledge of the mechanism of resistance to fungicides coupled with a clear understanding of their mechanism of action is a powerful tool in combating and preventing resistance problems in practice. Simply offering a fungicide to farmers and resistance managers with a different mechanism of action is neither acceptable or necessary if resistance is based on mechanisms independent of the mode-of-action or dependant on specific types of chemistry. A key advantage in managing resistance problems lies in the availability of several types of fungicides to control a single pathogen. The decreasing number of fungicide options is a growing concern as there is a continued loss of the older fungicides with multiple sites of action and no resistance problems. It is therefore critically important for resistance managers to introduce new compounds with a prudent resistance strategy and for industry and academic institutions to continue to support fundamental research on mechanisms of fungicidal action and resistance.

REFERENCES

1. Delp, C.J., Benzimidazole and related fungicides. In Modern Selective Fungicides. ed. H. Lyr. VEB Gustav Fischer Verlag, Jena and Logman Group UK Ltd., London, 1987, pp 233-244.

2. Delp, D.J., and Klopping, H. L., Performance attributes of a new fungicide and ovicide candidate. Plant Dis. Rep. 1968, **52**, 95-99.

3. Clemons, G. P., and Sisler, H. D., Localization of the site of action of a fungitoxic benomyl derivative. Pestic. Biochem. Physiol., 1971, **1**, 32-42.

4. Davidse, L. C., Antimitotic activity of methyl benzimidazole-2-yl carbamate (MBC) in *Aspergillus nidulans*. Pestic. Biochem. Physiol., 1976, **3**, 317-325.

5. Hammerschlag, R. S. and Sisler, H. D., Benomyl and methyl-2-benzimidazole carbamate (MBC): Biochemical, cytological and chemical aspects of toxicity to *Ustilago maydis* and *Saccharomyces cerevisiae*. Pestic. Biochem. Physiol., 1973, **3**, 42-54.

6. Jung, M. K., Dunne, OP. W., Suen, I.H., and Oakley, B. R., Sequence alterations in β-tubulin mutations of *A. nidulans*. J. Cell Biol., 1987, **105**, (Abstract 1560).

7. Orbach, M.J., Porro, E. B., and Yanofsky, C., Cloning and characterization of the gene for β-tubulin from a benomyl-resistant mutant of *Neurospora crassa* and its use as a dominant selectable marker. Mol. Cell. Biol., 1986, **6**, 2452-2461.

225

8. Thomas, J. H., Neff, N. F., and Botstein, D., Isolation and characterization of mutantions in the β-tubulin gene of *Saccharomyces cerevisiae*. Genetics, 1985, 111, 715-734.

9. Davidse, L. C., Benzimidazole fungicides: Mechanism of action and biological impact. Ann. Rev. Phytopathol., 1986, 24, 43-65.

10. Leroux, P. and Gredt M., Phenomena of negative cross-resistance in *Botrytis cinerea* Perg. between benzimidazole fungicides and carbamate herbicides. Phytiatrie-Phytopharmacie, 1979, 28, 79-86.

11. Kato, T., Suzuki, K., Takahashi, J., and Kamoshita, K., Negatively correlated cross-resistance between benzimidazole fungicides and methyl N-(3,5-dichlorophenyl) carbamate. J. Pesticide. Sci., 1984, 9, 489-495.

12. Demakopoulou, M. G., and Georgopolous, S. G., Sensitivity to N-phenylcarbamates as related to benzimidazole resistance in *Cercospora beticola*. Proceeding of the 6th IUPAC Congress of Plant Protection., 1987, 3E-05

13. Leroux, P., and Cavalier N., Characteristics of *Pseudocercosporella herpotrichoides* strains resistant to benzimidazole and thiophanate fungicides. La Defense de Vegetaux, 1983, 222, 231-238.

14. Schwinn, F. J,. and Staub, T., Phenylamides and other fungicides against Oomyctes. In Modern Selective Fungicides. ed. H. Lyr. VEB Gustav Fischer Verlag, Jena and Logman Group UK Ltd., London, 1987, pp 259-273.

15. Kerkenaar, A., On the antifungal mode of action of metalaxyl, an inhibitor of nucleic acid synthesis in *Pythium splendens*. Pestic. Biochem. Physol., 1981, 16, 1-13.

16. Davidse, L.C., Hofman, A. E., and Velthuis , G.C.M., Specific interference of metalaxyl with endogenous RNA polymerase activity in isolated nuclei from *Phytophthora megasperma f. sp. medicaginis*. Exp. Mycol. 1983, 7, 344-361.

17. Davidse, L. C., Phenylamide fungicides: Mechanism of action and resistance. In Fungicide Resistance in North America. ed. C. Delp, APS Press, St. Paul, 1988, pp. 63-65.

18. von Schmeling, B., and Kulka, M., Systemic fungicidal activity of 1,4-oxathiin derivatives. Science, 1966, 152, 659-660.

19. White, G. A., Thorn G. D., and Georgopolous, S. G., Oxathiin carboxamides highly active against carboxin-resistant succinic dehydrogenase complexes from carboxin-selected mutants of *Ustilago maydis* and *Aspergillus nidulans*. Pestic. Biochem. and Physiol., 1978, 9, 165-182.

20. Schewe, T., and Lyr, H., Mechanism of action of carboxin fungicides and related compounds. In Modern Selective Fungicides. ed. H. Lyr. VEB Gustav Fischer Verlag, Jena and Logman Group UK Ltd., London, 1987, pp 133-142.

21. Kulka, M., and von Schmeling, B., Carboxin fungicides and related compounds. In Modern Selective Fungicides. ed. H. Lyr. VEB Gustav Fischer Verlag, Jena and Logman Group UK Ltd., London, 1987, pp 119-131.

22. Locke, T., Current incidence in the United Kingdom of fungicide resistance in pathogens of cereals. Proc of the British Crop Prot. Conf., 1986. **2**, 781-786.

23. Hori, M., Kakiki, K., and Misato, T., Further study on the relation of polyoxin structure to chitin synthetase inhibition. Agr Biol. Chem., 1974, **38**, 691-698.

24. Hori, M., Kakiki, K., Eguchi, J., and Misato, T., Studies on the mode of action of polyoxins. VI effect of polyoxin B on chitin synthesis in polyoxin-sensitive and resistant strains of *Alternaria kikuchiana*. J. Antibiot., 1974, **27**, 260-266.

25 Hori, M., Kakiki, K., and Misato, T., Antagonistic effect of dipeptides on the uptake of polyoxin A by *Alternaria kikuchinana*. J. Pesticide Sci., 1977, **2**, 139-149.

26. Akatsuka, T., Kodama, O., and Yamada, H., A novel mode of action of Kitazin P in *Pyricularia oryzae*. Agric. Biol Chem., 1977, **41**, 2111-2112.

27. Uesugi, Y. and Katagiri, M., Metabolism of a phosphorothiolate fungicide IBP by strains of *Pyricularia oryzae* with varied sensitivity. In Pesticide Chemistry, Human Welfare and the Environment, Vol. 3, Mode of Action, Metabolism and Toxicology. ed. J. Miyamoto and P.C. Kearney, Pergammon Press, New York, 1983, pp. 165-170.

28. de Waard, M. A. and Van Nistelrooy, J. G. M., Mechanism of resistance to pyrazophos in *Pyricularia oryzae*. Neth. J. Pl. Path., 1980, 86, 251-258.

29. de Waard, M. A., Mechanisms of action of the organophosphorus fungicide pyrazophos. Meded. LandbHogesch. Wageningen, 1974, **74**(14), 1-98.

30. Brent, K. J. and Hollmon D. W., Risk of resistance against sterol biosynthesis inhibitiors in plant protection. In Sterol biosynthesis inhibitors : Pharmaceutical and agrochemical aspects. ed. D. Berg and M. Plempel., Ellis Horwood, Chichester, West Sussex, 1989, pp. 332-346.

31. Sisler, H. and Ragsdale, N. N., Biochemical and cellular aspects of the antifungal action of ergosterol biosynthesis inhibitors. In Mode of Action of Antifungal Agents. ed. A. P. J. Trinici and J. F. Ryley, Cambridge University Press, Cambridge, 1984, pp. 257-282.

32. Kerkenaar, A. Mechanism of action of morpholine fungicides. In Modern Selective Fungicides. ed. H. Lyr. VEB Gustav Fischer Verlag, Jena and Logman Group UK Ltd., London, 1987, pp. 159-171.

33. Scheinpflug, H. and Kuck, K. W., Sterol biosynthesis inhibiting piperazine, pyridine, pyrimidine and azole fungicides. In Modern Selective Fungicides. ed. H. Lyr. VEB Gustav Fischer Verlag, Jena and Logman Group UK Ltd., London, 1987, pp. 173-204.

34. de Waard, M. A., Kipp, E. M. C., Horn, N. M., and van Nistelrooy, J. G. M., Variation in sensitivity to fungicides which inhibit ergosterol biosynthesis in wheat powdery mildew. Neth. J. Plant Pathol., 1986, **92**, 21-32.

35. de Waard, M. A. and van Nistelrooy, J. G. M., An energy-dependent efflux mechanism for fenarimol in a wild-type strain and fenarimol-resistant mutants of *Aspergillus nidulans*. Pestic. Biochem. Physiol., 1980, **13**, 255-266.

36. de Waard, M. A. and van Nistelrooy, J. G. M., Accumulation of SBI fungicides in a wild-type and fenarimol-resistant isolates of *Penicillium italicum*. Pestic. Sci., 1988, **22**, 371-382.

37. Kalb, V. F., Loper, J. C., Dey, C. R., Woods, C.W., and Sutter, T. R., Isolation of a cytochrome P-450 structural gene from *Saccharomyces cerevisiae* . Gene, 1986, **45**, 237-242.

38. Henry, M. J. and Trivellas, A. E., Laboratory-induced fungicide resistance to benzimidazole and azole fungicides in *Cercospora beticola*. Pestic. Biochem. Physiol., 1989, **35**, 89-96.

39. Baloch, R. I., Mercer. E. I., Wiggins, T. E., and Baldwin, B. C., Where do morpholines inhibit sterol biosynthesis? Proc. Brit. Crop Prot. Conf., 1984, **3**, 893-898.

40. Schneegurt, M. and Henry, M. J., A comparison of the effects of fenpropidin and piperalin on sterol biosynthesis in *Ustilago maydis*. Pestic Biochem. Physiol., 1991, (submitted)

41. Ryder, N. S., Frank, I., and Dupont, M. C., Ergosterol biosynthesis inhibition by the thiocarbamate antifungal agents tolnaftate and tolciclate. Antimicrob. Agent. and Chemother., 1986, **29**, 858-861.

42. Ryder, N. S., Specific inhibition of fungal sterol biosynthesis by SF 86-327, a new allylamine antimycotic agent. Antimicrob. Agent. and Chemother., 1985, **27**, 252-259.

43. Orth, A. B., Henry, M. J. and Sisler, H. D., Mechanism of resistance to terbinafine in two isolates of *Ustilago maydis*. Pestic. Biochem. Physiol., 1990, **37**, 182-191.

VOLTAGE-DEPENDENT SODIUM CHANNELS IN SUSCEPTIBLE AND PYRETHROID-RESISTANT DROSOPHILA STRAINS

DAVID PAURON, MARCEL AMICHOT and JEAN-BAPTISTE BERGE
Laboratoire de Biologie des Invertébrés, 123 bd Francis
Meilland, B.P. 2078, 06606 Antibes Cedex, France

ABSTRACT

We report here the molecular and biochemical characterization of voltage-dependent sodium channels in two *Drosophila melanogaster* strains. One, Tübingen, is sensitive to pyrethroids such as deltamethrin and the other one, Tübingen DDT, is less susceptible to the "knock down" effect of those insecticides (kdr). We have found that the resistance of the Tübingen DDT flies is not due to a difference of metabolism of the pyrethroids nor to a difference in the sodium channel density between the two strains. On the contrary, binding experiments indicate that the apparent affinity of deltamethrin is shifted to lower values in the Tübingen DDT flies. Polyclonal antibodies directed against the two proteins which are supposed to act as sodium channels in Drosophila were used in western blotting experiments and labelled one glycoprotein of $Mr \approx 270,000$ on membranes of both Tübingen and Tübingen DDT strains. Partial sequencing of one of the putative sodium channel genes has revealed one mutation in the Tübingen DDT sequence that might be involved in the kd resistance.

INTRODUCTION

Voltage-sensitive sodium channels play a major role in nerve conduction both in vertebrates and in invertebrates. Several classes of toxins which specifically act at discrete receptor sites on the sodium channel and which modulate its activity in different manners have been identified (for a review, see 1).

One of these classes is the family of pyrethroid insecticides which have been shown to be highly toxic to many insects by acting on their nervous system (2). The identification of sodium channels as the primary target of action of pyrethroids and of DDT has been shown by electrophysiological studies (3-5), flux studies (6-7), and binding experiments performed both on vertebrates (8-9) and on houseflies (10).

In this study, we have analyzed the pharmacological properties of sodium channels in two strains of *Drosophila melanogaster*. One, Tübingen is susceptible to the action of pyrethroids and the other one is resistant to their action, resistance which is expressed *in vivo* by a delay to the "knock down" effect.

To go further on the elucidation of the resistance mechanism we then compared the size of sodium channels as well as the sequences of the alleles of one gene, *sch*, which is supposed to encode one type of sodium channel, in the two drosophila strains.

MATERIALS AND METHODS

Toxicity measurements
The Tübingen strain, a wild-type drosophila strain collected in the Tübingen area, Germany, in 1982 has been maintained in our laboratory under insecticide-free conditions. Tübingen DDT was derived from it after 50 generations selected with DDT at the LC_{50}. The tarsal contact method was used to measure the knock down effect of deltamethrin. Basically, 10 to 15 male flies were put in glass tubes (2.5 cm diameter, 10 cm high) pretreated with a solution of 10 mM deltamethrin in acetone. The kd_{50} values correspond to the time for half of the population to fall to the bottom of the tubes.

Metabolism assays
[14C] deltamethrin labelled on the acid or the alcohol moiety (2.19 GBq/mmol) were gifts of Roussel-Uclaf. Microsomal preparations of abdomens of adults flies from both strains and *in vitro* metabolism of labelled deltamethrin were achieved as reported earlier (11).

Binding assays
[3H]saxitoxin ([3H]STX) (1.33 TBq/mmol) and [3H]batrachotoxinin A 20-α-benzoate ([3H]BTX-B) (2.22 TBq/mmol) were purchased from Amersham and Du Pont de Nemours, respectively. Deltamethrin and Esbiol, a very potent isomer of bioallethrin which is a non cyano-type pyrethroid, were gifts of Mr. Demoute, Procida, France. Purified drosophila head membranes preparations, direct [3H]STX binding and competition studies using [3H]BTX-B and various concentrations of pyrethroids were performed as described elsewhere for housefly head membranes(10). For these last experiments, stock solutions of pyrethroids were prepared in dimethylsulfoxide and non specific binding was determined in parallel incubations containing 0.2 mM veratridine.

Production of antibodies

The 3-1 AL cDNA clone which encodes the third and fourth homology domains of the drosophila sch protein (12) was a gift of Dr. L. Salkoff(Washington University School of Medicine, St. Louis, USA). The P15 cDNA clone was given to us by Dr. M. Ramaswami (Caltech, Pasadena, USA) and corresponds to segments S3 to S6 of homology domain D of the drosophila para gene (13).

The SacI-EcoRI fragment of 3-1AL corresponding to part of the S5 region and the S6 region of homology domain D and the 3' end of *sch* (680 bp) was sequentially subcloned in M13 mp19 (Boehringer Mannheim), then in the SalI-EcoRI sites of pE 24 (a Bluescript plasmid with extra polylinker kindly provided by Dr. D. Fournier) and at last in the BamHI-PstI sites of the fusion vector pEX 2 (Genofit) which was then transformed in the E. coli strain pop 2136. P15 cDNA (640 bp) was directly inserted in the EcoRI site of pEX 2. Fusion proteins were produced according to Stanley and Luzio (14) and purified by precipitation in 4 M urea. Pellets were dissolved in buffer I (2% SDS, 9% glycerol, 75 mM Tris-HCl, pH 6.8, 2.5% ß-mercaptoethanol) and boiled for 5 min. Preparative SDS-PAGE (7.5% acrylamide) was performed according to Laemmli (15) and the bands corresponding to fusion proteins were cut from the gels. Rabbits were first injected with 600 µg of protein in complete Freund adjuvant. They were boosted with 300 µg of protein in incomplete Freund adjuvant every six weeks and bled every two weeks after the boosts.

Immunoblot assays

Samples of head membranes from Tübingen and Tübingen DDT flies(130 µg) were denaturated in buffer I and loaded onto a 6% polyacrylamide gel. Transfer of the proteins to nitrocellulose (Hybond C, Amersham) was achieved at 400 mA for 3 hours in a Transphor apparatus (Pharmacia). Blots were stained with Ponceau S (Sigma) to identify the molecular weight markers (BioRad) and rinsed in 1 x TBS (150 mM NaCl, 10 mM Tris-HCl, pH 7.4). They were autoclaved (120°C, 30 min.) and incubated with 1 x TBS/5% non fat milk/0.2% Tween 20 for 1 hour at RT. Incubation with pre-immune serums, anti-sch and anti-para antibodies (1:500 dilution in 1 x TBS/1% non fat milk) were done at 4°C under gentle agitation for 16 hours. To reduce non specific background, total pEX proteins produced from the vector alone were added to the incubations at a 1:100 dilution. Blots were washed with 1 x TBS (10 min), 3 x TBS (15 min), 1 x TBS (10 min). They were next incubated with 0.5 µg/ml protein A-peroxidase (Sigma) for 3 hours at RT. 4-chloro-1-naphthol was combined with N,N-dimethylphenylenediamine (Sigma) to detect the labelled bands as recommended by Conyers and Kidwell (16). The apparent molecular weight values which appear in the text are the means from three independent experiments.

Isolation of genomic DNA clones homologous to sch in Tübingen and Tübingen DDT

Genomic libraries were constructed in the EMBL3 vector and screened using *sch* genomic (3-1 clone) and cDNA clones (3-1AL clone) as probes. Hybridizations were carried out at 65°C in 6 x SSPE (20 x SSPE contains 3M NaCl, 0.176M NaH$_2$PO$_4$, 20 mM EDTA, pH 7.4), 5 x Denhardt's solution (100 x solution is 2% bovine serum

albumin, 2% polyvinylpyrrolidone 360, 2% ficoll 400), 75 µg/ml
denatured salmon sperm DNA. Final washings were performed in
0.2 x SSPE, 0.2 % SDS at 65°C. After two screens, five Tübingen
clones and seven Tübingen DDT clones were selected and sequenced
using the sequenase system (United States Biochemical).

RESULTS and DISCUSSION

Toxicology and metabolism of deltamethrin

The Tübingen DDT (TDDT) strain was originally derived from
Tübingen(Tüb) using DDT as the pressure agent, but flies of this
strain were also resistant to deltamethrin. For a concentration
of deltamethrin of 10 mM, the time necessary to observe the
knock down effect (kd_{50}) is 11.63±0.97 min for Tübingen flies and
24.75±2.69 min for Tübingen DDT flies, a ratio of about 2.13.

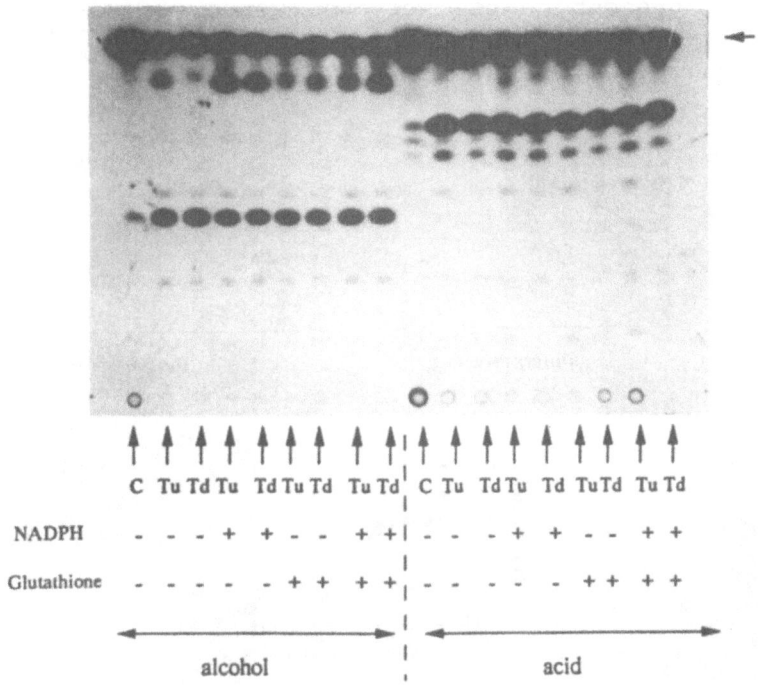

Figure 1. In vitro metabolism of deltamethrin. This picture
presents an autoradiogram of the TLC of deltamethrin metabolites
obtained with microsomes prepared from Tübingen (Tu) and
TübingenDDT (Td). Incubations were performed in the absence (-)
or in the presence (+) of glutathione (1 mM) and/or NADPH (1 mM)
and contained 1.1 10^5dpm of [^{14}C]deltamethrin labelled in the
acid or in the alcohol. The same amount of radiolabelled
deltamethrin was spotted on the two control lanes (C). The arrow
on the top right indicates the migration of deltamethrin.

To ensure that this difference in toxicity was not due to a difference of metabolism of the insecticide, we compared the pattern of metabolites obtained by incubating deltamethrin, radiolabelled in the alcohol or in the acid moiety, with microsomes prepared from abdomens of Tüb and TDDT flies. As illustrated in figure 1, no enhancement of the metabolic activity was observed in TDDT microsomes in standard conditions or when cofactors such as glutathione (for glutathione transferases) or NADPH (for mixed function oxidases) were added to the incubation medium. As observed in most drosophila strains, the major metabolic activity is due to esterases, and both glutathione transferases and mixed function oxidases play only a minor role in the degradation of this insecticide.

Binding Assays

As metabolism of deltamethrin cannot account for the resistance observed in TDDT, other mechanisms involving the target of pyrethroids had to be studied. As some authors have reported reductions in the number of sodium channels in relation to pyrethroid resistance in insects (17), we have titrated sodium channels present in the nervous system of flies of the two strains using saxitoxin (STX), a very potent toxin which is commonly used to determine the density of sodium channels.

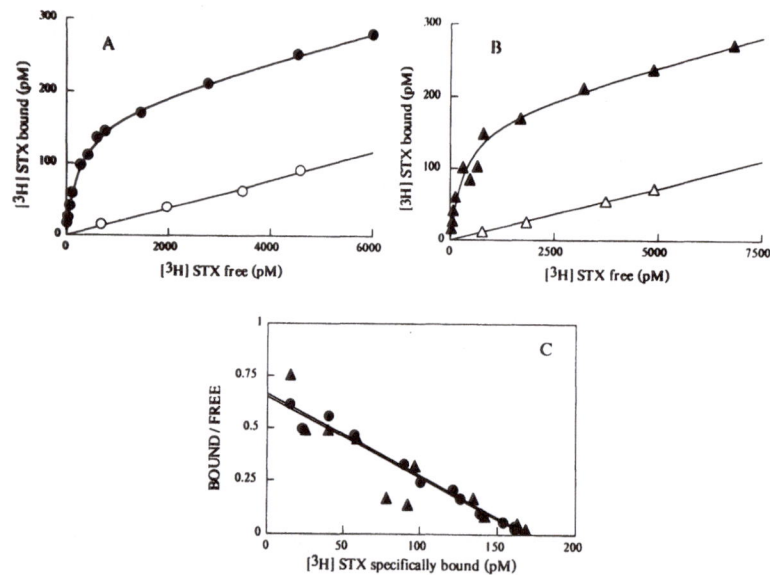

Figure 2. Direct binding of [3H]STX to purified head membranes of Tübingen and Tübingen DDT flies. Membranes (0.4 mg/ml of protein) were incubated with increasing concentrations of [3H]STX for one hour at 4°C. A. Total binding (●) and non specific binding (O, measured in the presence of 1 µM tetrodotoxin) of tritiated saxitoxin to Tüb membranes. B. Total (▲) and non specific binding (△) measured on TDDT membranes. C. Scatchard plots of specific bindings.

Figure 2 shows the results of the binding of tritiated saxitoxin to purified head membranes from Tüb and TDDT flies. Figure 2A presents the binding of increasing concentrations of [3H]STX to a fixed amount of Tüb head membranes measured in the absence or in the presence of 1 µM unlabelled STX (non specific binding). The results of such experiments performed on TDDT membranes is shown on figure 2B. The Scatchard plots (figure 2C) of the specific binding, which is the difference between total and non specific binding is linear for both experiments, indicating that the toxins bind to a single class of receptor sites in both cases. From these plots were calculated the values for the affinity of the tritiated toxin for its receptor, K_d, and the maximal binding capacity, B_{max}. These values were found similar for the two strains, namely 0.25±0.05 nM and 440±20 fmol/mg of protein for Tüb and 0.26±0.07 nM and 416±25 fmol/mg for TDDT. These results lead to the conclusion that the kd resistance mechanism we have studied is not related to a change in sodium channel density.

We then investigated the hypothesis of a modification of the affinity of the pyrethroids for their binding site on the sodium channel. As direct pyrethroid binding experiments were unsuccessful, we decided to exploit previous reports about a synergistic effect between pyrethroids and another effector of sodium channels, batrachotoxin (6,8-10).

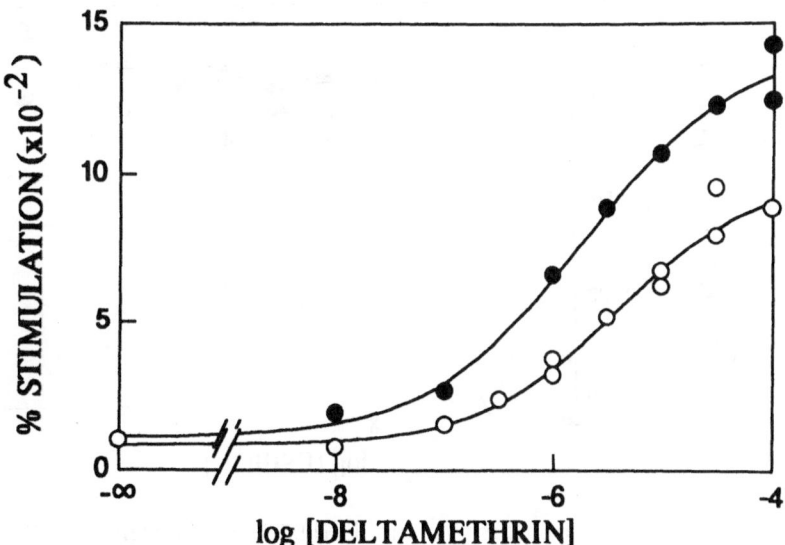

Figure 3. Stimulation of [3H]BTX-B binding by deltamethrin. Results are presented as percentage of enhancement of [3H]BTX-B specific binding on Tüb (●) or TDDT (○) membranes. Each incubation was done at 22°C for one hour and contained 1.5 mg/ml of protein and 7 nM of [3H]BTX-B. Non specific binding was measured in parallel incubations containing 0.2 mM veratridine.

Thus, we performed on the same membrane preparations the binding of [³H]BTX-B, a potent derivative of batrachotoxin in the absence or the presence of pyrethroids. Specific binding of [³H]BTX-B to Tüb and TDDT membranes measured in the absence of any insecticide were identical (data not shown). This indicates that the BTX-B binding site is not affected in the resistant strain. Then, incubations were done in the presence of increasing concentrations of deltamethrin. As shown in figure 3, there is a synergistic effect of deltamethrin on [³H]BTX-B binding to Tüb membranes in a dose-dependent manner. This stimulation reaches a plateau value for a concentration of 30 µM with a maximal increase of about 12-fold. The apparent affinity of deltamethrin determined at half-maximal stimulation, $K_{0.5}$ is about 0.8 µM. In contrast, this stimulatory effect is only partially observed on TDDT membranes. There is a shift in the affinity of deltamethrin for its site as the value of $K_{0.5}$ is 5.5 µM, 7 times higher as compared to Tüb.These data suggest that the knock down resistance mechanism in TDDT is related to a change of the pyrethroid binding site or to its linkage to the batrachotoxin binding site on the sodium channel.
The same kind of experiment was done with esbiol instead of deltamethrin to see if this synergistic effect was common to all insecticides of the pyrethroid family.

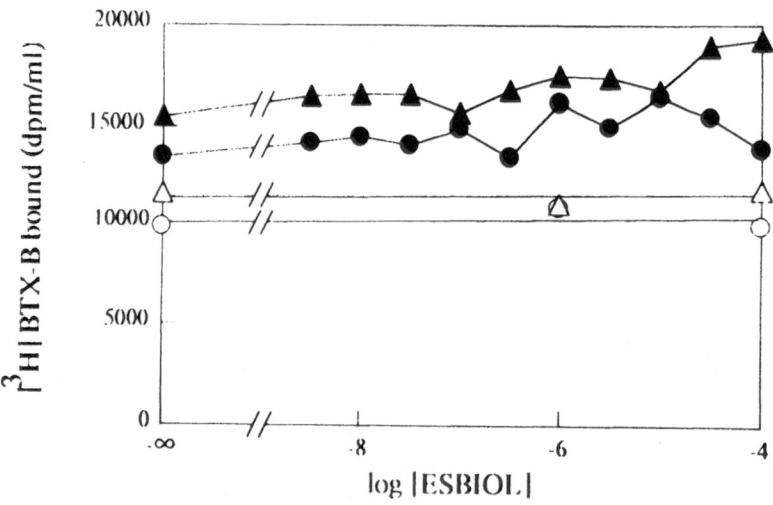

Figure 4. Effects of Esbiol on [³H]BTX-B binding to Drosophila head membranes. Incubation conditions were identical as for figure 3. This graph presents the data obtained for [³H]BTX-B total (●) and non specific (O) binding to Tüb membranes and to TDDT membranes (▲, △).

As seen on figure 4, increasing concentrations of esbiol have almost no effect on BTX-B binding. Moreover, no significant difference is observed when the binding is done on Tüb or TDDT membranes. These results are in agreement with previous report from Jacques et al. (6) who showed that some pyrethroids were unable to stimulate the $^{22}Na^+$ uptake on cell cultures even in the presence of veratridine which has the same binding site as batrachotoxin.

Immunological characterization of sodium channel proteins in Tübingen and Tübingen DDT
As mutations on the sodium channel proteins in TDDT flies seem to be involved in the resistance mechanism, it was interesting to compare the size and the structure of these proteins in the two strains. Two fusion proteins were designed to yield antisera directed against polypeptides produced by the two genes supposed to code for voltage-sensitive sodium channels in *Drosophila melanogaster*, *sch* and *para*. The first construct contains 229 amino acids of the *sch* sequence (numbered 1452-1680 in ref.18) including part of the S5 region, the entire S6 region of homology domain D, and the C-terminal end of *sch*. The P15 fusion protein contains a 213 amino acids-long sequence which covers part of the S3, and S4, S5 and S6 regions of homology domain D from *para*. Polyclonal antibodies were prepared as described in "Materials and Methods". In all experiments discussed below, we will refer to anti-sch and anti-para for the polyclonal antibodies obtained from the injection to rabbits of the 3-1AL and P15 purified hybrid proteins, respectively.

These antibodies were used at first on immunoblots performed on purified head membranes. Samples of these preparations were denatured in disulfide bond-reducing conditions and the proteins were separated by SDS-PAGE (6% acrylamide) then transferred to nitrocellulose. Results of probing experiments of these blots with anti-sch antibodies are illustrated in figure 5. This antibody specifically labels one band of Mr≈270,000 for both strains. The diffuse aspect of the bands is typical of a high percentage of glycosylation. We constantly found that the polypeptide identified in Tüb was identical in size to the one identified in TDDT. The same pattern was obtained when anti-para antibodies were used instead of anti-sch. This indicate that mutation(s) which might be present on TDDT sodium channels are small enough not to be detected by western blotting experiments. Moreover, these results are in good agreement with a recent report in which a Mr≈260,000 was assessed for the sodium channel in the central nervous system of another diptera, <u>Sarcophaga falculata</u>, by immunoprecipitation with anti-rat brain sodium channel antibodies (19).

To go further on the characterization of the proteins coded by the two genes, both antibodies were incubated with adult Tüb and TDDT cryosections. No qualitative or quantitative difference was noted between the two strains. With these results, we may now exclude the hypothesis of differential pattern of localizations to explain the resistance mechanism.

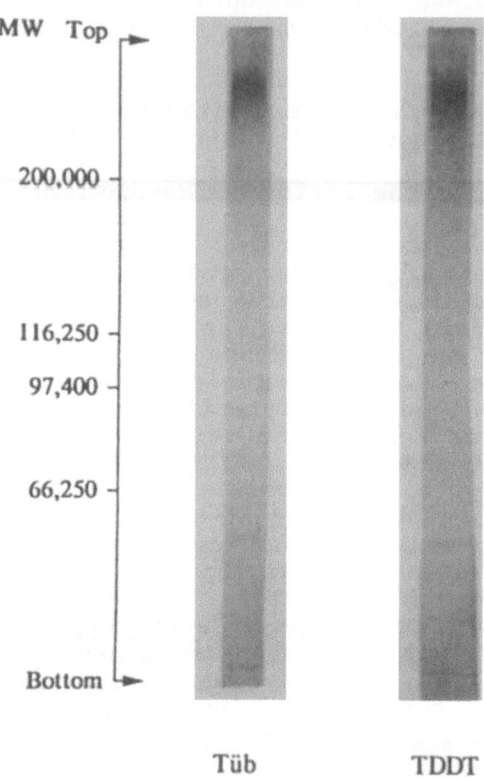

Figure 5. Identification of sch product in Tübingen and Tübingen DDT head membranes. The antibody exclusively labels one band of Mr≈270,000. No band is observed when pre-immune serum was used instead of immune sera. Molecular weight markers are myosin (200,000), E. coli ß-galactosidase (116,250), rabbit muscle phosphorylase b (97,400), and bovine serum albumin (66,200).

Cloning and sequencing of the sch allele in Tübingen and Tübingen DDT.

We have performed genetic experiments on Tüb and TDDT strains which strongly suggest that the kdr factor is on the second chromosome as is the *sch* gene (12), and not on the X chromosome which carries the *para* gene (20) (data not shown). This is why *sch* clones kindly provided by L. Salkoff were used to screen genomic libraries to clone *sch* alleles from Tüb and TDDT. We have sequenced the same coding regions as published in Salkoff et al. (18).

By comparing the sequences, we only found one non-silent mutation which is presented in figure 6. This mutation is located in the peptide sequence between segments S5 and S6 of the third homology domain. It consists in the replacement of the aspartic acid residue (D) present in the sequence of sch in Oregon R and Tüb strains in position 1172 by an asparagine (N) in TDDT. It might be important to note that this mutation is located very close to a potential glycosylation site. Another potentially important factor is that this linker section is no longer considered as an extracellular portion, but would appear to constitute the pore of the channel if recent models drawn for the *Shaker*-type potassium channels (21-23) can be applied to sodium channels.

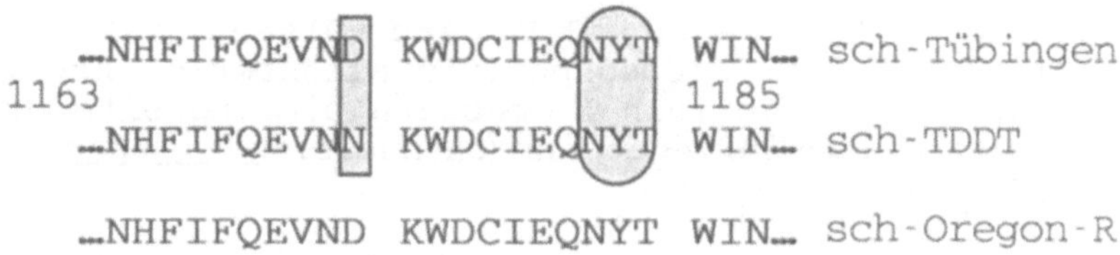

...NHFIFQEVND KWDCIEQNYT WIN... sch-Tübingen
1163   1185
...NHFIFQEVNN KWDCIEQNYT WIN... sch-TDDT

...NHFIFQEVND KWDCIEQNYT WIN... sch-Oregon-R

Figure 6. Deduced amino acid sequence of the linker region between segments S5 and S6 of homology domain C using the one-letter code. The shaded boxes correspond to the mutation between Tüb and TDDT sequences (D→N) and the potential glycosylation sequence (NYT). These two sequences are compared to the sch sequence from Oregon-R reported in ref.18.

Work is now in progress to determine if this mutation is indeed involved in the kd resistance mechanism and if other mutation(s) might also act to produce such a phenomenon.

REFERENCES

1. Catterall, W.A., Molecular properties of voltage-sensitive sodium channels. Ann. Rev. Biochem., 1986, 55, 953-85.

2. Sawicki, R.M., Resistance to pyrethroid insecticides in arthropods. In Progress in Pesticide Biochemistry and Toxicology, volume 5: Insecticides, ed. D.H. Hutson & T.R. Roberts, John Wiley & Sons, Chichester, 1985, pp. 143-92.

3. Lund, A.E. and Narahashi, T. Kinetics of sodium channel modification by the insecticide tetramethrin in squid axon membranes. J. Pharmacol. Exp. Therap., 1981, 219, 464-73.

4. Vijverberg, H.P.M., Van der Zalm, J.M. and Van der Bercken, J. Similar mode of action of pyrethroids and DDT on sodium channel gating in myelinated nerves. Nature, 1982, 295, 601-3.

5. Scott, J.G., Pyrethroid insecticides. In ISI Atlas of Science: Pharmacology, 1988, pp. 125-8.

238

6. Jacques, Y., Romey, G., Cavey, M.T., Kartalovski, B. and Lazdunski, M. Interaction of pyrethroids with the Na$^+$ channel in mammalian neuronal cells in culture. Biochim. Biophys. Acta, 1980, 600, 882-97.

7. Holan, G., Frelin, C. and Lazdunski, M. Selectivity of action between pyrethroids and combined DDT-pyrethroid insecticides on Na$^+$ influx into mammalian neuroblastoma. In Experientia 41, Birkhäuser Verlag, Basel, 1985, pp. 520-2.

8. Brown, G.B., Gaupp, J.E. and Olsen, R.W. Pyrethroid insecticides: Stereospecific allosteric interaction with the batrachotoxinin-A benzoate binding site of mammalian voltage-sensitive sodium channels. Mol. Pharmacol., 1988, 34, 54-9.

9. Lombet, A., Mourre, C. and Lazdunski, M. Interaction of insecticides of the pyrethroid family with specific binding sites on the voltage-dependent sodium channel from mammalian brain. Brain Res., 1988, 459, 44-53.

10. Pauron, D., Barhanin, J., Amichot, M., Pralavorio, M., Berge, J.B. and Lazdunski, M. Pyrethroid receptor in the insect Na$^+$ channel: Alteration of its properties in pyrethroid-resistant flies. Biochemistry, 1989, 28, 1673-1677.

11. Cuany, A., Pralavorio, M., Pauron, D., Bergé, J.B., Fournier, D., Blais, C., Lafont, R., Salaün, J.P., Weissbart, D., Larroque, C. and Lange, R. Characterization of microsomal oxidative activities in a wild-type and in a DDT resistant strain of Drosophila melanogaster. Pestic. Biochem. Physiol., 1990, 37, 293-302.

12. Salkoff, L., Butler, A., Wei, A., Scavarda, N., Giffen, K., Ifune, C., Goodman, R., and Mandel, G. Genomic organization and deduced amino acid sequence of a putative sodium channel gene in Drosophila. Science, 1987, 237, 744-9.

13. Ramaswami, M., and Tanouye, M.A. Two sodium-channel genes in Drosophila : Implications for channel diversity. Proc. Natl. Acad. Sci. USA, 1989, 86, 2079-82.

14. Stanley, K.K., and Luzio, J.P. Construction of a new family of high efficiency bacterial expression vectors: identification of cDNA clones coding for human liver proteins. EMBO J., 1984, 3, 1429-34.

15. Laemmli, U.K. Cleavage of structural proteins during the assembly of the head of bacteriophage T4. Nature, 1970, 227, 680-5.

16. Conyers, S.M. and Kidwell, D.A. Chromogenic substrates for horseradish peroxidase. Anal. Biochem., 1991, 192, 207-11.

17. Rossignol, D.P. Reduction in number of nerve membrane sodium channels in pyrethroid resistant house flies. Pestic. Biochem. Physiol., 1988, 32, 146-52.

18. Salkoff, L., Butler, A., Scavarda, N., and Wei, A. Nucleotide sequence of the putative sodium channel gene from Drosophila : the four homologous domains. Nucl. Acid Res.,1987, 15, 8569-72.

19. Gordon, D., Moskowitz, H., and Zlotkin, E. Sodium channel polypeptides in central nervous systems of various insects identified with site directed antibodies. Biochim. Biophys. Acta, 1990, 1026, 80-86.

20. Suzuki, D.T., Grigliatti, T. and Williamson, R. Temperature-sensitive mutants in *Drosophila melanogaster*: a mutation (*para*[ts]) causing reversible adult paralysis. Proc. Natl. Acad. Sci. USA, 1971, 68, 890-3.

21. Yool, A.J. and Schwarz, T.L. Alteration of ionic selectivity of a K^+ channel by mutation in the H5 region. Nature, 1991, 349, 700-4.

22. Yellen, G., Jurman, M.E., Abramson, T. and MacKinnon, R. Mutations affecting internal TEA blockade identify the probable pore-forming region of a K^+ channel. Science, 1991, 251, 939-42.

23. Hartmann, H.A., Kirsch, G.E., Drewe, J.A., Taglialatela, M., Joho, R.H. and Brown, A.M. Exchange of conduction pathways between two related K^+ channels. Science, 1991, 251, 942-4.

INSECTICIDE RESISTANCE BY GENE AMPLIFICATION IN *MYZUS PERSICAE*

LINDA M FIELD AND ALAN L DEVONSHIRE
AFRC Institute of Arable Crops Research
Rothamsted Experimental Station
Harpenden, Herts, AL5 2JQ

ABSTRACT

Gene amplification in *Myzus persicae* permits the synthesis of large quantities of esterase that cause resistance by detoxifying a wide range of insecticides. Two, slightly different forms of esterase can be elevated and the corresponding amplified genes show much homology, but differ at the 3′ end. Spontaneous loss of resistance in clonal populations occurs through reduced transcription of these amplified genes and this is associated with loss of methylation from the 5-methylcytosine (5mC) present at some sites in the expressed genes; three such sites have been mapped, two of which are in coding regions of the gene. Field populations of resistant *M. persicae* have been screened for the type of amplified esterase genes present and the occurrence of 5mC. Insecticide-resistant aphids of a closely related species, *Myzus nicotianae*, contain elevated esterase and amplified genes, indistinguishable from those of *M. persicae*.

GENE AMPLIFICATION AND DETOXIFICATION

The presence of multiple gene copies can facilitate the rapid synthesis of cellular components both constitutively, as for ribosomal DNA sequences present in all eukaryotes [1], or in a developmentally regulated way, as for the chorion genes responsible for eggshell proteins in insects [2]. In addition, exposure to toxic substances can result in the selection of amplified genes that confer resistance, either by increasing the synthesis of a target protein or by enhancing detoxication mechanisms [3]. Most studies on resistance by gene amplification have involved treatment of micro-organisms or cell cultures with toxic chemicals, a situation clearly similar to insecticide treatment of insect populations, and it was partly this analogy that led Devonshire & Sawicki [4] to propose that the increased amount of esterase responsible for insecticide resistance in the peach-potato aphid *M. persicae* might result from amplification of esterase genes. This paper will review the evidence for such amplification and discuss recent studies on the structure and expression of amplified esterase sequences.

INSECTICIDE RESISTANCE AND AMPLIFIED ESTERASE GENES IN
M. PERSICAE

Populations of *M. persicae* throughout the world have developed resistance to a wide range of insecticides. In the UK, glasshouse and field populations are sprayed intensively to prevent both direct damage and transmission of potato and sugar beet viral diseases, consequently selection of resistance is common. In all cases studied worldwide only one resistance mechanism has been found, the synthesis of high levels of one of two closely related, insecticide-degrading esterases, E4 or FE4. The presence of large amounts of E4 is associated with a particular (A1,3) chromosome translocation, whereas FE4, a slightly larger protein, is synthesized abundantly in resistant aphids of apparently normal karyotype [5].

An E4 cDNA probe, which can detect both E4 and FE4 DNA sequences has been used [6] to probe dot blots of DNA from resistant aphid clones, containing approximately 4-times (R_1), 16-times (R_2) and 64-times (R_3) the enzyme levels of susceptible aphids (Figure 1A). The increase in probe binding between S, R_1, R_2 and

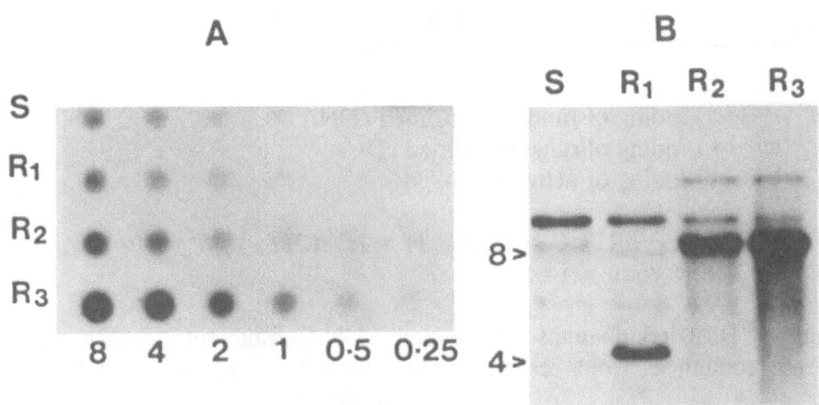

Figure 1. Binding of an E4 cDNA to DNA extracted from susceptible (S) and 3 resistant aphid clones (R_1, R_2 and R_3). A. Dot blots of denatured aphid DNA (8, 4, 2, 0.5, 0.25 µg) made up to 8 µg with non-specific (herring sperm) DNA. B. Southern blot of *Eco*RI-digested aphid DNA (10 µg). Arrows indicate sizes of major bands.

R_3 clearly demonstrated esterase gene amplification, although some, apparently non-specific binding, masked any serial increase in E4 DNA, especially for the less resistant types. However, this was shown more clearly by binding of the cDNA to Southern blots of *EcoR*1-digested aphid DNA (Figure 1B) which also showed that amplified esterase sequences are on a 4 kb fragment in resistant aphids that synthesize FE4 (R_1, Figure 1B) and on an 8 kb fragment in those with E4 (R_2 and R_3, Figure 1B). The correlation between fragment size and type of esterase was consistent for all resistant aphid clones studied [6].

The 8 kb and 4 kb *EcoRI* genomic fragments containing E4 and FE4 DNA sequences respectively have now been cloned and analysed by restriction mapping and binding of various probes (Figure 2). The binding of a cDNA probe made by random

A. 8 kb

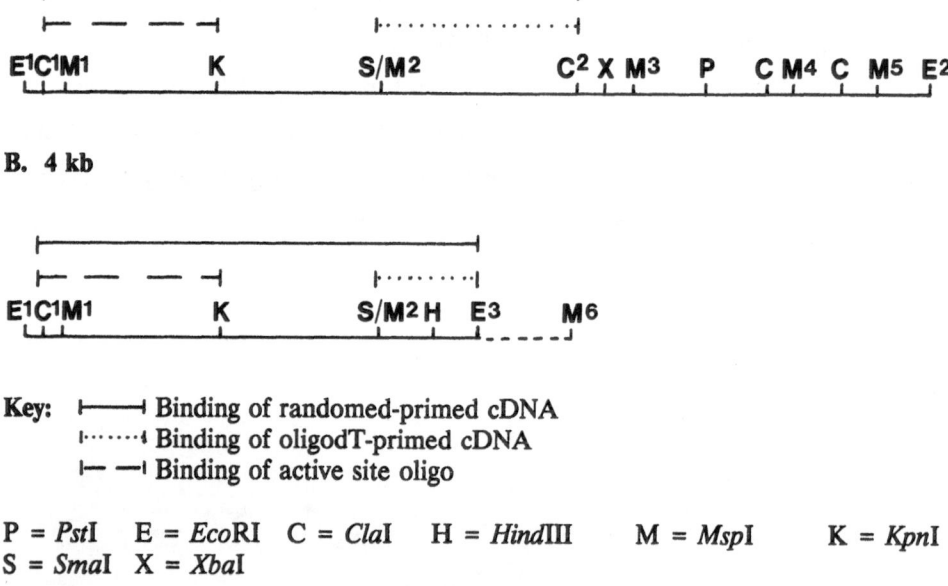

B. 4 kb

Key: ├────┤ Binding of randomed-primed cDNA
├·······┤ Binding of oligodT-primed cDNA
├── ──┤ Binding of active site oligo

P = *Pst*I E = *Eco*RI C = *Cla*I H = *Hind*III M = *Msp*I K = *Kpn*I
S = *Sma*I X = *Xba*I

Figure 2. Restriction maps of genomic DNA fragments containing amplified esterase sequences from resistant aphids with A. E4 enzyme and B.FE4 enzyme (see text for details).

priming of R_3 aphid polyA$^+$ RNA shows that coding sequences for E4 are contained between the 2 *Cla*I sites (C^1 and C^2) in the 8 kb fragment and that FE4 coding sequences are distributed throughout most of the 4 kb fragment; since the *Kpn*I site in both fragments is not in coding sequence (personal observation from sequence data) there must be at least one intron in both genes. The binding of cDNA made by oligodT priming of R_3 aphid polyA$^+$ RNA, and deliberately limited to <500 bases, shows that the E4 and FE4 genes are orientated 5' → 3' as shown in Figure 2B and that the 3' end of E4 lies between the *Sma*I and *Cla* I (C^2) sites, whilst that of FE4 is downstream of *Sma*I. Recent comparisons of E4 and FE4 cDNAs show that FE4 is approx 600 bases longer than E4, in line with the larger FE4 polypeptide thus, the 2 genes differ both qualitatively and quantitatively at the 3' end. By contrast the restriction maps show that upstream of the *Sma*I site the 2 genes are the same, including the region that contains the active site, as judged by the binding of a degenerate oligonucleotide (14-mer) complementary to a cyanogen bromide fragment close to the active site of E4 (Figure 2).

Current work on sequencing of a full length E4 cDNA has shown that the 5' end of E4 is upstream of the 8 kb fragment, further sequencing and isolation of regions flanking the genomic fragments should provide much more detail of the relative structure of the 2 genes. It is possible that the reduced size of the amplified E4 gene compared to FE4 is a consequence of a rearrangement within the gene caused by the chromosome translocation which is linked with the presence of E4.

STABILITY OF RESISTANCE AND EXPRESSION OF AMPLIFIED E4 GENES IN *M. PERSICAE* CLONES

Apparent instability is often a feature of insecticide resistance, especially when populations are heterogeneous, or rearing conditions do not prevent contamination. However, these factors were excluded for a detailed study of *M. persicae* in 1980, when Sawicki *et al* [7] reported that extremely resistant (R_3 level) aphid clones with the A1,3 chromsome translocation could spontaneously lose resistance and elevated E4. Later, when E4 cDNA was used to probe nucleic acids from such "revertant" clones, it was found that the high levels of E4 mRNA were lost but that the amplified E4 genes remained [8]. In addition, the amplified E4 sequences in revertants were still contained within an 8 kb *Eco*RI fragment and cloning and mapping of this fragment has shown it to be identical to its counterpart in resistant aphids. Thus the basis of reversion in *M. persicae* is a change in expression of the amplified E4 genes which occurs without any major DNA rearrangement. This contrasts with the analagous loss of resistance in mammalian cell cultures, where in the absence of selection, amplified genes are lost from the cell [3]. This can happen when the amplified genes are located on extrachromosomal elements, termed double minutes that lack a centromere and therefore segregate unequally at cell division. Cells with fewer amplified genes are then at a significant advantage in the rapidly reproducing culture. In cell lines with amplified genes integrated into the chromosomes, resistance is generally stable over long periods, even in the absence of selection [3]. However, there is at least one report of the loss of multi-drug resistance in a tumour cell line without loss of the amplified P-glycoprotein genes responsible for resistance [9].

The only difference so far observed [8] between the expressed and silent E4 sequences in *M. persicae* is a change in the presence of 5-methylcytosine (5 mC), a fifth base known to be present and affect gene transcription, in vertebrates [10]. The presence of 5 mC in esterase DNA sequences was demonstrated by comparing *Msp*I and *Hpa*II restriction digests probed with the E4 cDNA [8] as shown in Figure 3. Both

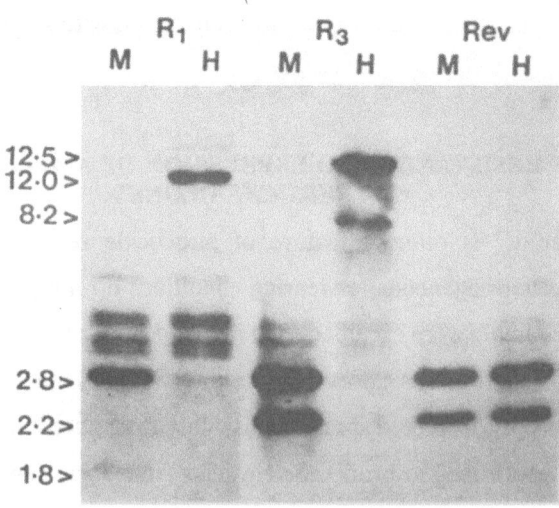

Figure 3. Binding of an E4 cDNA to a Southern blot of DNA extracted from resistant (R$_1$ and R$_3$) and revertant (Rev) aphid clones and digested with *Msp*I (M) or *Hpa*II (H) (See text for details). Arrows indicate sizes of major bands.

*Msp*I and *Hpa*II cut at the sequence CCGG but only *Msp*I cuts if the internal cytosine is methylated, so that a difference in banding pattern between the 2 enzyme digests indicates the presence of 5mC. DNA from both resistant clones in Figure 3 (R_1 with FE4 and R_3 with E4) is cut differently by *Msp*I and *Hpa*II whereas for the revertant clone (REV), the banding patterns are the same for both enzymes. Thus both types of esterase genes are expressed in the presence of DNA methylation, whereas lack of expression of E4 genes correlates with the absence of 5 mC. This is a very surprising result; in the vast majority of vertebrate systems studied so far the presence of 5mC is associated with lack of gene expression and demethylation causes genes to be turned on. The banding patterns in Figure 3 can be interpreted in the light of the known restriction maps of the amplified E4 and FE4 sequences (Figure 2). The 2.8 and 2.2 kb bands in *Msp*I digests of resistant aphids with E4 genes arise by cutting at the three *Msp*I sites, M^1, M^2 and M^3. The 2.2 kb band is replaced in the DNA from FE4-synthesizing aphids by a 1.8 kb fragment which must result from another *Msp*I site (M^6) 0.5 kb downstream from *Eco*RI (E^3). Since the 2.8 kb fragment (M^1/M^2) is probably identical for both E4 and FE4 genes, the E4 cDNA probe will have equal homology to this region and the difference in intensity between R1 and R3 reflects differences in gene copy number. On the other hand, the homology of the E4 cDNA to fragments downstream of M^2, where the 2 genes differ, should be greater for the E4 gene (i.e. the 2.2 kb M^2/M^3) than the FE4 (i.e. the 1.8 kb M^2/M^6), this is clearly visible from Figure 2. Thus the presence of E4 or FE4 in aphids can be identified from the *Msp*I digest patterns of their DNA. The presence of 5 mC in the FE4 gene (at M^1, M^2 and M^6) of R1 aphids and in the E4 gene (at M^1, M^2 and M^3) of R3 aphids results in the loss of the lower molecular weight fragments in *Hpa*II digests. These are replaced by 8.2 kb and 12.5 kb bands in R_3 and a 12 kb band in R_1 that must arise by *Hpa*II cutting at other unmethylated CCGG sites at least one of which must flank the cloned sequences. Clearly it is a loss of 5 mC at M^1, M^2 and M^3 which accompanies loss of resistance in *M. persicae*, both M^1 and M^2 are in coding regions of the E4 gene and studies are underway to look for further upstream *Msp*I sites which may, by comparison with vertebrate genes, be involved in the control of esterase gene expression.

AMPLIFIED ESTERASE GENES AND THEIR EXPRESSION IN FIELD POPULATIONS OF *M. PERSICAE*

The previous section discussed the loss of resistance in some *M. persicae* clones but to what extent does this occur in field populations? During 1990 aphids were collected from the field and individuals were used to set up "Miniclones" on single potato leaves in Blackman boxes [11]. Each first generation of c20 aphids produced parthenogenetically was collected and their DNA extracted and analysed by probing *Msp*I and *Hpa*II digests with a mixture of cloned E4 and FE4 genomic sequences (fragments E^1/K and S/C^2 subcloned from the 8 kb fragment and E^1/K and S/E^3 from the 4 kb fragment, see Figure 2). Figure 4 shows a representative range of results obtained from 84 samples analyzed as summarised in Table 1 which also indicates the esterase content as judged by immunoassay [12]. An accurate assessment of the level of amplification was not possible from the Southern blots.

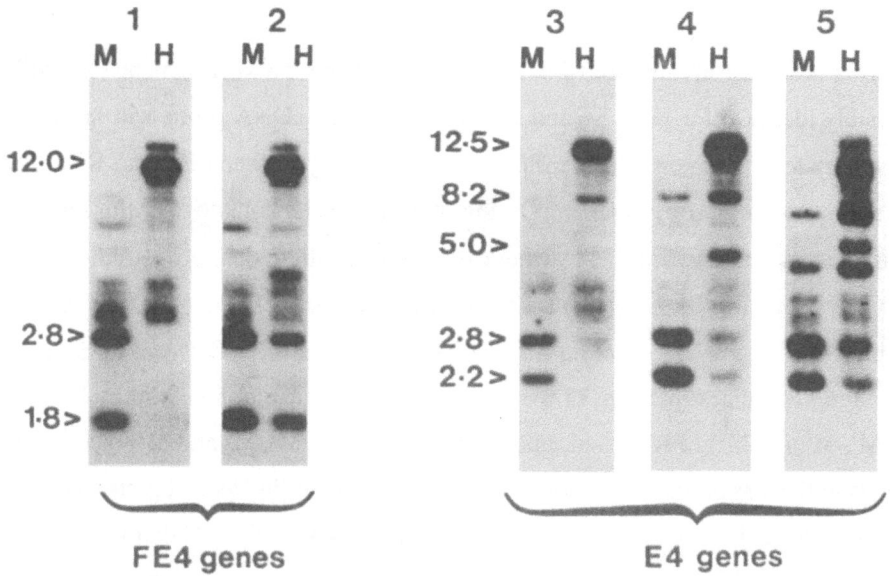

Figure 4. Binding of esterase genomic sequences to a Southern blot of DNA from aphid "Miniclones" digested with *Msp*I (M) or *Hpa*II (H) (see text for details). Arrows indicate sizes of major bands.

Most [75] of the samples had amplified esterase genes, 25 with E4 sequences and 50 with FE4; in no case were both types of sequence found in the same aphid. Most showed full methylation of the *Msp*I sites discussed above for both E4 and FE4 genes

TABLE 1.

Results of DNA probing and immunoassay of *M. persicae* Miniclones set up from field populations during 1990

| Gene | DNA Probing | | Immunoassay | | |
	Amplification	Methylation	$R_{2/3}$	R_1	S
E4	+	Full	21	1	0
	+	Partial	2*	1	0
FE4	+	Full	13	31	0
	+	Partial	1*	5	0
?	-	None	-	1	8

* Samples 2, 4 and 5 in Figure 4

(Table 1 and Figure 4, samples 1 and 3) but a few were deemed to have "partial" methylation as judged by the presence of either 2.8 and 2.2 or 2.8 and 1.8 kb fragments as well as the larger fragments in *Hpa*II digests (Figure 4 samples 2, 4 and 5). There were no examples where both *Msp*I and *Hpa*II digests were identical i.e. complete loss of 5 mC as seen in the revertant aphid clone (Figure 3). It is interesting that in the field samples, partial methylation of esterase sequences was seen in aphids with both E4 and FE4 genes although instability of resistance has never been reported in aphids which have FE4 enzyme. In general the aphid samples with partial methylation showed lower levels of esterase (as judged by immunoassay) than would be anticipated by the amount of probe binding, indicating that partial loss of 5 mC correlates with a reduction in esterase gene expression, but this could not be quantified accurately. Recent refinements in extraction of aphid DNA, which allow analysis of individual insects should permit a more extensive study during 1991 in which esterase gene copy number will be quantified, to establish any correlation between copy number, esterase level and degree of methylation. If reversion occurs only from extremely resistant (R_3) aphids, it should become more frequent as such variants become more common in field populations.

Table 1 also shows the results for 9 samples where there was no esterase gene amplification and only low (mostly susceptible) levels of enzyme. It has not been possible to establish which esterase gene is present in these aphids, since although the 2.8 kb *Msp*I fragment is seen, neither the 2.2 nor the 1.8 kb bands are detected, it is possible therefore that the 3′ end of the original esterase gene in susceptible insects is slightly different from E4 and FE4, and that these alleles are generated during the amplification process. This is in accord with the ambiguity often seen for susceptible aphids on enzyme electrophoresis (Moores & Devonshire, unpublished).

A tobacco adapted form of *M. persicae* recently classified as *M. nicotianae* [13] has become resistant to insecticides by an increase in carboxylesterase activity [14]. We have examined both fundatrigeniae of *M. nicotianae* from peach and summer populations from tobacco. Electrophoretic studies showed that the esterases present are indistinguishable from those of very resistant *M. persicae* of normal karyotype [5] and DNA analyses gave restriction digests consistent with the presence of methylated amplified FE4 genes [15]. This identity of resistance genes between the 2 extremely closely related species, could have arisen either by the adaptation to tobacco of resistant aphids or by the same gene amplification event occurring independantly in both species.

CONCLUSION

The data presented indicate that amplification of esterase genes in *M. persicae* populations is very conservative, giving rise to only 2 types of amplified gene even in the form of this aphid adapted to tobacco. This parallels the situation in the *Culex pipiens* complex, where only one pattern of amplification is seen for the B2 esterase gene [18]. Refinements in DNA preparation and analyses hold promise to provide a better insight of the population genetics of amplified esterase genes and the association between DNA methylation and gene expression. More fundamental studies should provide information on how methylation controls esterase gene transcription.

ACKNOWLEDGEMENTS

We thank Sara Smith and Mary Stribley for rearing and biochemical characterisation of aphid clones.

REFERENCES

1. Long, E.O. and Dawid, I.B., Repeated genes in eukaryotes. *Ann. Rev. Biochem.*, 1980, **49**, 727-64.

2. Spradling, A.C. and Mahowald, A.P., Amplification of genes for chorion proteins during oogenesis in *Drosophila melanogaster*. *Proc. Natl. Acad. Sci. USA*, 1980, **77**, 1096-100.

3. Devonshire, A.L. and Field, L.M., Gene amplification and insecticide resistance. *Ann. Rev. Ent.*, 1991, **36**, 1-23.

4. Devonshire, A.L. and Sawicki, R.M., Insecticide-resistant *Myzus persicae* as an example of evolution by gene duplication. *Nature*, 1979, **270**, 140-1.

5. Devonshire, A.L., Insecticide resistance in *Myzus persicae*: From field to gene and back again. *Pestic. Sci.*, 1989, **26**, 375-82.

6. Field, L.M., Devonshire, A.L. and Forde, B.G., Molecular evidence that insecticide resistance in peach-potato aphids (*Myzus persicae* Sulz.) results from amplification of an esterase gene. *Biochem. J.*, 1988, **251**, 309-12.

7. Sawicki, R.M., Devonshire, A.L., Payne, R.W. and Petzing, S.M., Stability of insecticide resistance in the peach-potato aphid, *Myzus persicae* (Sulzer). *Pestic. Sci.*, 1980, **11**, 33-42.

8. Field, L.M., Devonshire, A.L., ffrench-Constant, R.H. and Forde, B.G., Changes in DNA methylation are associated with loss of insecticide resistance in the peach-potato aphid *Myzus persicae* (Sulz.). *FEBS Letts.*, 1989, **243**, 323-7.

9. Sugimoto, Y., Roninson, I.B. and Tsurou, T. Decreased expression of the amplified *mdr*1 gene in revertants of multidrug-resistant human myelogenous leukemia K562 occurs without loss of amplified DNA. *Mol. Cell Biol.*, 1987, **7**, 4549-52.

10. Holliday, R., A different kind of inheritance. *Sci. Amer.*, 1989 (June) 40-8.

11. Blackman, R.L., Rearing and handling aphids. In *World Crop Pests, Aphids their Biology, Natural Enemies and Control.* Vol B., Eds. A.K. Minks and P. Harrewijn, Elsevier Science Publishers, The Netherlands, 1988, 59-68.

12. Devonshire, A.L., Moores, G.D. and ffrench-Constant, R.H., Detection of insecticide resistance by immunological estimation of carboxylesterase activity in *Myzus persicae* (Sulzer) and cross reaction of the antiserum with *Phorodon humuli* (Schrank) (Hemiptera: Aphididae). *Bull. ent. Res.*, 1986, **76**, 97-107.

13. Blackman, R.L., Morphological discrimination of a tobacco-feeding form from *Myzus persicae* (Sulzer) (Hemiptera: Aphididae) and a key to New World *Myzus (Nectarosiphon)* species. *Bull. ent. Res.*, 1987, **77**, 713-30.

14. Abdel-Aal, Y.A.I., Wolff, M.A., Roe, R.M. and Lampeut, E.P., Aphid carboxylesterases: Biochemical aspects and importance in the diagnosis of insecticide resistance. *Pestic. Biochem. Physiol*, 1990, **38**, 255-66.

15. Field, L.M. and Devonshire, A.L., Esterase genes conferring insecticide resistance in aphids. *Proc. Am. Chem. Soc.* New York 1991 (Eds. C.A. Mullin and J.G. Scott (In preparation).

16. Raymond, M., Callaghan, A., Fort, P. and Pasteur, N., Worldwide migration of amplified insecticide resistance genes in mosquitoes. *Nature*, 1991, **350**, 151-3.

BIOCHEMICAL MECHANISMS OF RESISTANCE TO PHOTOSYSTEM II HERBICIDES

JACK J.S. VAN RENSEN AND OSCAR J. DE VOS
Laboratory of Plant Physiological Research,
Wageningen Agricultural University
Gen. Foulkesweg 72, 6703 BW Wageningen, The Netherlands

ABSTRACT

Photosystem II herbicides bind to the D1 protein of the reaction centre of photosystem II. Resistance to these herbicides in plants is confined almost exclusively to the triazine group and involves an altered D1 protein: at site 264 the serine of the wild-type is replaced by a glycine.
The electron flow rate between the primary electron acceptor of photosystem II and the plastoquinone pool is about three times slower in triazine-resistant plants, but the overall electron transport rate is not different from that in sensitive plants. When plants are grown at a low light intensity, there is no difference in photosynthetic capacity and in fitness. However, when grown at a high light intensity, resistant plants have a lower photosynthetic capacity and are less fit. This difference is ascribed to a higher sensitivity of resistant plants to photoinhibition.
Measurements of the reversible binding and release kinetics revealed that the binding kinetics are not much different in sensitive and resistant plants. Triazine resistance appears to originate from a significant increase of the release kinetics.

INTRODUCTION

Triazine resistance was noticed for the first time when Ryan (1) reported in 1970 that atrazine and simazine no longer controlled *Senecio vulgaris* at a nursery where triazine herbicides had been used annually for a long time. Since then this resistance has spread all over the world.

Searching for the mechanism of triazine resistance, no differences between resistant and sensitive biotypes were found in uptake, transport and biodegradation. After Radosevich and De Villiers (2) found that atrazine did not inhibit the Hill reaction in isolated chloroplasts from resistant plants, Pfister and Arntzen (3) demonstrated absence of binding of ^{14}C-labelled atrazine to resistant chloroplasts.

Souza Machado *et al.* (4) reported that atrazine resistance in *Brassica campestris* was only inherited through the female parent. Later, it was shown that the resistance trait was controlled by cytoplasmatic inheritance via chloroplast DNA.

PHOTOSYNTHETIC ELECTRON TRANSPORT; THE D1 PROTEIN

Within the process of photosynthesis herbicides do not inhibit CO_2-reduction, but interfere exclusively with electron flow. A scheme of photosynthetic electron transport is presented in Fig. 1. Herbicides act either at photosystem I (PSI), *e.g.*, paraquat, or at photosystem II (PSII). The latter photosystem II herbicides interrupt electron flow between the primary quinone electron acceptor of photosystem II, Q_A, and the plastoquinone pool.

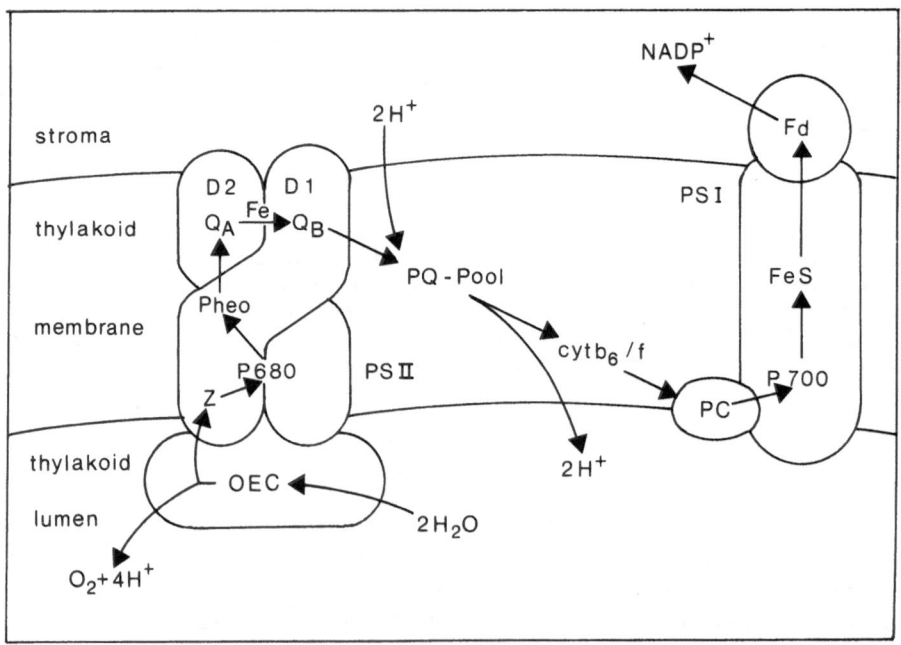

Figure 1. Electron transport pathway in photosynthesis.

Of special interest for understanding the mechanism of resistance to photosystem II herbicides is the PS II reaction centre, which can be considered as a water-plastoquinone oxido-reductase. Consuming four photons it transfers four electrons from two molecules of water to plastoquinone producing molecular oxygen and two molecules of doubly reduced plastoquinone. Our current conception about the many aspects of PS II activity involving protein and cofactor interactions, was greatly stimulated by the crystallisation and X-ray diffraction analysis of the reaction centre from *Rhodopseudomonas viridis* (5,6). It was realised that there are significant structural and functional homologies between the D1 and D2 proteins in PS II and the L (low molecular weight) and the M (medium molecular weight) subunit of this photosynthetic purple bacterium (7,8). The D1 and the D2 proteins appear to be the basic polypeptides of the reaction centre of PS II. Both have a molecular weight of 32 kDa. The D2 protein contains the binding site for Q_A, while the D1 protein binds Q_B.

The D1 protein is not only the binding site of Q_B, but also of the various classes of photosystem II herbicides. This protein appears to include five membrane spanning helices (7). The binding sites for Q_B and the herbicides are suggested to be located between the 4th and the 5th helix. The amino acid residues forming the binding niche of Q_B and the herbicides are known (9), but an exact steric model of the shape of the pocket has been hard to establish.

Triazine resistance could be correlated with an alteration of the D1 protein. After it was found that the chloroplast *psb*A gene codes for the D1 protein, Hirschberg *et al.* (10) demonstrated a change from adenine in the sensitive to guanine in the resistant biotype. This leads to a substitution of serine for glycine at position 264 within the D1 protein. Many triazine-resistant plants have been analyzed; in all cases triazine resistance is correlated with this same mutation (11).

Triazine resistance initiated the study of the molecular biology of the D1 protein. Many mutants of algae and bacteria resistant to various photosystem II herbicides have been obtained by selection at a low concentration of a herbicide, or by site directed mutagenesis. The matter was recently reviewed by Trebst (12). All mutations are located between amino acids 211 and 275 of the D1 protein. This area includes the hydrophobic transmembrane helices IV and V and the parallel helix connecting these on the matrix side of the D1 protein.

INHIBITION OF ELECTRON TRANSPORT BY PHOTOSYSTEM II HERBICIDES

Research on the action of herbicides inhibiting photosynthesis has yielded much detailed information about their mechanism of action. Furthermore, our understanding of the photosynthetic process has been greatly enhanced by the use of these chemicals as specific inhibitors (13).

The photosystem II herbicides can be divided into the diuron-type herbicides and the phenol-type chemicals (14, 15). The diuron-type group include amongst others the urea and the triazine herbicides; the phenol-type ones include, e.g., DNOC (4,6-dinitro-o-cresol), dinoseb and ioxynil. The structures of diuron (DCMU), atrazine and ioxynil are presented in Fig. 2.

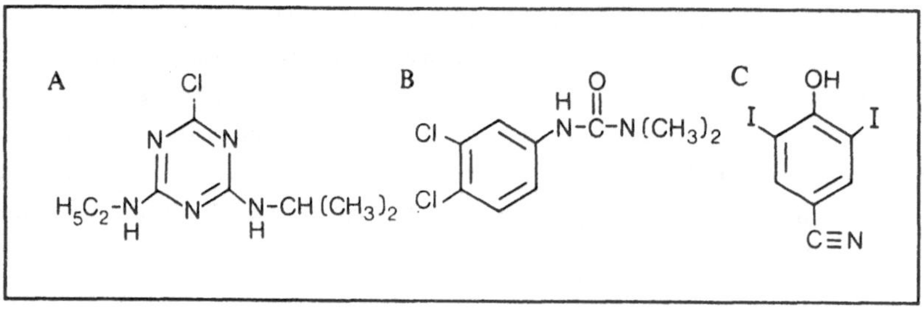

Figure 2. Structures of some PS II herbicides; A = atrazine, B = DCMU, C = ioxynil.

Trebst (16) listed the many differences between the two groups of herbicides. Recently Wildner et al. (17), using five mutants of Chlamydomonas reinhardtii and a large number of PS II inhibitors, confirmed the division into two families; their data were further used to refine the folding model of the D1 protein and the orientation of the inhibitors in their binding site.

Triazine resistance is specific for all chemicals of the triazine group; the concentration of triazine herbicide causing 50% inhibition of the Hill reaction in isolated chloroplasts from resistant biotypes is often about 1000 times higher than in chloroplasts from the sensitive plants. There is little difference in sensitivity for urea herbicides, while resistant biotypes are somewhat more sensitive to phenol-type herbicides (negative cross-resistance).

ARE TRIAZINE-RESISTANT PLANTS LESS FIT?

The alteration of the D1 protein does not only result in triazine resistance, but also in a change in the equilibrium of the $Q_A^- Q_B$ to $Q_A Q_B^-$ reaction. This is assumed to be due to a large decrease in the rate of electron flow from Q_A^- to Q_B. Jansen and Pfister (18) measured the rate of

Q_A^- oxidation in 28 different plant species and found an average half-time of 314 ± 46 μs. In five different triazine-resistant plants this average was 946 ± 100 μs. This three-fold decrease in rate is less than previously reported. These authors suggested that the changed kinetics of the Q_A/Q_B reaction does not simply decrease primary photosynthetic efficiency via a direct effect on photosynthetic electron flow, but that the mutation in the D1 protein affects also some other functional aspects of the PS II complex important for regulation of photosynthesis and biomass production, e.g., the turnover of the D1 protein.

Triazine resistance has been transferred to several crop plants: rapeseed (*Brassica napus* L.), chinese cabbage (*Brassica campestris* L.) and foxtail millet (*Setaria italica* L.). Unfortunately, a significant reduction in yield accompanies the resistance trait in most species studied. Field studies of resistant rapeseed showed decreased growth and crop yield. The above-mentioned slower rate of electron flow between Q_A and Q_B has been suggested as the cause of the reduction in photon yield, lowered maximum photosynthesis, decreased growth and ecological fitness (e.g., 19, 20).

An influence of the impaired electron transport between Q_A and Q_B on the overall electron transport was first questioned by Jansen et al. (21). The lower rate of electron flow from Q_A to Q_B in the resistant biotype is still about 20 times faster than the oxidation of reduced plastoquinone. The latter reaction remains the rate limiting step having a half-time of about 20 ms. These authors demonstrated that the electron transport between water and plastoquinone has a lower rate and a lower quantum yield in isolated chloroplasts of resistant plants. However, no significant differences were found for the rate and quantum yield of whole chain electron transport.

It was recently suggested that the differences between resistant and sensitive plants occur only when grown at high light intensity and are much less when grown at a low light intensity (22, 23 and 24). Hart and Stemler (22) compared triazine-resistant and susceptible *B. napus* plants, grown under low photon flux density. They found that the slow electron transfer from the primary to secondary quinone electron acceptors of photosystem II was still present in the resistant line but photon yield and light-saturated oxygen evolution were similar in the two *B. napus* lines. Van Rensen et al. (23) measured growth and quantum yield of triazine-resistant and sensitive *Chenopodium album* plants grown at low (20 W m^{-2}) and at high (150 W m^{-2}) light intensity (Table 1).

At the lower light intensity the dry matter production of the resistant and the sensitive plants is almost similar; at the higher light intensity the resistant biotype shows a significantly lower productivity. The quantum yield, which was measured in chloroplasts, isolated from these plants, is almost similar for the resistant and the sensitive plants when

TABLE 1

Dry weight production and quantum yield of resistant (R) and sensitive (S) *Chenopodium album* plants grown at light intensities of 20 and 150 W m^{-2} (R20, S20, R150 and S150)

	R20	S20	R150	S150
dry weight (g)	0.09 ± 0.02	0.11 ± 0.03	1.14 ± 0.10	1.65 ± 0.21
quantum yield	3.6 ± 0.4	3.8 ± 0.2	5.3 ± 0.3	7.3 ± 0.5

grown at low light intensity, but at the higher light intensity the quantum yield of the resistant biotype is substantially lower.

It was suggested the differences which occur at high light intensity were caused by a greater sensitivity of the resistant plants to photoinhibition (23, 24). Van Rensen *et al.* (23) measured photoinhibition in resistant and sensitive biotypes of *C. album* grown at low and high light intensity (Table 2). They treated leaves during 15 min at 5 °C while one half of the leaf was kept in the dark, and the other half was exposed to light at an intensity of 800 W m^{-2} (photoinhibitory treatment, PIT). The resulting photoinhibition was measured by Fv/Fm (variable chlorophyll *a* fluorescence over maximum fluorescence).

TABLE 2

Photoinhibition measured as Fv/Fm of resistant (R) and sensitive (S) leaves of *Chenopodium album*, grown at 20 and at 150 W m^{-2}. The leaves were kept at 5 °C during 15 min. One half of the leaf was darkened (dark); the other half illuminated with 800 W m^{-2} (photoinhibitory treatment, PIT)

	R20	S20	R150	S150
Dark (a)	0.81 ± 0.01	0.84 ± 0.01	0.80 ± 0.01	0.85 ± 0.01
PIT (b)	0.51 ± 0.03	0.58 ± 0.04	0.58 ± 0.02	0.69 ± 0.04
(a)-(b)=(c)	0.30	0.26	0.22	0.16
(c)as % of(a)	37%	31%	28%	18%

Although plants suffer less from photoinhibition when grown at high light intensity compared when grown at low light, it appears that the resistant biotype is more sensitive to photoinhibition than the sensitive one; the difference is less when grown at low light.

It may be concluded that resistant plants are indeed less fit than sensitive biotypes; however, only at high light intensities. This may be explained by the phenomenon that in resistant plants Q_A remains longer in the Q_A^- state because of the lower electron flow rate between Q_A and Q_B. This leads to less charge separation and to more photoinhibition. In this way the resistant plants have lower productivity, especially at higher light intensities.

TRIAZINE RESISTANCE IS CAUSED BY INCREASED TRIAZINE HERBICIDE RELEASE KINETICS

Interaction of diuron-type herbicides with plastoquinone was first proposed by Van Rensen (25). It is now widely accepted that the mechanism of action of these herbicides is a displacement of Q_B from its binding site at the D1 protein. The binding of the herbicides appears to be competitive with plastoquinone. It was expected that the inhibitors stay rather long at their binding site at the D1 protein preventing access of Q_B. The inhibitory herbicides cannot be reduced by Q_A and electron transport beyond this point is thereby prevented.

The Q_B-quinone is not permanently bound to the D1 protein. In its oxidized and reduced state, Q_B and Q_BH_2, it is only weakly bound and can easily leave its binding site; the semiquinone Q_B^- is more tightly bound. Also herbicides bind and release at their binding site on the D1 protein. These exchange parameters can be measured using a method, initiated by Vermaas *et al.* (26) and adapted by Naber (11). It is based on the flash-induced oxygen evolution patterns of isolated broken chloroplasts, which are measured in the absence and in the presence of herbicides. The exchange parameters are obtained by fitting experimental data to those calculated with a kinetic model. This model is derived from the following equations:

$$Sn.Q_A.Q_B + I \overset{E_1}{\underset{E_3}{\rightleftharpoons}} Sn.Q_A.I + Q_B$$

$$Sn.(Q_A.Q_B)^- + I \overset{E_2}{\underset{E_4}{\rightleftharpoons}} Sn.(Q_A.I)^- + Q_B$$

In these equations, Sn (where $n = 0,1,2,3$) represents the redox state of the oxygen evolving complex. In the presence of slowly exchanging herbicides, having residence times on the D1 protein of the same order of magnitude as the duration of the flash train or longer, the oscillation is hardly damped compared to the control. In this case only the amplitude of the signal is diminished. However, when herbicide exchange is occurring with the same or higher frequency than the firing of the flashes, the damping of the oscillation is considerably stronger. This is caused by the fact that reaction centres are blocked for a certain time span, and start making turnovers at the moment the herbicide is displaced by a PQ-molecule. Thus, centres can get out of phase with each other, and produce oxygen at different flashes. By comparing flash patterns obtained with different flash frequences and herbicide concentrations, the exchange parameters E_1 to E_4 can be calculated (Naber and Van Rensen, 27).

TABLE 3

Values for the atrazine exchange parameters E_1 to E_4 measured in chloroplasts isolated from resistant (R) and sensitive (S) biotypes of *Chenopodium album*; E_1 and E_2: $\mu M^{-1}.s^{-1}$; E_3 and E_4 : s^{-1}.

	E_1	E_2	E_3	E_4
C. album S	0.1	0.02	0.11	0.04
C. album R	0.05	0.002	15	2.25

As is shown in Table 3, in the resistant biotype the binding parameters E_1 and E_2 are only decreased by 2- and 10-fold, respectively. However, the release parameters E_3 and E_4 are 136 and 56 times higher in the resistant plants, compared with those in the sensitive ones. This indicates that triazine resistance originates from a significant increase of the release kinetics.

This can be explained at the molecular level. The "on" kinetics of a compound to the binding environment are determined principally by the accessibility of the niche to the compound. This is determined by the chemical structure of the herbicide, especially its molecular dimensions, charges and hydrophobicity. These properties are, of course, the same when atrazine is added to resistant or to sensitive chloroplasts. A very slight change of hydrophobicity of the binding pocket can be expected as a result of the serine to glycine substitution at position 264 in the triazine-resistant biotype. However, the atrazine molecule cannot be stabilised in its binding environment in the mutant protein, probably because the serine-OH group provides an important H-bonding possibility in the sensitive biotype. The result is a decrease in herbicidal activity of 2 to 3 orders of magnitude.

REFERENCES

1. Ryan, G.F., Resistance of common groundsel to simazine and atrazine. Weed Sci., 1970, 18, 614-616.

2. Radosevich, S.R. and De Villiers, O.T., Studies on the mechanism of s-triazine resistance in common groundsel. Weed Sci., 1976, 24, 229-232.

3. Pfister, K. and Arntzen, C.J., The mode of action of photosystem II-specific inhibitors in herbicide-resistant weed biotypes. Z. Naturforsch., 1979, 34c, 996-1009.

4. Souza Machado, V. Bandeen, J.D., Stephenson, G.R. and Lavigne, P., Uniparental inheritance of chloroplast atrazine tolerance in Brassica campestris. Can. J. Plant Sci., 58, 977-981.

5. Deisenhofer, J., Epp, O., Miki, K., Huber, R. and Michel, H., X-ray structure analysis of a membrane protein complex. Electron density map at 3 Å resolution and a model of the chromatophores of the photosynthetic reaction center from Rhodopseudomonas viridis. J. Mol. Biol., 1984, 180, 385-398.

6. Deisenhofer, J., Epp, O., Miki, K., Huber, R. and Michel, H., X-ray structure analysis at 3 Å resolution of a membrane protein complex: Folding of the protein subunits in the photosynthetic reaction center from Rps. viridis. Nature, 1985, 318, 618-624.

7. Trebst, A., The topology of the plastoquinone and herbicide binding peptides of photosystem II in the thylakoid membrane. Z. Naturforsch., 1986, 41c, 240-245.

8. Barber, J., Photosynthetic reaction centers: a common link. TIBS, 1987, 12, 321-326.

9. Trebst, A., The three-dimensional structure of the herbicide binding niche on the reaction center polypeptides of photosystem II. Z. Naturforsch., 1987, 42c, 742-750.

10. Hirschberg, J., Bleeker, A., Kyle, D.J., McIntosh, L. and Arntzen, C.J., The molecular basis of triazine-herbicide resistance in higher plant chloroplasts. Z. Naturforsch., 1984, 39c, 412-420.

11. Naber, J.D., Molecular aspects of herbicide binding in chloroplasts. PhD thesis, Agricultural University Wageningen, 1989, pp. 1-116.

12. Trebst, A., The molecular .basis of resistance of photosystem II herbicides. In Proc. 11th Long Ashton Int. Symposium "Herbicide Resistance". ed. J.C. Caseley, Butterworths Publishers, London, 1991, pp. 52-61.

13. Van Rensen, J.J.S., Herbicides interacting with photosystem II. In Herbicides and Plant Metabolism, ed. A.D. Dodge, Cambridge University Press, Cambridge, 1989, pp. 21-36.

14. Van Rensen, J.J.S., Van der Vet, W. and Van Vliet, W.P.A., Inhibition and uncoupling of electron transport in isolated chloroplasts by the herbicide 4,6-dinotro-o-cresol. Photochem. Photobiol., 1977, 25, 579-583.

15. Trebst, A. and Draber, W., Structure activity correlations of recent herbicides in photosynthetic reactions. In Advances in Pesticide Science. part 2, ed. H. Geissbühler, Pergamon Press, Oxford, 1979, pp. 223-234.

16. Trebst, A., The three-dimensional structure of the herbicide binding niche on the reaction center polypeptides of photosystem II. Z. Naturforsch., 1987, 42c, 742-750.

17. Wildner, G.F., Heisterkamp, U. and Trebst, A., Herbicide cross-resistance and mutations of the psbA gene in Chlamydomonas reinhardtii. Z. Naturforsch., 1990, 45c, 1142-1150.

18. Jansen, M.A.K. and Pfister, K., Conserved kinetics at the reducing side of reaction-center II in photosynthetic organisms; changed kinetics in triazine-resistant weeds. Z. Naturforsch., 1990, 45c, 441-445.

19. Holt, J.S., Stemler, A. and Radosevich, S.R., Differential light responses of photosynthesis by triazine-resistant and triazine-susceptible Senecio vulgaris biotypes. Plant Physiol., 1981, 67, 744-748.

20. Ireland, C.R., Telfer, A., Covello, P.S., Baker, N.R. and Barber, J., Studies on the limitation to photosynthesis in leaves of the atrazine-resistant mutant of *Senecio vulgaris* L. Planta, 1988, **173**, 459-467.

21. Jansen, M.A.K., Hobé, J.H., Wesselius, J.C. and Van Rensen, J.J.S., Comparison of photosynthetic activity and growth performance in triazine-resistant and susceptible biotypes of *Chenopodium album*. Physiol. Vég., 1986, **24**, 475-484.

22. Hart, J.J. and Stemler, A., Similar photosynthetic performance in low light-grown isonuclear triazine-resistant and -susceptible *Brassica napus* L. Plant Physiol., 1990, **94**, 1295-1300.

23. Van Rensen, J.J.S., Curwiel, V.B. and De Vos, O.J., The effect of light intensity on growth, quantum yield and photoinhibition of triazine-resistant and susceptible biotypes of *Chenopodium album*. EBEC REPORTS, 1990, **6**, 46.

24. Hart, J.J. and Stemler, A., High light-induced reduction and low light-enhanced recovery of photon yield in triazine-resistant *Brassica napus* L. Plant Physiol., 1990, **94**, 1301-1307.

25. Van Rensen, J.J.S., Action of some herbicides in photosynthesis of *Scenedesmus*, as studied by their effects on oxygen evolution and cyclic photophosphorylation. Meded. Landbouwhogeschool Wageningen, 1971, **71-9**, 1-80.

26. Vermaas, W.F.J., Dohnt, G, and Renger, G., Binding and release kinetics of inhibitors of Q_A^- oxidation in thylakoid membranes. Biochim. Biophys. Acta, 1984, **765**, 74-83.

27. Naber, J.D. and Van Rensen, J.J.S., Activity of photosystem II herbicides is related with their residence times at the D1 protein. Z. Naturforsch., 1991, **46c**, 49-52.

MOLECULAR BIOLOGY AND DIAGNOSTICS IN DISEASE CONTROL

D. W. HOLLOMON
Department of Agricultural Sciences,
University of Bristol,
Institute of Arable Crops Research,
Long Ashton Research Station,
Bristol BS18 9AF

ABSTRACT

Visual diagnosis of fungal diseases is often inaccurate, whilst detecting fungicide resistance using bioassays may be too slow to allow effective anti-resistance strategies to be implemented in time. Immunology and DNA probe technology offer ways to overcome some of these limitations, and are currently subject of much active research. Monoclonal and polyclonal antibodies have both been used in specific disease diagnostic systems and offer either rapid (10 min) on farm methods, or more sophisticated and potentially quantitative laboratory based techniques. Less progress has been made in using immunology to detect fungicide resistance, although single DNA base changes connected with resistance have already been detected using appropriate oligonucleotide probes. The effects of these techniques are unclear, but are likely to improve fungicide effectiveness. Less desirable consequences may arise through the survival of fewer wild-type individuals which might otherwise compete with fungicide resistant strains in the absence of fungicides. Improvements in resistance monitoring may be needed to match improved disease control measures, and earlier detection of resistance should ensure that effective anti-resistance strategies can be introduced before it is too late.

INTRODUCTION

Biotechnology has generated a wide variety of immunological and nucleic acid diagnostic techniques in human and veterinary medicine. Progress towards the detection and quantification of fungal diseases has been much slower for several reasons [1], but a number of recent reports [2-6] suggest significant progress is now being made. Precise identification and quantification of disease may soon be available to optimise control measures for a number of important plant pathogens. Immunoassays are ideally suited for pathogen detection [7], either as sensitive and quantitative Enzyme Linked Immunoabsorbent Assays (ELISA), or more rapid "user friendly" qualitative diagnostics. Where monoclonal and polyclonal antibodies lack the required antigenic specificity needed to produce a useful diagnostic, DNA probe technology linked to fast evolving techniques based on the Polymerase Chain Reaction (PCR) [8], offer alternative approaches. So too do microscopically based systems [9], which allow detection of several diseases on the one sample, although quantification through counts of fruiting bodies may not always be easy using unskilled labour. This paper reviews some recent developments in the detection of both fungal diseases and fungicide resistance, and explores their likely impact on both the development of fungicide resistance, and strategies to combat it.

IMMUNODIAGNOSTICS FOR DISEASE DETECTION

Grey Mould
Small molecular weight antigenic glycolipids produced by *Botrytis cinerea*, and specific polyclonal antisera, form the basis of a non-competitive indirect enzyme immunoassay able to quantify grey mould infection levels below 1% harvested berries [10]. A carbohydrate antigen forms the basis of another equally sensitive Double Antibody Sandwich (DAS)-ELISA developed by DuPont (USA) [11]. Designed to speed up inspection methods to determine whether to accept or reject grapes prior to pressing, these diagnostics may be of limited value in determining thresholds for fungicide treatments, since in many regions *B. cinerea* is ubiquitous, and sprays to control grey mould are routine.

Damping-off of Seedlings
Both *Phytophthora megasperma* f.sp. *glycinea* and *Rhizoctonia solani* cause indistinguishable damping-off symptoms in soybeans, although different fungicides must be used to control each disease. Specific antibodies adsorbed onto a membrane and formatted as a DAS-ELISA provide a rapid (10 min) "on site" identification test to distinguish these two pathogens, allowing the correct fungicide to be used [2]. Use of these kits to improve fungicide timing for *Rhizoctonia* control in turf grass was apparently less successful, and hardly detected the pathogen prior to symptom expression. Whether essentially qualitative results will be a feature of all "on site" rapid diagnostic systems is not clear.

Septoria Diseases of Wheat

Septoria tritici and *S. nodorum* cause damaging diseases of wheat, and in the case of *S. nodorum* barley too. Although the symptoms and biology of these two diseases are different, a three week latent period prior to symptom expression coupled with limited curative action by most fungicides, suggests that presymptomatic quantitative detection of *S. tritici* at least, would be a valuable tool used in conjunction with existing forecasting systems [12]. Monoclonal immunoassays specific for either *S. tritici* or *S. nodorum* have recently been developed [3, 13] which do not cross-react with other cereal pathogens. Formatted either as a DAS-ELISA using a secondary polyclonal antibody conjugated with peroxidase, or as an "on site" kit, the two Septoria species can now be diagnosed accurately. However, at this stage it is not clear just how quantitative these assays are, and whether presymptomatic detection is at all possible. Clearly this is crucial if they are to be useful in helping to decide when to apply a fungicide. A lack of presymptomatic detection may reflect the antigen source, which was germinating conidia for *S. tritici*, and so the antigen may be less abundant during mycelial spread within the host tissue. Polyclonal immunoassays for both Septorias have recently been described [14] which do not cross-react with a large number of phytopathogenic fungi and, at least, for *S. nodorum* is sufficiently sensitive to detect the disease three days after inoculation.

Cereal eyespot

Eyespot (*Pseudocercosporella herpotrichoides*) is a troublesome, stem-base, disease of wheat, barley and rye. Although text-book pictures of eyespot, sharp eyespot (*Rhizoctonia cerealis*) and Fusarium distinguish these diseases of the cereal stem-base complex, in practice identification by visual symptomology is far from easy. Mixed infections occur, but are difficult to detect against a general blackening of the stem base caused by Fusarium, or perhaps other factors. Not surprisingly, a "threshold" value based on visual assessment of 20% stem infection, has not always proved a safe criterion on which to decide to apply an eyespot fungicide. Consequently, the effectiveness of disease control measures are hampered by accurate diagnosis and quantification at GS.31, when fungicides are often applied for maximum control.

A variety of immunogens have been used to raise antisera to the eyespot pathogen, although none of these antisera distinguish between the wheat (*P. herpotrichoides* var *herpotrichoides*) [15] and rye (*var. acuformis*) pathotypes. Polyclonal antisera were raised in rabbits to macerated mycelium

[4] or dialysed culture filtrate [16]; mouse monoclonal antibodies were raised to surface washings of mycelial fragments [1]. Although protocols are not always well described, limited cross adsorption to remove unwanted cross-reactivity was used by DuPont (USA) to produce antisera [16]. All eyespot diagnostic kits developed so far use some form of DAS-ELISA and, as yet, no user friendly formats have been resourced for commercial use. The capture antibody is bound to the micro-titre plate well prior to despatch, and the user simply washes away stabilizer and adds antigen sample. The secondary antibody may be conjugated with either alkaline phosphatase or peroxidase, or involve a biotin-avidin system to enhance sensitivity. At least three standard eyespot antigen preparations should be included in each assay, and chosen to give a straight-line relationship from which antigen levels in unknown samples can be calculated. Ninety-six wells are available in each micro-titre plate, but when standards are included and dilutions, there is seldom scope to include more than 50 samples for testing. Eyespot antigens appear quite stable and extracts can be made by macerating stem bases, and freezing at -20°C until required. When thawed and mixed, samples can be added to each well and the assay should be completed in under three hours.

TABLE 1
Cereal diseases and eyespot immunoassay

Disease	Absorbance units	Eyespot antigen units/plant
Uninfected	0.016	<2
Eyespot (Severe)	1.150*	35,000
Eyespot (Moderate)	1.450	670
Eyespot (Presymptomatic)	0.804	51
Sharp eyespot (*R. cerealis*)	0.030	<2
Take-all (*G. graminis*)	0.025	<2
Root rot (*Fusarium spp.*)	0.040	<2
Yellow rust (*Puccinia striiformis*)	0.025	<2
Brown rust (*P. recondita*)	0.001	<2
Powdery mildew (*E. graminis*)	0.050	<2
Leaf blotch (*Septoria tritici*)	0.073	<2
Glume blotch (*Septoria nodorum*)	0.033	<2

* Sample diluted 1:729   Data from [5].

Cross reactivity with other stem base pathogens seems a problem with some monoclonal antisera [1], whereas eyespot polyclonal systems show little or no cross-reactivity [Table 1] against major cereal pathogens. Eyespot can be detected pre-symptomatically, and antigen levels generally correlate well with the amount of disease present [Fig. 1]. ELISA also showed that the fungicide flusilazole arrests eyespot development rapidly, although the stable antigen remains until leaf sheaths are lost from the plant [Fig. 2].

DNA PROBE TECHNOLOGY FOR DISEASE DIAGNOSIS

Take-all

The soil borne ascomycete *Gaeumannomyces graminis* f.sp. *tritici* causes an important root disease of wheat and barley. Infected plants are stunted and carry sterile ears, although these symptoms may be caused by other micro-organisms, or even adverse environmental factors. Take-all can be diagnosed visually, but can be confused with non-pathogenic formae speciales of *G. graminis*, and its definitive identification requires time consuming microscopic examination, coupled with culturing on selective media.

A cloned 4.3 Kb mitochondrial DNA fragment of *G. graminis* f.sp. *tritici* hybridized specifically to genomic DNA extracted from cultures of the pathogen [6]. Schesser *et al.* [8] explored the possibility of using infected plant material for diagnosis and the Polymerase Chain Reaction (PCR), to amplify the region of genomic DNA of interest. Primers (25 mer) derived from sequence analysis of the mitochondrial DNA probes, were used to amplify the corresponding region of genomic *G. graminis* f.sp. *tritici* DNA, allowing seedlings grown under sterile conditions, but otherwise infected with the take-all pathogen, to be distinguished from seedlings inoculated with *Fusarium culmorum*, *Rhizoctonia* sp, *Cochliobolus sativus* and *Trichoderma* sp. Presumably, other primers will be needed to diagnose differences within species and identify *G. graminis* f.sp. which are non-pathogenic. Extension of this work to amplification of DNA from the roots of field grown plants, contaminated no doubt with soil, will certainly require rigorous control of PCR conditions to avoid artefacts. Although amplified sequences may be visualised in gels stained with ethidium bromide, routine detection based perhaps on a dot-blot system, will require use of non-radioactive probes. Considerable progress is currently being made toward replacement of ^{32}P with more user-friendly detection systems, and this should not be a problem in the future.

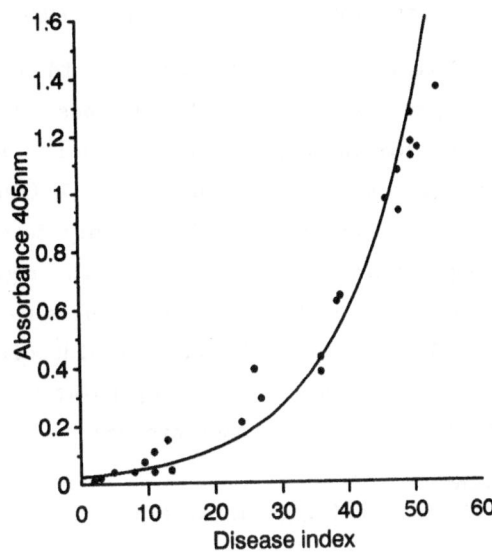

Figure 1. Correlation between ELISA values and eyespot disease index derived from visual assessments. From [4].

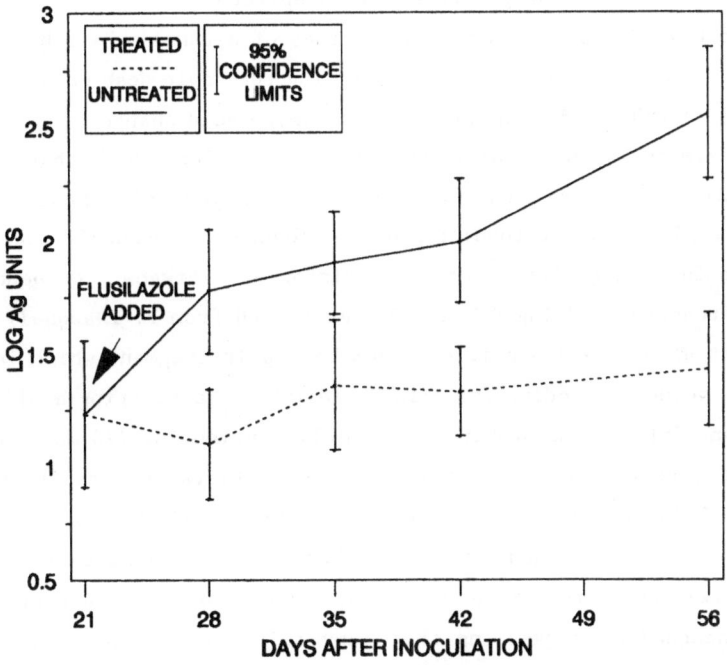

Figure 2. Antigen levels and eyespot control after treatment with flusilazole. Data from [5].

DETECTION OF FUNGICIDE RESISTANCE

Current methods for detecting and monitoring fungicide resistance rely solely on bioassay techniques, which identify various phenotypic levels of resistance (See Brent, this volume). Bioassays are frequently time consuming and imprecise, and allow analysis of only a small fraction of a pathogen population. Consequently resistance is often only detected when it is too late to adopt strategies which might avoid a loss in fungicide effectiveness. Biochemically based methods should allow more samples to be tested and detect resistance earlier. In addition, they will provide direct evidence of the mechanisms of resistance present in pathogen populations, and should provide a better understanding of how resistance evolves.

Carbendazim Resistance

Carbendazim and related fungicides inhibit the polymerisation of α and β tubulins to form microtubules. These fungicides normally bind to ß-tubulin but in resistant strains an altered ß-tubulin no longer binds these fungicides [17, 18]. Attempts to use immunology to detect carbendazim resistance in *B. cinerea* were unsuccessful [19] and it seems the amino acid changes involved do not generate unique epitopes.

Both wild-type and resistant alleles of ß-tubulin have been cloned and sequenced from a number of fungi, including the apple scab pathogen (*Venturia inaequalis*) [20]. In each case a single amino acid change is correlated with carbendazim resistance, and is the result of a single base change in the DNA (Table 2). Although a number of different amino acid changes occur in a limited region of the ß-tubulin protein, changes at codon 198 seem common and can confer negative cross resistance between carbendazim and phenylcarbamates. Using PCR and DNA extracted from *V. inaequalis* cultures, Koenraadt *et al.* [21] amplified a 408 base-pair fragment which included the critical sequences coding for amino acid 198. Transferring amplified DNA to a membrane filter, and probing with an oligonucleotide (20 mer) specific for a single base change, allowed the correct resistance classification of at least 80 *V.inaequalis* isolates obtained from around the world. Despite the several possible base changes identified from studies on laboratory resistant mutants [22], very few probes were needed to survey all these field isolates of *V. inaequalis*, suggesting that many amino acid changes result in poor fitness. Since ß-tubulin genes are highly conserved across many organisms, and the same amino-acid changes occur in carbendazim resistant strains of many plant pathogens, the same oligonucleotide probes may be used to identify carbendazim resistance in different fungi. Indeed, the probes used to identify carbendazim resistance in *V. inaequalis* have correctly identified resistance in *Penicillium* sp. and *Monilinia fructicola* [23]. It is now important to extend these techniques so that DNA can be amplified from diseased lesions to avoid the need to isolate each pathogen.

TABLE 2
Amino acid substitutions and carbendazim resistance in *Venturia inaequalis*.

	Carbendazim	Phenylcarbamate	codon 198	codon 200
Wild-type	S	R	glu	phe
VHR	R	S	ala	phe
HR	R	R	lys	phe
MR⁻	R	S	gly	phe
MR⁺	R	R	glu	tyr

Data from [20].

IMPLICATIONS FOR FUNGICIDE RESISTANCE

Recent research on diagnostic methods is likely to improve the accuracy of disease identification. Both ELISA and oligonucleotide probe techniques offer the potential for quantitative disease measurements, more accurate forecasting systems and better timing of fungicide applications. Determination of application thresholds as antigen levels will require several years field experimentation; are likely to vary on a regional basis and depend on cultivar use. One aim of diagnostic methods is to help maintain pathogen populations at lower levels than at present which should reduce the likelihood of resistant mutants emerging. Any reduction in the number of fungicide applications will also have a beneficial effect on resistance. However, more effective fungicide use could lead to fewer wild-type individuals surviving treatment, and able to compete with resistant strains during periods when fungicides are not used. Furthermore, if pre-symptomatic detection encourages earlier treatment of smaller pathogen populations, genetic drift may become a more important factor in the spread of fungicide resistance.

Improved methods to detect fungicide resistance may well be needed to complement these improvements in disease detection. Earlier and better detection of resistance will not only allow anti-resistance strategies to be implemented where available, but their effectiveness can be monitored. Precise information on resistance mechanisms operating in natural pathogen populations will allow predictions of the risk of resistance to be updated and modified by further research if necessary. Where several mechanisms of resistance are known to exist, available cross-resistance data might identify a fungicide likely to be the most effective against the resistance mechanism prominent in field populations.

REFERENCES

1. Dewey, F.M., Development of immunological diagnostic assays for fungal plant pathogens. *Proc. Brighton Crop Prot. Conf. - Pests and Diseases,* 1988, 777-785.

2. Miller, S.A., Rittenburg, J.H., Peterson, F.P. and Grothaus, G.D., Application of rapid, field usable immunoassays for the diagnosis and monitoring of fungal pathogens in plants. *Proc. Brighton Crop Prot. Conf. - Pests and Diseases,* 1988, 795-803.

3. Petersen, F.P., Rittenburg, J.H., Miller, S.A. and Grothaus, G.D., Development of monoclonal antibody-based immunoassays for detection and differentiation of *Septoria nodorum* and *S. tritici* in wheat. *Proc. Brighton Crop Prot. Conf. - Pests and Diseases,* 1990, 751-756.

4. Unger, J.G. and Wolfe, G., Detection of *Pseudocercosporella herpotrichoides* (From) Deighton in wheat by indirect ELISA, *J. Phytopath.,* 1988, **122,** 281-286.

5. Smith, C.M., Saunders, D.W., Allison, D.A., Johnson, L.E.B., Labit, B., Kendall, S.A. and Hollomon, D.W., Immunodiagnostic assay for cereal eyespot: novel technology for disease detection. *Proc. Brighton Crop Prot. Conf. - Pests and Diseases,* 1990, 763-770.

6. Hensen, J.M., DNA Probe for identification of the take-all fungus *Gaeumannomyces graminis, Appl. Environ. Microbiol.,* 1989, **55,** 284-288.

7. Miller, S.A. and Martin, R.R., Molecular diagnosis of plant disease. *Ann. Rev. Phytopath.* 1988, **26,** 409-432.

8. Schesser, K., Luder, A. and Henson, J.M., Use of polymerase chain reaction to detect the take-all fungus, *Gaeumannomyces graminis,* in infected wheat plants. *Appl. Environ. Microbiol.,* **57,** 553-556.

9. Vereet, J.A. and Hoffmann, G.M., Threshold based control of wheat diseases using the BAYER cereal diagnosis system after Verreet/Hoffmann. *Proc. Brighton Crop Prot. Conf. - Pests and Diseases,* 1990, 745-750.

10. Richer, R.W., Marois, J.J., Dlott, J.W., Bostock, R.M. and Morrison, J.C., Immunodetection and quantification of *Botrytis cinerea* on harvested wine grapes. *Phytopath*. 1991; 81, 404-411.

11. Cagnieul, P. and Majarian, W., Diagnostic kit for grape grey mold. *ANPP - Troisiéme Conference Internationale sur les Maladies des Plantes, Bordeaux*, 1991, 555-560.

12. Shaw, M.W. and Royle, D.J., Saving *Septoria* fungicide sprays: the use of disease forecasts. *Proc. Brit. Crop Prot. Conf. - Pests and Diseases*, 1986, 1193-1200.

13. Mittermeier, L., Dercks, W., West, S.J.E. and Miller, S.A., Field results with a diagnostic system for the identification of *Septoria nodorum* and *Septoria tritici*. *Proc. Brighton Crop Prot. Conf. - Pests and Diseases*, 1990, 757-762.

14. Cagnieul, P., Joerger, M. and Hirata, L., Septoria diagnolab kit - diagnostic kit for wheat septoria (*Septoria nodorum* and *Septoria tritici*). *ANPP - Troisiéme Conference Internationale sur les Maladies des Plantes, Bordeaux*, 1991, 539-545.

15. Nirenberg, H.I., Differenzierung der Erreger der Halmbruchkrankeit. I. Morphologie. *J. Pl. Dis. Prot*., 1981, 88, 241-248.

16. Saunders, D., Feindt, D., Johnson, L., Smith, C. and Stave, J., A diagnostic immunoassay to detect areal eyespot (footrot) disease in wheat. *Phytopath*. 1990, 80, Abst. 38.

17. Davidse, L.C. and Flack, W., Interaction of thiabendazole with fungal tubulin. *Biochem. Biophys. Acta*., 1978, 543, 82-90.

18. Oakley, B.R. and Morris, N.R., A ß-tubulin mutation in *Aspergillus nidulans* that blocks microtubule function. *Cell*, 1981, 24, 837-845.

19. Groves, J.D., Fox, R. and Baldwin, B.C., Development of an immunodiagnostic for MBC resistance in *Botrytis*. *Proc. Brighton Crop Prot. Conf. - Pests and Diseases*, 1988, 415-420.

20. Koenraadt, H., Somerville, S.C. and Jones, A.L., Molecular characterisation of the beta-tubulin gene from benomyl-sensitive and benomyl-resistant field strains of *Venturia inaequalis*. *Phytopath*. 1990, 80, Abstr. 288.

21. Koenraadt, H., Jones, A.L. and Somerville, S.C., Use of allele-specific oligonucleotide probes to characterise resistance to benomyl in field strains of *Venturia inaequalis*. *Phytopath*. 1991, 81, Abstr. 163.

22. Fujimura, M., Oeda, K., Inoue, H. and Kato, T., Mechanism of action of N-phenylcarbamates in benzimidazole-resistant *Neurospora* strains. In: *Managing Resistance to Agrochemicals: From Fundamental Research to Practical Strategies*, Eds. M.B. Green, H.M. Le Baron and W.K. Moberg. ACS Symposium Series 421, Washington DC. 224-236.

23. Jones, A.L., (personal communication. 1991).

INSECT RESISTANCE TO BIOTECHNOLOGY PRODUCTS: AN OVERVIEW OF RESEARCH AND POSSIBLE MANAGEMENT STRATEGIES

Pamela G. Marrone and Susan C. MacIntosh
Entotech, Inc.
(A Novo Nordisk Subsidiary)
1497 Drew Avenue, Davis, CA USA 95616

ABSTRACT

Products based on *Bacillus thuringiensis* (B.t.) are the most successful biological pesticides used in agriculture, forestry and public health. Despite their success, usage in comparison to chemical presticides has been small. Several developments have led to rapidly increasing usage of B.t. products. Recombinant DNA technology has demonstrated the potential to significantly improve B.t. efficacy and persistence, which will lead to even wider use of B.t.-based biotechnology products. During the past 30 years of use of *Bacillus thuringiensis* as a microbial insecticide, there have been very few documented cases of B.t. resistance development in actual field use. In these situations, the insects were under repeated, intensive use of B.t. The likelihood for B.t. resistance development as a result of increased selection pressure on the various target pest populations can be minimized if sound integrated pest management and resistance management strategies are practiced.

INTRODUCTION

Insects have shown a remarkable ability to develop resistance to insecticidal chemicals. Over 500 species of arthropod are now reported to resist insecticides, and this resistance can extend to all known classes of insecticidal chemistry [1]. The development of resistance is one of the most critical issues facing growers and pest managers today. The large development costs, estimated to be between $20-40 million, for registration of a new chemical pesticide highlights the critical need for new and existing pesticides to be managed wisely.

Products based on *Bacillus thuringiensis* are the most successful biological pesticides used in agriculture, forestry and public health. Despite their success, usage in comparison to chemical pesticides has been small, accounting for only about 1% of the world insecticide market. Several developments, such as use of B.t. as a resistance management tool and increased environmental concerns, have led to rapidly increasing usage of B.t. products.

During the past 30 years of use of *Bacillus thuringiensis* as a microbial insecticide, there have been few documented cases of B.t. resistance development. In these situations, (Indian meal moth in stored grain and diamondback moth in certain vegetables) the insects were under continuous or intensive B.t. use. A large number of companies and universities have undertaken research programs to manipulate B.t. protein endotoxin genes with the aim of developing vastly improved insect control products. Some microbial products under development are plant endophytes engineered to contain certain B.t. protein genes (Crop Genetics International) and B.t. protein genes engineered into *Pseudomonas* reportedly with longer residual activity than conventional B.t. products (Mycogen). Several companies have transferred B.t. genes into crop plants such as cotton, potato, and tomato (Monsanto, Plant Genetics Systems, Agracetus, Calgene, Lubrizol). B.t. products based on recombinant DNA technology have the potential to improve B.t. efficacy and persistence, which will lead to wider use of B.t. based biotechnology products. Resistance development is *the* critical issue in implementation of these new products.

This paper provides an overview of reported incidences of B.t. resistance, laboratory research, cross-resistance, genetic, and biochemical mechanisms and possible management strategies to maximize B.t. efficacy.

FIELD RESISTANCE DEVELOPMENT TO B.T.

The Indian meal moth, *Plodia interpunctella*, a pest in grain bins, was the first insect with documented resistance to B.t. Kinsinger and McGaughey [2] reported up to a 42-fold difference in susceptibility to Dipel® between laboratory and field strains. In a subsequent study [3], populations surveyed over a five state area showed a moderate reduction in susceptibility to Dipel®. Laboratory selection of a colony derived from a field population resulted in a 27-fold decrease in susceptibility after two generations and a 97-fold decrease after 15 generations. It was concluded that the decrease in susceptibility was true resistance development, as evidenced by the lack of significant difference in slopes of the dose mortality lines [4].

Until very recently, the development of resistance by the Indian meal moth was considered unique due to the prolonged, intimate association between the pest and the B.t. product. Tabashnik et al. [5] documented the first case of B.t. resistance development by a field crop pest, the diamondback moth, *Plutella xylostella*. On one particular farm in Hawaii, diamondback moths on watercress were treated with B.t. 50-100 times from 1978-1982 and 15 times again in 1989-90. Resistance ratios of insects from this and another heavily sprayed field ranged from 20-33. Subsequent laboratory selection of insects from one of these watercress farms resulted in 150-190 fold resistance after five generations and 430-820 fold resistance after nine generations of selection. Field selected resistance declined slowly in the absence of B.t. treatments [6].

Shelton [7] surveyed eleven populations of diamondback moths from six states and Indonesia. High levels of resistance to Dipel® and Javelin®, two commercial B.t. *kurstaki* products tested with leaf dip bioassays, revealed extremely high resistance ratios (240 and 470) in Florida and New York populations. Some of the resistance in New York insects could be traced to diamondback moths on cabbage transplants from Florida.

The variation in susceptibility of a forest pest, the gypsy moth, *Lymantria dispar* to B.t. (HD-1 strain) was studied to determine the potential for resistance development [8]. There were significant differences in the susceptibility of four field populations and also families within each population. The authors concluded that the variation in susceptibility to B.t. is based on vigor differences in growth and developmental capability. In addition, they concluded that resistance genotypes will be favored if these genotypes are also more vigorous.

LABORATORY SELECTION

Several researchers have attempted selection of resistance to B.t. in laboratory experiments with varying results. Unsuccessful attempts to develop resistance to B.t. in the laboratory have been summarized by Georghiou [9] and Briese [10]. Since those reviews, there have been a number of successful selection experiments. As discussed earlier, both Tabashnik et al. [6] and McGaughey [4] increased the resistance through laboratory selections in colonies derived from field populations. McGaughey and Beeman [11] reported that five Indian meal moth colonies derived from populations in midwestern grain bins developed resistance at different rates. The highest resistance was 250-fold at the F36, and the lowest 15-fold at the F39.

Stone et al. [12] reported the first incidence of laboratory selection to a field crop pest. *Heliothis virescens* was exposed to increasing doses of a recombinant *Pseudomonas fluorescens*, which expressed the CryIA(b) gene product [13]. Three fold resistance developed after three generations of selection at the LC$_{50-80}$. At the F14, resistance was approximately 24-fold. In subsequent studies, from the F14-18, the strain was subcultured on Dipel®, and from the F18-F22 was selected on the recombinant *Pseudomonas*. The strain had approximately 75-fold resistance to the *Pseudomonas*, 57-fold resistance to Dipel® (which contains CryIA(a), CryIA(b), CryIA(c), and CryIIA proteins), 71-fold resistance to the CryIA(b) purified protein, and 16-fold resistance to CryIA(c) [14].

Another field crop insect that has since been shown to develop resistance under laboratory selection is the Colorado potato beetle, *Leptinotarsa decemlineata*. The selected colony, which originated from field populations sprayed with M-One®, was selected in the laboratory for 13 generations on M-One® treated foliage. Bioassays showed 67-fold resistance at the F10 generation [15].

Several studies have been carried out to determine the potential for mosquitoes to develop resistance to B.t. *israelensis* (B.t.i.), resulting in small decreases in susceptibility [16-19]

It has been hypothesized that the reason for limited B.t.i. resistance development in mosquitoes is the presence of multiple protein toxin genes. Dai and Gill [20] selected *C. quinquefasciatus* for 20 generations on a preparation containing B.t.i. spores and endotoxin, and then subcultured the mosquitoes on purified 72 kDa protein endotoxin. The mosquitoes developed a 70-fold resistance to purified 72 kDa protein, but only a 3-fold resistance to the spore/endotoxin preparation.

CROSS RESISTANCE OF B.T. TO CHEMICALS AND OTHER B.T. STRAINS

B.t. has a unique mode of action and therefore it is not surprising that to date there have been no reported cases of cross resistance to insecticidal chemicals. However, there have been reports of cross resistance among B.t. strains or proteins. Knowledge on cross resistance among B.t. strains and proteins will be critical when devising resistance management strategies.

The cross resistance to Dipel® of 57 strains active against *P. interpunctella* was determined [21]. Twenty-one strains were equally effective on Dipel® resistant and Dipel® sensitive insects. The insects were also resistant to *E. coli*-cloned CryIA(a-c) proteins. The researchers concluded that resistance in *P. interpunctella* was limited to the HD-1 type strains with CryIA(a-c) proteins [21,22].

H. virescens, selected to a *Pseudomonas* containing CryIA(b) protein gene and Dipel® had varying levels of cross resistance (2.4-57 fold) to all of the approximately 20 different B.t. strains from subspecies *aizawai, colmeri, darmstadiensis, entomocidus, kurstaki, thuringiensis* and others [23].

GENETICS OF RESISTANCE

The tactics used to manage resistance will depend in part on the genetic basis of the resistance. Roush and McKenzie [24] concluded that laboratory strains of insects, initiated with a small sample of the total species genetic variability, usually develop polygenic resistance mechanisms. If studies are carried out on laboratory colonies, caution must be taken because they may or may not reflect field situations.

Resistance to *Pseudomonas* expressing a B.t. CryIA(b) protein in a laboratory strain of *H. virescens* was found to be autosomally inherited, incompletely dominant, and controlled by several genetic factors [25]. Resistance of 67-fold in one selected line declined to 13-fold by the fifth generation without selection. Follow-up research [26] studied the inheritance of *H. virescens* to the 63 kDa activated fragment of the delta-endotoxin protein.

Resistance (139-fold) was controlled by several genetic factors, was partially recessive, and control was predominantly autosomal, with some sex-linked factors involved. Resistance to Dipel® and a recombinant *Pseudomonas* expressing CryIA(b) protein was unstable in the absence of selection pressure. In contrast, resistance to HD-73, which contains a different protein, CryIA(c), was stable over seven generations without selection.

McGaughey [4] and McGaughey and Beeman [11] studied the genetics of B.t. (Dipel®) resistance in *P. interpunctella* with a field history of prior B.t. exposure. They concluded that resistance is recessive, but variable in recessiveness and suggested that it is controlled by a single major factor. After three generations without selection, resistance in one strain declined from 63-70 fold to 2-4 fold, suggesting that the strain still contained susceptible genotypes and had not reached a resistance plateau.

MECHANISM OF RESISTANCE

The mechanism by which an insect evolves resistance to a particular toxin is unavoidably tied to the toxin mode-of-action. The majority of published data examines the mechanism of resistance to chemical insecticides. Genetic, biochemical, physiological, and ecological elements are a few of the many interdependent factors influencing the onset of resistance [27-28].

Since development of resistance to B.t. has only recently been documented in the field, reports on the mechanism of resistance have focused entirely on laboratory selected insect colonies. One study has been carried on a field population of diamondback moth [29]. Significant variability was found in wild populations of Indian meal moth [3,4], tobacco budworm [12], and gypsy moth [8].

B.t. has a distinct yet complex mode-of-action: once the B.t. protoxin is ingested, it is first proteolyzed to an activated toxin protein [30-32] that diffuses through the peritrophic membrane and binds to high affinity receptors present on the midgut epithelium. The gut becomes paralyzed and the larvae stops feeding [33,34]. Once the activated toxin binds [35-39,24], the punitive receptor/toxin interaction becomes irreversible and the toxin is believed to insert into the membrane causing a lesion or pore to form [40-42]. The pore formation disrupts the inward K^+ gradient [43] leading to microvilli swelling and destruction [44-46]. If enough midgut cells are destroyed the larvae will die.

McGaughey's [3,4] studies of B.t. susceptibility in wild populations of Indian meal moth led to the laboratory selection of a colony collected from stored grain bins. The resultant resistant colony had a resistance ratio of >200 fold when compared to a susceptible laboratory strain [47]. Two possible resistance mechanisms were investigated, proteolytic processing of B.t. proteins and specific binding of B.t. toxin proteins. Post-binding events, such as membrane insertion and pore formation were not evaluated.

The proteolytic digestion of the full-length B.t. protoxins (130-140 kDa) to the trypsin stable proteins (55-70 kDa) is necessary for insecticidal activity. A mode of resistance could involve the decrease in pH or proteolytic properties of the insect midgut. Since the peritrophic membrane acts as a molecular sieve, the intact crystal proteins would be prevented from interacting with the midgut epithelium and be excreted. In the case of Indian meal moth, no differences in midgut pH or midgut proteinase patterns were found [48].

Additional studies focused on the specific binding of CryIA(b) and CryIC toxin proteins to isolated brush border membrane vesicles (BBMV) from the resistant and susceptible Indian meal moth strains [47]. Resistance ratio of sensitivity to the CryIA(b) toxin was >300-fold while the sensitivity to CryIC was actually enhanced in the resistant insects by 3.6-fold. The binding parameters correlated with the observed differences in toxicity. The binding affinity of CryIA(b) was significantly reduced, 50-fold, in the resistant Indian meal moth strain whereas the binding site concentration was virtually unchanged. Although the binding affinities of the CryIC toxin were identical between the resistant and susceptible insect strains, the binding site concentration was increased by a factor of three for resistant Indian meal moth. Therefore, the resistance development in this laboratory selected Indian meal moth was accompanied by two distinct molecular changes in the specific binding of two different toxin proteins, CryIA(b) and CryIC.

The tobacco budworm, the first field insect to be selected for resistance to B.t. [12], displayed a mechanism of resistance also related to specific binding of the toxin proteins. The proteolytic composition of the gut juice and proteolytic processing of the toxin proteins, CryIA(b) and CryIA(c), were unchanged in the resistant insect strain [14]. The resistant ratio of sensitivity of CryIA(c) and CryIA(b) was 16- and 70-fold, respectively, towards the resistant insects but only a 4- and 2-fold reduction in receptor affinity was observed [14]. Surprisingly, this loss of affinity was accompanied by an increase in binding site concentration for both toxins, 4-fold for CryIA(c) and 6-fold for CryIA(b). Competition experiments revealed differential receptor affinities in the resistant tobacco budworm. CryIA(b) was much less efficient in competing with CryIA(c) in the resistant than the susceptible strain.

In the first report of mechanism of resistance in field populations, Ferré et al. (1991) determined that 200-fold resistance to Dipel by diamondback moths from the Philippines could be explained by a complete lack of binding of CryIA(b) protein in the midgut. However, binding to CryIB and CryIC proteins, which are not present in Dipel, was similar in resistant and susceptible insects. Heterologous competition experiments indicate that diamondback moth has three distinct high-affinity binding sites (CryIA(b), CryIB, and CryIC).

The mechanism of resistance of all insects studied to date was influenced by alterations in specific toxin binding and not tied to the proteolytic processing of the protoxin proteins. Receptor binding of a variety of B.t. proteins have been analyzed from BBMV of a number of lepidopteran species [14,36-40]. In most cases, insect sensitivity correlated with the presence of high affinity binding sites. Two studies found an inverse correlation of binding with toxicity, CryIC in tobacco budworm [37] and CryIA(b) and CryIA(c) in gypsy moth [39].

The complexity of the resistance mechanism and, of course, the mode-of-action is evident from studies of the resistant insects. In Indian meal moth and tobacco budworm at least two different molecular changes occurred, affecting binding affinity and binding site concentration. Other aspects of the possible mechanisms of resistance remain to be explored, these include post-binding events such as membrane insertion and pore formation and behavior studies. These types of studies are difficult to interpret and in the case of pore formation, can only be evaluated indirectly by analyzing amino acid transport [43,49]. Since the properties and characterization of the punitive B.t. receptor is still elusive [50,51], the physiological function of the receptor also remains known. Therefore, it is not surprising to find that the resistant insects have developed multiple modifications to overcome the B.t. effects.

STRATEGIES FOR DELAYING RESISTANCE DEVELOPMENT

The successful use of new B.t.-based biotechnology products depends on development and implementation of sound management strategies. In the past, resistance management to chemical insecticides has occurred after resistance reached crisis proportions. With current B.t. products, resistance has developed in fields under continuous, intensive use of B.t. This provides a lesson for future management and use of new B.t.-based biotechnology products. At present, there is an opportunity to develop strategies in advance of marketing products such as first generation, single gene transgenic plants, which are thought by many experts to have a high risk for resistance development [52,53].

The B.t. Management Working Group, a consortium of biotechnology and agrochemical companies, was formed in 1988 to address the issue of B.t. resistance development [54]. The goals of this group are to determine the potential for resistance development in key pests, determine the mechanisms for resistance, develop resistance management strategies, and serve as a resource on B.t. resistance. Over a two year period, the group provided funds to five research groups. The purpose of the grants is to develop a base of information necessary to develop strategies for maximizing efficacy of B.t. products.

Experts agree that the most effective way to manage resistance is to prevent its occurrence and avoid pesticide applications [6,56]. One obvious approach is alternation across time of B.t. products with other pesticides of different modes of action. The basic principle behind alternation is that fitness disadvantages will cause the decline of resistance in the absence of the selecting pesticide [56]. Alternation across generations is thought to be considerably more effective than a programmed rotation on a weekly or five-day schedule. For B.t. products, a alternation strategy across generations would only be practical with microbial products, rather than transgenic plants expressing B.t. protein the entire crop season.

Another possible strategy, for both microbials and transgenic plants is mixtures. Currently, microbial B.t. products are primarily applied in mixtures with chemical insecticides. However, there are no studies that demonstrate that this will delay resistance. In fact, it is possible that it may actually enhance resistance development. Roush [58]) provides some advice on the choice between mixtures and alternation. In order for mixtures to be very effective for delaying resistance, initial frequencies of resistance should be low, the fraction escaping treatment should be high relative to dominance and linkage, the pesticides must be nearly 100% effective against treated susceptible heterozygotes, and equal in persistence. If all of these conditions are met, then mixtures are the choice over alternation. In reality, it is very difficult to meet these conditions and rotation across generations may be the wisest choice.

Another possible strategy is the use of mosaics. The number of B.t. expressing or B.t.-treated plants in a field would be diluted with non-expressing or untreated plants. Gould [56,57] indicates that a pest will evolve much more slowly to a mixture of plants of 80% with resistance factors and 20% with no resistance factors. He cites the example of the Hessian fly, where an 80%/20% mixture will take the flies five times longer to adapt than to a widely used cultivar with two resistance factors. He is skeptical that this strategy would be practical to those developing B.t.-expressing transgenic plants, because it is easier to develop genetically homogeneous cultivars than mixtures, and farmers will buy from the company offering the most toxic plants [52].

Roush [58] states that alternation is sometimes much better and certainly never worse than mosaics of more than one pesticide. His model assumes completely random mating for the mosaic. Where mating is not completely random, mosaics will be as durable as alternation. When resistance is completely recessive to both pesticides and there are no fitness disadvantages, alternation and mosaics are equal in delaying resistance.

An obvious strategy for transgenic plants is the incorporation of more than one gene into the crop plant. This strategy would only be effective if the two genes were of unique modes of action. Unfortunately, insecticidal genes available for incorporation into plants are currently limited to B.t. and protease inhibitors [58-60].

A number of companies developing transgenic plants view the incorporation of two or more B.t. genes into the crop plant as the most promising strategy for delaying resistance. The idea behind the strategy is that a particular B.t. protein is active at one midgut binding site, while a different protein is active at another binding site. The combination of the two in the plant should delay resistance when compared to a plant expressing only one protein.

Although a seemingly simple solution, the situation is more complicated than it appears. The model assumes that protein activity in the gut and development of resistance is primarily due to binding in the midgut. Wolfersberger [39] found that B.t. activity in gypsy moth was not always explained by binding. Likewise, MacIntosh et al. [14] concluded that while binding was extremely important in explaining B.t. resistance in *H. virescens*, post-binding events may be even more important. Resistance could develop to both proteins if post binding events or some aspect of the B.t "receptor" share a commonality in physiological function. In fact, we now know that field insects can develop resistance to products with multiple genes, such as Dipel®. Tabashnik et al. [6] state that the use of mixtures of toxin genes will not necessarily halt the development of resistance. Clearly more knowledge of the resistance mechanism and B.t. mode of action is required.

A long term approach for B.t. expressing transgenic plants is tissue and temporal specific expression of the B.t. protein. The protein will be expressed only when and where the insect feeds, thus limiting selection pressure. Gould [52] suggests that if a B.t. protein were expressed only in the cotton boll tissues, selection would be discontinuous because only one of three to four generations of bollworms feeds on the fruiting structures. Scientists have succeeded in transferring a proteinase inhibitor gene from potato into tobacco. The protein was only expressed in response to wounding [60].

278

CONCLUSIONS

Due to environmental concerns, the use of microbial products will continue to increase. Biotechnology can provide exciting new tools in the quest for new, environmentally sound pest control methods. As new B.t.-based biotechnology products are developed, the challenge to use these products in a sound manner that maximizes their field life becomes critical. Pest control specialists view resistance as the single most important issue in the development of single gene B.t.-based products, such as first generation, single gene transgenic plants. We now have the opportunity to demonstrate the lessons of past use of chemical insecticides by developing sound strategies for using B.t.-based biotechnology products in pest management systems in advance of their actual commercialization and marketing.

REFERENCES

1. Georghiou, G.P. and Lagunes, A., The occurrence of resistance to pesticides: cases of resistance reported worldwide through 1988. F.A.O., Rome, 1988, 325.

2. Kinsinger, R.A. and McGaughey, W.H., Susceptibility of populations of Indianmeal moth and almond moth to *Bacillus thuringiensis*, J. Econ. Entomol., 1979, **72**, 346-349.

3. McGaughey, W.H., Evaluation of *Bacillus thuringiensis* for controlling Indianmeal moths (Lepidoptera: Pyralidae) in farm grain bins and elevator silos, J. Econ. Entomol., 1985a, **78**, 1089-1094.

4. McGaughey, W.H., Insect resistance to the biological insecticide *Bacillus thuringiensis*, Science, 1985b, **229**, 193-195.

5. Tabashnik, B.E., Cushing, N.L., Finson, N., and Johnson, M., Field development of resistance to *Bacillus thuringiensis* in diamondback moth (Lepidoptera:Plutellidae), J.Econ. Entomol., 1990, **83**, 1671-1676.

6. Tabashnik, B.E., Finson, N., and Johnson, M., Managing resistance to *Bacillus thuringiensis*: Lessons from the Diamondback moth (Lepidoptera: Plutellidae), J. Econ. Entomol., 1991, **84**, 49-55.

7. Shelton, A.M. and Wyman, J.A., Insecticide resistance of diamondback moth (Lepidoptera: Plutellidae) in North America, presented at the International Diamondback Workshop, Tainan, Taiwan, December 10-14, 1990.

8. Rossiter, M., Yendol, W.G., and Dubois, N.R., Resistance to *Bacillus thuringiensis* in gypsy moth (Lepidoptera: Lymantriidae): Genetic and Environmental Causes, J. Econ. Entomol., 1990, **83**, 2211-2218.

9. Georghiou, G.P., Implications of potential resistance to biopesticides. In Biotechnology, Biological Pesticides, and Novel Plant Pest Resistance for Insect Pest Management, eds. D.W. Roberts and R.R. Granadas, Boyce Thompson Institute, Ithaca, NY, 1988, pp. 137-146.

10. Briese, D.T., Resistance of insect species to microbial pathogens. In Pathogenesis of Invertebrate Microbial Diseases, ed. E. Davidson, Allenheld, Osmun, Totowa, N. J., 1981, pp. 511-545.

11. McGaughey, W.H. and Beeman, R.W., Resistance to *Bacillus thuringiensis* in

colonies of the Indianmeal moth and almond moth (Lep.:Pyralidae), J. Econ. Entomol., 1988, **81**, 28-33.

12. Stone, T.B., Sims, S.R., and Marrone, P.G., Selection of tobacco budworm to a genetically engineered *Pseudomonas fluorescens* containing the delta-endotoxin of *Bacillus thuringiensis* subsp. *kurstaki,* J. Invert. Path., 1989, **53**, 228-234.

13. Watrud, L.S., Perlak, F.J., Tran, M.-T., Kusano, K., Mayer, E.J., Miller-Wideman, M.A., Obukowicz, M.G., Nelson, D.R., Kreitenger, J.P., and Kaufman, R.J., Cloning of the *Bacillus thuringiensis* subsp. *kurstaki* delta-endotoxin gene into *Pseudomonas fluorescens*: Molecular biology and ecology of an engineered microbial pesticide. In Engineered Organisms in the Environment, eds. H.O. Halverson, D. Pramer, and M. Rogul, Amer. Soc. Microbiol., Washington, 1985, pp. 40-46.

14. MacIntosh, S.C., Stone, T.B., Jokerst, R.S., and Fuchs, R.L., Binding of *Bacillus thuringiensis* proteins to a laboratory selected line of *Heliothis virescens*, Proc. Nat. Acad. Sci. (in press).

15. Miller, D.L., Rahardja, U., and Whalon, M.E., Development of a strain of Colorado potato beetle resistant to the delta-endotoxin of B.t. Pest Resistance Management, WRCC Newsletter, 1990, **2**, 25.

16. Georghiou, G.P., Insecticide resistance in mosquitoes: research on new chemicals and techniques for management, Ann. Rept., Mosquito Control Research, University of California, 1984, 97-99.

17. Georghiou, G.P. and Vazquez-Garcia, M., Assessing the potential for development of resistance to *Bacillus thuringiensis* subsp. *israelensis* toxin (*Bti*) by mosquitoes, Ann. Rept., Mosquito Control Research, University of California, 1982, 80-81.

18. Goldman, I.F., Arnold, J., and Carlton, B.C., Selection of resistance to *Bacillus thuringiensis* subsp. *israelensis*, in field and laboratory populations of *Aedes aegypti*. J. Invert. Path., 1986, **47**, 317-324.

19. Gharib, A.H. and Szalay-Marzso, L., Selection for Resistance to *Bacillus thuringiensis* serotype H-14 in a laboratory strain of *Aedes aegypti* L. In Fundamental and Applied Aspects of Invertebrate Pathology, eds. R.A. Samson, J.M. Vlak, and D. Peters, Found. Fourth Int. Cooloq. Invert. Pathol., Wageningen, Netherlands, 1986, p. 37.

20. Dai, S.-M., and Gill, S., Development of resistance to the 72kDa toxin of *Bacillus thuringiensis* in *Culex quinquefasciatus*. In Proceedings of International Symposium of Molecular Insect Science, October 22-27, 1989, Tucson, AZ.

21. McGaughey, W.H., and Johnson, D.E., Toxicity of different serotypes of *Bacillus thuringiensis* to resistant and susceptible Indianmeal moths (Lepidoptera: Pyralidae), J. Econ. Entomol., 1987, **80**, 1122-1126.

22. Han, E.-S., McGaughey, W.H., Johnson, D.E., Dunn, P., and Aronson, A.I., Characterization of *Bacillus thuringiensis* isolates effective on resistant Indian meal moth. In Genetics and Biotechnology of the Bacilli, Vol. 2, eds. A.T. Ganesan and J.A. Hoc, Academic Press, NY, 1988, pp. 233-238.

23. Stone, T.B., Sims, S.R., MacIntosh, S.C., Marrone, P.G., Armbruster, B.A., and Fuchs, R.L., Insect resistance to Bacillus thuringiensis delta-endotoxins, presented at International Symposium of Molecular Insect Science, October 22-27, 1989, Tucson, AZ.

24. Roush, R.T. and McKenzie, J.A., Ecological Genetics of insecticide and acaracide resistance, <u>Ann Rev. Entomol.</u>, 1987, **32**, 361-380.

25. Sims, S.R. and Stone, T.B., Genetic basis of tobacco budworm resistance to an engineered *Pseudomonas fluorescens* expressing the delta endotoxin of *Bacillus thuringiensis kurstaki*, <u>J. Invert. Path.</u>, 1991, **57**, 206-210.

26. Sims, S.R. and Stone, T.B., (unpublished),

27. Wilkinson, C. F., Role of mixed-function oxidases in insecticide resistance. In <u>Pest resistance to pesticides</u>, eds. G.P. Georghiou, and T. Saito, Plenum Press, New York, 1983, p. 175.

28. Knipple, D.C., Bloomquist, J.R., and Soderlund, D.M., Molecular genetic approach to the study of target-site resistance to pyrethroids and DDT in insects. In <u>Biotechnology for crop protection</u>, eds. P.A. Hedin, J.J. Menn, and R.M. Hollingsworth, Amer. Chem. Soc., Washington, D.C., 1988, pp. 199-214.

29. Ferré, J., Real, M.D., Van Rie, J., Jansens, S., and Peferoen, M. (1991) Resistance to *Bacillus thuringiensis* bioinsecticides in a field population of *Plutella xylostella* is free to change in midgut membrane receptor, <u>Proc. Natl. Acad. Sci.</u>, **88**, 5119-5123..

30. Fast, P.G., Bacteria: the crystal toxin of *Bacillus thuringiensis*. In <u>Microbial control of pests and plant diseases, 1970-1980</u>, ed. H.D. Burges, Academic Press, Inc., London, 1981, pp. 223-248.

31. Fast, P.G., *Bacillus thuringiensis* parasporal toxin: aspects of chemistry and mode-of-action, <u>Toxicon</u>, Suppl., 1983, **3**, 123-125.

32. Huber, H.E. and Lüthy, P., *Bacillus thuringiensis* delta-endotoxin composition and activation. In <u>Pathogenesis of Invertebrate Microbial Diseases</u>, ed. E.W. Davidson, Allenheld, Osmun., Totowa, NJ., 1981, pp. 209-234.

33. Dulmage, H.T., Graham, H.M., and Martinez, E., Interactions between the Tobacco budworm, *Heliothis virescens*, and the delta-endotoxin produced by the HD-1 isolate of *Bacillus thuringiensis* var. *kurstaki*: relationship between length of exposure to the toxin and survival, <u>J. Invert. Path.</u>, 1978, **32**, 40-50.

34. Salma, H.S. and Sharaby, A., Histopathological changes in *Heliothis armigera* infected with *Bacillus thuringiensis* as detected by electron microscopy, <u>Insect Sci. Applic.</u>, 1985, **6**, 503-511.

35. Hofmann, C., Lüthy, P., Hütter, R., and Pliska, V., Binding of the delta-endotoxin from *Bacillus thuringiensis* to brush-border membrane vesicles of the cabbage butterfly (*Pieris brassicae*), <u>Eur. J. Biochem.</u>, 1988a, **173**, 85-91.

36. Hofmann, C., Vanderbruggen, H., Höfte, H., Van Rie, J., Jansens, S., and Van Mellaert, H., Specificity of *Bacillus thuringiensis* d-endotoxins is correlated with the presence of high-affinity binding sites in the brush border membrane of target insect midguts, <u>Proc. Natl. Acad. Sci.</u> USA, 1988b, **85**, 7844-7848.

37. Van Rie, J., Jansens, S., Höfte, H., Degheele, D., and Van Mellaert, H., Specificity of *Bacillus thuringiensis* delta-endotoxins: Importance of specific receptors on the brush border membrane of the midgut of target insects, <u>Eur. J. Biochem</u>, 1989, 186, 239-247.

38. Van Rie, J., Jansens, S., Höfte, H., Degheele, D., and Van Mellaert, H., Receptors on the brush border membrane of the insect midgut as determinants of the specificity of *Bacillus thuringiensis* delta-endotoxins, Appl. Environ. Microbiol., 1990a, **56**, 1378-1385.

39. Wolfersberger, M.G., The toxicity of two *Bacillus thuringiensis* delta-endotoxins to gypsy moth larvae is inversely related to the affinity of binding sites on midgut brush border membranes for the toxins, Experientia, 1990, **46**, 475-477.

40. Knowles, B.H. and Ellar, D.J., Characterization and partial purification of a plasma membrane receptor for *Bacillus thuringiensis* var. *kurstaki* lepidopteran-specific delta-endotoxin, J. Cell Sci., 1986, **83**, 89-101.

41. Ellar, D.J., Knowles, B.H., Haider, M.Z., and Drobniewski, F.A., Investigation of the specificity, cytotoxic mechanisms and relatedness of *Bacillus thuringiensis* insecticidal d-endotoxins from different pathotypes. In Bacterial protein toxins, eds. P. Falmagne, F.J. Fehrenbach, J. Jeljaszewics, and M. Thelestam, Gustav Fischer, New York, 1986, pp. 41.

42. Hendrick, K., De Loof, A., and Van Mellaert, H., Effects of *Bacillus thuringiensis* delta-endotoxin on the permeability of brush border membrane vesicles from tobacco budworm (*Manduca sexta*) midgut, Comp. Biochem. Physiol., 1990, **95C**, 241-245.

43. Sacchi, V.F., Parenti, P., Hanozet, G.M., Giordana, B., Lüthy, P., and Wolfersberger, M.G., *Bacillus thuringiensis* toxin inhibits K+-gradient-dependent amino acid transport across the brush border membrane of *Pieris brassicae* midgut cells, FEBS Let., 1986, **204**, 213-218.

44. Lüthy, P., Jaquet, F., Huber-Lukac, H. E., and Huber-Lukac, M., Physiology of the delta-endotoxin of *Bacillus thuringiensis* including the ultrastructure and histopathological studies. In Basic Biology of Microbial Larvicides of Vectors of Human Diseases, ed. F. Michal, UNDP/World Bank/WHO, Geneva, Switzerland, 1982, pp. 29-36.

45. Percy, J. and Fast, P.G., *Bacillus thuringiensis* crystal toxin: ultrastructural studies of its effect on silkworm midgut cells, J. Invert. Path., 1983, **41**, 86-98.

46. deLello, E., Hanton, W.K., Bishoff, S.T., and Misch, D.W., Histopathological effects of *Bacillus thuringiensis* on the midgut of Tobacco hornworm larvae (*Manduca sexta*): low doses compared with fasting, J. Invert. Path., 1984, **43**, 169-181.

47. Van Rie, J., McGaughey, W.H., Johnson, D.E., Barnett, B.D., and Van Mellaert, H., Mechanism of insect resistance to the microbial insecticide *Bacillus thuringiensis*, Science, 1990b, **247**, 72-74.

48. Johnson, D.E., Brookhart, G.L., Kramer, K.J., Barnett, B.D., and McGaughey, W.H., Resistance to *Bacillus thuringiensis* by the Indian mealmoth, *Plodia interpunctella*: comparison of midgut proteinases from susceptible and resistant larvae, J. Invert. Pathol., 1990, **55**, 235-244.

49. Wolfersberger, M., Lüthy, P., Maurer, A., Parenti, P., Sacchi, F. V., Giordana, B., and Hanozet, G. M., Preparation and partial characterization of amino acid transporting brush border membrane vesicles from the larval midgut of the cabbage butterfly (*Pieris brassicae*), Comp. Biochem. Physiol., 1987, **86A**, 301-308.

50. Ellar, D.J., Investigation of the molecular basis of *Bacillus thuringiensis* delta-endotoxin specificity and toxicity, presented at the Soc. Invert. Path., College Park,

MD., August 20 to 24, 1989.

51. Muthukumar, G. and Nickerson, K.W., The glycoprotein toxin of *Bacillus thuringiensis* subsp. *israelensis* indicates a lectinlike receptor in the larval mosquito gut, Appl. Environ. Microbiol., 1987, **53**, 2650-2655.

52. Gould, F., Evolutionary Biology and Genetically Engineered Crops, BioScience, 1988, **38**, 26-32.

53. Raffa, K.F., Genetic engineering of trees to enhance resistance to insects, BioScience, 1989, **39**, 524-533.

54. Marrone, P.G., B.t. Working Group. Pesticide Resistance Management, WRCC Newsletter, 1989, **1**, p. 13.

55. Roush, R., Designing resistance management programs: how can you choose?, Pestic. Sci., 1989, **26**, 423-441.

56. Gould, F., Simulation models for predicting durability of insect resistant germplasm: a deterministic diploid, two-locus model, Environ. Entomol., 1986a, 15, 1-10.

57. Gould, F., Simulation models for predicting durability of insect resistant germplasm: Hessian fly (Diptera: Cecidomyiidae)-resistant winter wheat, Environ. Entomol., 1986b, **15**, 11-23.

58. Hilder, V.A., Gatehouse, A.M.R., Sheerman, S.E., Barker, R.F., and Boulter, D., A novel mechanism of insect resistance engineered into tobacco, Nature, 1987, **330**, 160-163.

59. MacIntosh, S.C., Kishore, G.M., Perlak, F.J., Marrone, P.G., Stone, T.B., Sims, S.R., and Fuchs, R.L., Potentiation of *Bacillus thuringiensis* insecticidal activity by serine protease inhibitors, J. Agric. Food Chem., 1990, **38**, 1145-1151.

60. Sanchez-Serrano, J.J., Keil, M., O'Connor, A., Schell, J., and Willmitzer, L., Wound induced expression of a potato proteinase inhibitor II gene in transgenic tobacco plants, EMBO J., 1987, **6**, 303-306.

THE NEEDS FOR NEW HERBICIDE-RESISTANT CROPS

JONATHAN GRESSEL

Plant Genetics, Weizmann Institute of Science, Rehovot, Israel

ABSTRACT

The technologies of isolating genes conferring resistance and engineering them into crops have been worked out, and a number of field trials are under way. There seem to be few new initiatives. There are three major threats to the world food supply that are not being addressed where engineering will help: Parasitic weeds, for which there are no selective herbicides, halve yields in the third world. Engineering certain target-site resistances into the crops should allow selective control of these parasites; (2) Weeds have evolved cross resistances to all previously usable wheat selective herbicides, which can be alleviated by engineering new resistances into wheat; (3) Major maize herbicides in the triazine and chloroacetamide groups have been banned in various countries. Triazine-resistant weeds are also problems on 3 million hectares. Thus, new selectivities may have to be engineered into maize.

INTRODUCTION

The biotechnological tools of plant tissue culture selection of mutants, protoplast fusion, gene excision, cloning and insertion into plants, as well as regeneration, have allowed the generation of a number of crops with new genes for herbicide resistance. Much had to be learnt about basic plant physiology to ensure that genes were expressed when, where and how needed to give a sufficient margin of crop resistance with a minimum loss in crop yield. Most of this research has been done by the private sector or under contract to chemical and biotechnology companies, and lately in conjunction with the seed companies that will market the final product. This required a choice of target herbicides, crops and markets that could provide the direct and immediate market return in herbicide and seed sales. The private sector is now having second thoughts about conferring resistance; the herbicides become off-patent by the time resistant-crops can be marketed, and only small market niches are viewed as being available.

There has been little effort to meet potential needs that are likely, but not guaranteed, to arise in a few years, nor to meet needs of the third world agriculture. Some of these needs are with crops such as wheat, where seed sales are limited because of farmer saving, and with crops where the herbicides must be inexpensive in unsubsidized agriculture. Fulfilling third world agricultural needs has little economic incentive for the private sector. The guantlet to meet these needs has not been raised by the public sector, even though it should be easy to do so now with the available gene constructs. For these reasons, progress in herbicide-resistant crops will be first summarized, and then the needs this author feels are most likely to be acute will be described.

PROGRESS

Target-Site Resistances
Many of the selected resistances involve the specific target enzymes affected by herbicides. Such resistances must be viewed with care as losses in target enzyme efficiency can reduce crop yields; the change in enzyme conformation that precludes herbicide binding may reduce binding of the natural substrates. Additionally, if it is easy to get target-site resistant mutations in crops or model plants such as *Arabidopsis*, then it may be easy to inadvertantly select for resistance in weed populations.

Photosystem II Herbicides: Target site resistance to triazines and related herbicides has evolved throughout the world [1] and has also been selected for in tissue culture of tobacco [2] and potato [3]. Resistance inherited on the *psb*A gene of the chloroplast genome has been transferred to potato by protoplast fusion, with the nuclear genome functionally excluded, from a closely related weed [4]. A complicated series of crosses with a weed have allowed this resistance to be transferred to rape [5], and transfer to further rape varieties has been either by cross-breeding [6] or protoplast fusion [7]. Protoplast fusion precludes the need for eight generations of backcrosses to reach a variety.

There is a considerable yield loss, (20-30%), with all triazine resistant material, which could not be overcome by further crossing [6]. This still allows economic use of the newly resistant varieties when there are highly competitive weeds that cannot be controlled in rape by other herbicides [8]. Of the many mutation sites found in algae and higher plants [9], only one ($ser_{264} -> gly_{264}$) has appeared in weeds and has been used in crops. Recently, thr_{264} [10] and asn_{264}[11] mutants have been found in tobacco, which may be less deleterious. Care must be exercised with triazine-resistant crops as there are 3 million hectares of newly evolved triazine-resistant weeds [1], and atrazine has been banned in some European countries after its appearance in water.

EPSP-Synthase: Glyphosate resistance has been transferred to a number of crops, utilizing a modified enolphosphate-shikimate phosphate (EPSP) synthase gene. Highly sophisticated modifications of the bacterial genes were needed to transfer of the enzyme from the cytoplasm to chloroplasts, and to cause high expression rates in growing tissue, where the herbicide becomes localized [12]. Glyphosate is an excellent, general, broad-spectrum post emergence herbicide with negligible residual effects in the soil. Farmers prefer high residual pre-emergence herbicides giving season long weed control, but this leads to resistance and leaching into ground water.

Glyphosate resistance has been engineered into tobacco, tomato, and rape [12], but there have yet to be reports of yield trials to ascertain if the modified enzyme lowers yields. There may also be problems of residues of glyphosate in crops, as the herbicide is not degraded. Efforts are indeed being made to find, clone and engineer genes into plants that will degrade this herbicide.

Acetolactate synthase (ALS) : It has been very easy to select for resistance to the sulfonylurea, imidazolinone and triazolopyrimidine herbicides inhibiting this key enzyme in the synthesis of branch-chained amino acids. Target site mutants have been found in bacteria, yeast [13], plant cell cultures [14,15], crops in field [16], as well as in *Arabidopsis* [17]. The problem is that resistance to these herbicides has readily evolved in weeds in the field [18], but only to the members of these classes that have long residual activities in soil. There are many sites on the ALS gene that can mutate and confer resistance [19]. The mutations all result in a large degree of cross-resistance to the ALS inhibitors despite their different chemistry. Chemists have developed ALS inhibiting herbicides that are degraded by most major crops. Their wide use may lead to more cases of resistance, especially as the long residual ones have not been withdrawn. Resistant tobacco, sugar-beet, cotton and other crops have been obtained by genetic engineering [e.g. 20,21], maize and cotton by selection/breeding [14, 22], and recently in maize inbreds by direct selection for heterozygotic offspring after mutagenizing pollen [22a]. This latter technique saves considerable breeding time. Field tests on resistant sugar beets and cotton and selected soybeans and maize are under way. Losses of yield due to using this gene have not been reported.

Acetyl CoA Carboxylase: The aryloxyphenoxy propionate (-fop) and cyclohexanedione (-dim) herbicides affect this enzyme in most graminae but not in other plant species, giving them great utility as grass killers in dicots [23]. Of these only diclofop-methyl is specifically degraded in wheat, allowing its use to control other graminae [24]. Target-site resistance was selected for in embryogenic maize tissue cultures [25] and the lines given to a seed company for transfer into commercial varieties. Care must be taken with use of these herbicides as two major weeds have already evolved target site resistance [26].

Cellulose biosynthesis: Mutant *Arabidopsis* strains have been isolated that have target site resistance to isoxaben [27] and dichlobenil [28], herbicides inhibiting cellulose biosynthesis. There is no cross resistance between the two herbicides on the different mutants suggesting different enzymes in the pathway, or distant sites on the same enzyme. Presumably efforts are underway to isolate the genes for further engineering.

Phytoene desaturase: A mutant gene was isolated and cloned from cyanobacteria that upon transformation into other cyanobacteria confers resistance to the pyridazinone and other herbicides inhibiting this target [29]. Despite widespread use of norflurazon in cotton, resistant weeds have not evolved. The cyanobacterial gene conferred norflurazon resistance on higher plants. Most of the phytoene disaturase inhibitors are non-selective, and engineered resistance can greatly change the market.

Dihydropterate synthase: A bacterial mutant gene conferring resistance to sulfonamide antibiotics was modified and engineered into tobacco, conferring cross resistance to asulam [30]. If plants are resistant to field rates, this could bring new uses to a not widely used herbicide.

Herbicide Degradations

Enzymes that degrade herbicides are less likely to depress yields, if the requirements for energy and co-substrates are easily met. It is hard to find and isolate enzymes causing degradation of herbicides, and there is always the possibility that such an enzyme will degrade important metabolites in the crops. Still, there has been some excellent progress in such systems.

Acetyl transferase: An acetyl CoA utilizing acetyltransferase that acetylates glufosinate and bialaphos has been isolated from various streptomyces and transferred into plants, with great success [31]. Field trials with high rates of these glutamine synthase inhibitors were without loss of transgenic-crop yields. Despite various efforts, it has not been possible to find target site resistance to these inhibitors [32], suggesting that they may be relatively resistance proof, especially as they are low residue, post emergence herbicides. The transferase gene has been introduced into alfalfa (lucerne) [33] and sugar beets [34].

Hydrolases: Hydrolases are excellent enzymes for genetic engineering of resistance as they have no energy requirement, and should have little load on the transgenic crop. A bacterial hydrolase has been isolated that cleaves phenmedipham [35], a PSII inhibiting herbicide used only in beets. The enzyme has no cofactor requirements. The cloned gene has been transformed into plants. Thus herbicide group is used on only limited areas and engineered crops could allow displacement of other, too widely used herbicides.

A highly efficient nitrilase has been isolated from soil bacteria. It cleaves the -CN group from bromoxynil and ioxynil at sufficient rates that only limited expression is required in the transgenic tobacco, tomato and cotton that have been regenerated [36]. This tobacco has been field-tested in France [37], and the cotton at 30 sites in 11 U.S. states [38]. Bromoxynil is probably not resistance-prone as it both inhibits photosystem II on the quinone binding site binding and it also uncouples photophosphorylation from electron transport, although not as efficiently as its analog ioxynil [39]. The likelihood of two simultaneous target site mutations in the same weed is negligible [40].

Monooxygenases: Bacterial monoxygenases that degrade herbicides have been widely studied [41]. They are typically soluble NADPH dependent cytochrome P-450 monooxygenases. The same group of enzymes in plants is usually membrane associated. A highly specific bacterial monooxygenase was isolated and engineered into tobacco conferring 2,4-D resistance [42, 43]. The monooxygenase was so specific that it did not oxidize the closely related phenoxy herbicide MCPA [42]. Constructs have been used to engineer *Arabidopsis* potato and cotton resistant to 2, 4-D [35]. The gene is now available for general use [35].

N-glucosyltransferases: O-glucosyltransferases often conjugate glucose from UDPG to herbicides or herbicide metabolites, but the glucose can be cleaved by ubiquitous β-glucosidases. Conversely, N-glucosyl-transferases irreversibly conjugate glucose to various herbicides. They would be a good target enzyme because cells do not lack the UDPG needed for conjugation (unlike glutathione with glutathione transferases). A metribuzin degrading N-glucosyl transferase activity conferring resistance to some soybean and tomato varieties has been well characterized [cf. 44]. The gene would have been a good candidate if triazine-resistant weeds were not cross-tolerant to metribuzin.

Pyridate is a proherbicide, which at least at the chloroplast thylakoid level inhibits photosystem II better in triazine-resistant weeds than related sensitive biotypes [45]. Efforts made to engineer the pyridate N-glucosyltransferase activity degrading the herbicide have stopped [46].

Others: A microbial dehalogenase gene has been engineered into tobacco conferring resistance to dalapon [47], but there seems to have been no follow up, despite the apparent utility of such a gene, as dalapon might well control grasses that have evolved herbicide resistances. Dalapon was withdrawn from U.S. use.

NEEDS

It is apparent that many gene constructs and technologies are available that confer resistance. Most have been engineered into the non-edible weed that is considered exceedingly dangerous to human health; tobacco. The time has come to allocate resources into learning how to insert these constructs into crops for situations where they are needed to enlarge food supplies on this limited planet. Three illustrative cases are described below:

Parasitic Weeds

Many crops are attacked by higher plant weeds that attach themselves to the growing plant; usually cease photosynthesis in cases where there is some initially, and then suck photosynthate, minerals and water from the plants [48]. The most common on field crops are: dodders (*Cuscuta* spp); broom rapes (*Orobanche* spp) [49]; and witchweeds (*Striga* spp) [50]. Mistletoes and other species attack forest trees. The most devastating of these species are the broom-rapes [49] and the witchweeds [50]. The areas infested are vast and ever growing, especially in third world countries. For example, a major survey of 180,000 km^2 in Nigeria showed 70% of fields to be infested with *Striga* seeds [51]. The areas are rapidly spreading due to agricultural practices: hand and mechanical weeding of other species leaves attached parasites as well as perennial weeds such as nutsedges, which are also an ever growing problem; lack of rotation or fallowing, which is supportive of the parasites; and the lack of highly selective chemical control mechanisms. Third world agriculture has learnt to use dangerous insecticides, but more than 60% of third world farmers' time is spent in weeding, usually when it is too late, and weeding leaves the parasites. The witchweeds affect the grains of 300 million people in sub-Sahara Africa [51], as well as in Asia. Yields are reduced 50%, more in drought years. The broomrapes devastate the legume and vegetable crops in much of north Africa and elsewhere, severely limiting the locally produced protein content of the diet of the inhabitants [49]. Clearly the broomrapes and witchweeds have a greater deleterious effect than AIDS on the health and welfare of these third world areas. Crop yields could be doubled without added fertilizer or irrigation, if these weeds could be effectively controlled.

There are few herbicides that can control these parasites with a sufficient margin of selectivity that will allow their use in third world situations. There *are* herbicides that can control the parasites, but they usually control the crops or are too costly [49, 50, 52]. Thus, there is a need to genetically engineer resistance into *many* third world crops to preserve genetic diversity and prevent too much monoculture, while controlling the weeds. It is probably necessary that target site resistances should be the type engineered for parasite control; it would not be helpful to have the crop degrade the herbicide before it reaches the parasite. It will not help to engineer resistance to herbicides solely affecting photosynthesis, as parasitic weeds are not dependent on performing this function. There is evidence that these organisms synthesize their own amino acids from mineral nitrogen and plant-derived carbohydrate [53, 54]. Herbicides inhibiting amino acid biosyntheses should and do kill these parasites [e.g. 49, 52]. These include glyphosate [49, 52, 55] (inhibiting aromatic amino acid syntheses), sulfonylureas, imidazolinones and triazolopyrimidines (inhibiting branch-chain amino acid synthesis) [49, 52]. It is not clear whether glufosinate inhibits the glutamine synthase of these parasites, as different isozymes predominate in the parasites than in crop leaves [56]. Glyphosate may be especially useful as it inhibits the biosynthesis of many shikimate derived phenolics. It has been suggested that the glyphosate-resistant crops being engineered should be tested with broomrapes, and that more crops should be engineered [49]. Glyphosate should be cost-effective candidate, as very low rates, (a tenth to a twentieth the rates used to control annual weeds) are sufficient to kill parasites. Thus the cost of herbicide would be negligible. Glyphosate is translocated from the crop into the parasite [56], a valuable trait as the surface areas of the parasites are small compared to the crops. Glyphosate is used to selectively control some parasites in crops [49, 55], but the margin of safety is very low, and crop damage can all too easily occur. If the crops would be engineered for high levels glyphosate resistance, it may also be able to selectively control the perennial nutsedges (*Cyperus*), species that cannot be manually controlled in the third world.

Maize with resistance to ALS inhibitors [15] should be tested for selective control of witchweed. The single dominant gene conferring target site resistance could easily be crossed into third world varieties, and these modified varieties could be released in just a few years. Such breeding can be directly performed in the target countries. These parasites are also highly sensitive to the soil applied carbamate and dinitroaniline herbicides [49, 58] that inhibit tubulin binding. Target site resistance has evolved to these herbicides in various grass weeds, including *Setaria viridis* [59], which crosses with the *S. italia* a crop used in Africa [60]. This resistance could be transferred genetically to this crop and could be selected for at the whole plant level with other grain crops, in local agricultural stations.

This scientist finds it reprehensible and irresponsible that groups purporting to be interested in helping humanity can, after reporting vis a vis biotechnologically derived herbicide-resistant crops, recommend that "Third World countries should be fully informed of potential negative impacts of herbicide-tolerant crops and the FAO of the UN be urged to develop restrictions on the export of herbicide-tolerant plants" [61]. Their report did not even allude to the effect of parasitic and other weeds on crop yields of the third world. It is to be hoped that they will reassess and support introducing such simple methods of overcoming the scourge of parasitic while doubling yields.

edict or resistance? At what cost will other available herbicides provide the necessary control? How quickly will maize-weeds become resistant to ALS inhibitors? Should engineering be considered? A recent review of genetic engineering of herbicide resistance in crops held in the U.S. Corn-Belt did not address these questions [68], even though it was brought up in the discussions. Edicts come suddenly, as the farmer is a small pawn in a political scene laden with emotions about pesticides. New answers for maize must be available.

CONCLUDING REMARKS

Other reasons for resistant-crops.
A broader spectrum of resistances in crops can increase the use of a broader spectrum of herbicides. This will decrease the potential problems from any given herbicide. Herbicides with less environmental impact could then be used on more crops. This requires much forethought in choice of candidate herbicides.

Hybrid vigor is a property that increases yield. Hybrids are expensive to produce and in crops such as wheat there is a problem of pollination. The pollen-parent must be planted as close as possible to the females and must be harvested separately. If the female had a dominant resistance to a herbicide that would still kill post anthesis, susceptible males could be interplanted with females in one sowing and then be culled by herbicide after pollination [65]. It would be much cheaper to produce hybrids this way and the hybrids would have two positive economic traits; higher yield and herbicide resistance.

Who should be engineering the crops?
Engineering has largely been the realm of the private sector with avoidance in the public sector, even in maize [68] and wheat growing areas. The parasitic weed and wheat cases cited above may not be touched by the private sector, as there is no perceived profit incentive from the third world or from wheat seed. The problems in maize do not seem sufficiently immediate to provide the profit motive. Growers cannot do the genetic engineering themselves. The public sector must get involved. The research and development costs may not be *that* expensive if the private sector makes the currently available gene constructs accessible to the public sector. The public sector should also be looking for genes for resistance to inexpensive ecologically-sound herbicides that have been avoided by the private sector. The genes are there, and they should be put into crops the farmer needs, if the farmer is to supply us with sustenance.

ACKNOWLEDGEMENTS

This article is dedicated to the memory of Roman Sawicki, a scientist and humanist who worried about pest control problems in the third as well as first world. The many researchers who supplied unpublished material are gratefully thanked. The author holds the Gilbert de Botton Chair of Plant Sciences.

REFERENCES

1. Holt, J.S. and LeBaron, H.M., Significance and distribution of herbicide resistance. *Weed Technol.*, 1990, 4, 141-149.

2. Rey, P., Eymery, F. and Peltier, G., Atrazine and diuron resistant plants from photoautotrophic protoplast-derived cultures of *Nicotiana plumbaginifolia*. *Plant Cell Rep.*, 1990, **9**, 241-244.

3. Smeda, R.J., Hasegawa, P.M. and Weller, S.C., Mechanism(s) of tolerance to atrazine in photo-autotrophic potato cells. *Weed Sci Soc. Amer. Abstr.*, 1989, **29**, 164.

4. Gressel, J., Method of Producing Herbicide Resistant Plant Varieties and Plants Produced Thereby. *United States Patent*, 1990, 4,900,676.

5. Souza-Machado, V., Inheritance and breeding potential of triazine tolerance and resistance in plants. In *Herbicide Resistance in Plants*, eds. H.M. LeBaron and J. Gressel, New York, Wiley, 1982, pp. 257-274.

6. Beversdorf, W.D., Hume, D.J. and Donnelly-Vanderloo, J.J., Agronomic performance of triazine-resistant and susceptible reciprocal spring Canola hybrids. *Crop Sci.*, 1988, **28**, 932-934.

7. Thomzik, J.E. and Hain, R., Introduction of metribuzin resistance into German winter oilseed rape of double-low quality. *Pflanzenschutz-Nachrichten Bayer*, 1990, **43**, 61-87.

8. Forcella, F., Herbicide-resistant crops: yield penalties and weed thresholds for oilseed rape (*Brassica napus* L.). *Weed Res.*, 1987, **27**, 31-34.

9. J.J.S. Van Rensen, this volume.

10. Sigematsu, Y., Sato, F. and Yamada, Y., The mechanism of herbicide resistance in tobacco cells with a new mutation in the Q_B protein. *Plant Physiol.*, **89**, 986-992.

11. Pay, A., Smith, M.A., Nagy, F. and Marton, L., Sequence of the *psbA* gene from wild type and triazine-resistant *Nicotiana plumbaginifolia*. *Nucleic Acids Res.*, 1988, **16**, 8176.

12. Kishore, G.M., EPSP-synthase - from biochemistry to engineering of glyphosate tolerance. In *Biotechnology in Crop Protection*, eds. J.J. Menn, R.M. Hollingsworth and P.A. Hedin, American Chemical Society Symposia No. 379, Washington, DC., pp. 37-48.

13. Falco, S.C., Dumas, K.S. and McDevitt, R.E., Molecular genetic analysis of sulfonylurea herbicide action resistance in yeast. In *Molecular Form and Function of the Plant Genome*, ed. L.Van Vloten-Doting, Plenum, New York, 1985, pp. 467-478.

14. Mazur, B.J. and Falco, S.C., The development of herbicide resistant crops. *Annu. Rev. Plant Physiol. Plant Mol. Biol.*, 1989, **40**, 441-470.

15. Anderson, P.C. and Georgeson, M., Herbicide-tolerant mutants of corn. *Genome*, 1989, **31**, 994-99.

16. Stannard, M.E. and Fay, P.K., Selection of alfalfa seedlings for tolerance to chlorsulfuron. *Weed Sci. Soc. of Amer. Abst.*, 1987, **27**, 61.

17. Haughn, G.W. and Somerville, C.R., A mutation causing imidazolinone resistance maps to the *Csr1* locus of *Arabidopsis thaliana*. *Plant Physiol.*, **92**, 1081-1085.

18. Saari, L.L., Cotterman, J.C. and Primiani, M.M., Mechanism of sulfonylurea herbicide resistance in the broadleaf weed, *Kochia scoparia*. *Plant Physiol.*, 1990, **93**, 55-61.

19. Schloss, J.V., Ciskanik, L.M. and Van Dyk, D.E., Origin of the herbicide binding site of acetolactate synthase. *Nature*, 1988, **331**, 360-362.

20. Gabard, J.M., Charest, P.J., Iyer, V. N. and Miki, B.L., Cross-resistance to short residual sulfonylurea herbicides in transgenic tobacco plants. *Plant Physiol.*, 1989, **91**, 574-580.

21. Miki, B.l., Labbe, H., Hattori, J., Ouellet, T., Gabard, J., Sunohara, G., Charest, P.J. and Iyer, V.N., Transformation of *Brassica napus* canola cultivars with *Arabidopsis thaliana* acetohydroxyacid synthase genes and analysis of herbicide resistance. *Theor. Appl. Genet.*, 1990, **80**, 449-458.

22. Subramanian, M.V., Hung, H-Y., Dias, J.M., Miner, V.W., Butler, J.H. and Jachetta, J.J., Properties of mutant acetolactate synthases resistant to triazolopyrimidine sulfonanilide. *Plant Physiol.*, 1990, **94**, 239-244.

22a. Dunwell, J.M., personal communication, 1991.

23. Secor, J. and Cseke, C., Inhibition of acetyl-CoA carboxylase activity by haloxyfop and tralkoxydim. *Plant Physiol.*, 1988, **86**, 10-12.

24. Jacobson, A., Shimabukuro, R.J. and McMichael, C., Response of wheat and oat seedlings to root-applied diclofop-methyl and 2,4-dichlorophenoxyacetic acid. *Pestic. Biochem. Physiol.*, 1985, **24**, 61-67.
25. Gronwald, J.W., Parker, W.B., Somers, D.A., Wyse, D.L., Gengenbach, B.G., Selection for tolerance to graminicide herbicides in maize tissue culture. *Brighton Crop Protection Conference -Weeds*, 1989, pp. 1217-1224.
26. Gronwald, J.W., Eberlein, C.B., Betts, K.J., Rosow, K. M., Ehlke, N.J. and Wyse, D.L., Diclofop resistance in a biotype of Italian rye-grass. *Plant Physiol.*, 1989, **89S**, 115,
27. Heim, D.R., Roberts, J.L., Pike, P.D. and I.M. Larrinua, A second locus *Ixr* B1 in *Arabidopsis thaliana* that confers resistance to the herbicide isoxaben. *Plant Physiol.*, 1990, **92**, 858-861.
28. Heim, D.R., Bjelk, L.A.ˆand I.M. Larrinua, Isolation and characterization of a dichlobenil resistant mutant of *Arabidopsis thaliana*. *Weed Sci. Soc. Amer. Abst.*, 1991, **31**, 76.
29. Chamovitz, D., Pecker, I., Sandmann, G., Böger. P. and Hirschberg, J. Cloning a gene coding for norflurazon resistance in Cyanobacteria. *Z. Naturforsch.*, 1990, **45c**, 482-486.
30. Guerineau, F., Brooks, L., Meadows, J., Lucy, A., Robinson, C., and Mullineaux, P., Sulfonamide resistance gene for plant transformation. *Plant Mol. Biol.* ,1990, **15**, 127-136.
31. De Greef, W., Delon, R., De Block, M., Leemans, J. and Botterman, J., Evaluation of herbicide resistance in transgenic crops under field conditions. *Bio/Technology*, 1989, **7**, 61-64.
32. Donn, G., Tischer, E., Smith, J.A. and Goodman, H. M., Herbicide resistant alfalfa cells; an example of gene amplification in plants. *J. Mol. Appl. Gen.*, 1984, **2**, 621-635.
33. D'Hulluin, K., Botterman, J. and De Greef, W., Engineering of herbicide-resistant alfalfa and evaluation under field conditions. *Crop Sci.*, 1989, **30**, 866-871.
34. Botterman, J., personal communication, 1991.
35. Streber, W. R,. and Pohlenz, H.B., personal communication, 1991.
36. Stalker, D.M., McBride, K.E. and Malyj, L.D., Herbicide resistance in transgenic plants expressing a bacterial detoxification gene. *Science*, 1988, **242**, 419-423.
37. Pelissier, B., Delon, R., Lutz, J.P., Borrod, G., Spicca, G., Leroux, B., Sailland, A., Lebrun, M., Bouchefra, O., Pallett, K. and Freyssinet, G., Use of bromoxynil for weed control on transgenic tobacco fields, *Proc. Corestra Meeting*, Thessalaniki, 1990. (in press).
38. Stalker, D.M. personal communication, 1991.
39. Sanders, G.E., Cobb, A.H. and Pallett, K.E., Physiological changes in *Matricaria inodora* following ioxynil and bromoxynil treatment. *Z. Naturforsch.*, 1983, **39c**, 505-509.
40. Gressel, J. and Segel, L. A., Interrelating factors controlling the rate of appearance of resistance: the outlook to the future. In *Herbicide Resistance in Plants*, eds. H.M. LeBaron and J. Gressel, Wiley, New York, 1982, pp. 325-347.
41. O'Keefe, D.P., Romesser, J.A. and Leto, K.J., Plant and bacterial cytochromes P-450: involvement in herbicide metabolism. In *Phytochemical Effects of Environmental compounds*, eds. J.A. Saunders, L.K. Channing and E.E. Conn, Plenum, New York, 1987, pp. 151-173.
42. Streber, W.R. and Willmitzer, L., Transgenic tobacco plants expressing a bacterial detoxifying enzyme are resistant to 2,4-D. *Bio/Technology.*, 1989, **7**, 811-816.
43. Lyon, B.R., Llewellyn, D.J., Huppatz, J.L., Dennis, E.S. and Peacock, W.J., Expression of a bacterial gene in transgenic tobacco plants confers resistance to the herbicide 2,4-dichlorophenoxyacetic acid. *Plant Mol. Biol.*, 1989, **13**, 533-540.
44. Davis, D.G., Olson, P.A., Swanson, H.R. and Frear, D.S., Metabolism of the herbicide metribuzin by an *N*-glucosyltransferase from tomato cell cultures. *Plant Sci.*, 1991, **74**, 73-80.

45. Durner, J., Thiel, A. and Böger, P., Phenolic herbicides: correlation between lipophilicity and increased inhibitor sensitivity in thylakoids from higher plant mutants. Z. Naturforsch., 1986, 41c, 881-884.
46. Kroath, H., Susani, M. and Zohner, A., Genetic engineering of resistance to the phenylpyridazine herbicide, pyridate. Proc. EWRS Symp. Factors Affecting Herbicidal Activity and Selectivity, 1988, pp. 343-348.
47. F. Cannon, personal communication, 1988.
48. Steward, G.R. and Press, M.C., The physiology and biochemistry of parasitic angiosperms., Annu. Rev. Plant Physiol. Plant Mol. Biol., 1990, 41, 127-151.
49. Foy, C.L., Jain, R. and Jacobsohn, R., Recent approaches for chemical control of broomrape (Orobanche spp.). Rev. Weed Sci., 1989, 4, 123-152.
50. Steward, G., Witchweed: a parasitic weed of grain crops. Outlook Agric., 1990, 19, 115-117.
51. Hartman, G.L. and Tanimonure, O.A., Seed populations of Striga species in Nigeria. Plant Dis., 1991, 75, 494-496.
52. Parker, C., Protection of crops against parasitic weeds. Crop Prot., 1991, 10, 6-21.
53. McNally, S.F. and Steward, G.R., Inorganic nitrogen assimilation by parasitic angiosperms. In, Parasitic Flowering Plants. eds. C. H. Weber, and W. Forstreuter, Proc. of the 4th ISPFP, Marburg, 1987, pp. 539-546.
54. Wolf, S.J. and Timko, M.P., In vitro root culture: a novel approach to study the obligate parasite Striga asiatica (L.) Kuntze. Plant Sci., 1991, 73, 233-242.
55. Dawson, J.H., Dodder (Cuscuta spp.) control in newly seeded alfalfa (Medicago sativa) with glyphosate. Weed Technol., 1990, 4, 880-885.
56. McNally, S.F., Ortebamjo, T.O., Hirel, B. and Steward, G.R., Glutamine synthetase isoenzymes of Striga hermonthica and other angiosperm root parasites. J. Exp. Bot., 1983, 34, 610-619.
57. Liu, Z.Q. and Fer, A., Influence d'un parasite (Cuscuta lupuliformis Krock.) sur la redistribution de deux herbicides systemiques appliques sur une legumineuse (Phaseolus aureus Roxb.). C.R. Acad. Sci., Paris, 1990, 311, 333-339.
58. Dawson, J.H., Dodder (Cuscuta spp.) control with dinitroaniline herbicides in alfalfa (Medicago sativa). Weed Technol., 1990, 4, 341-348.
59. Morrison, I.N., Todd, B.G., and Nawolsky, K.M. Confirmation of trifluralin-resistant green foxtail (Setaria viridis) in Manitoba. Weed Technol., 1989, 3, 544-551.
60. Darmency, H. and Pernes, J., Use of wild Setaria viridis(L) Beauv. to improve triazine resistance in cultivated S. italica (L.) by hybridization. Weed Res., 1985, 25, 175-179.
61. Goldburg, R., Rissler, J., Shand, H. and Hassebrook, C., Biotechnology's Bitter Harvest, Environmental Defense Fund, New York, 1990, 73pp.
62. Moss, S.R. (1991) This volume.
63. Powles, S.B. (1991) This volume.
64. Gressel, J., Multiple resistances to wheat selective herbicides: new challenges to molecular biology, Oxford Surv. Plant Mol. Cell Biol., 1988, 5, 195-203.
65. Gressel, J., Wheat herbicides: The challenge of emerging resistance, Biotechnology Affiliates, Checkendon/Reading, U.K. 247 pp.
66. Frear, D.S., Swanson, H.R. and Thalacker, F.W., Induced microsomal oxidation of diclofop, triasulfuron, chlorsulfuron and linuron in wheat. (submitted).
67. Christopher, J.T., Powles, S.B., Liljegren, D.R. and Holtum, J.A.M., Cross resistance to herbicides in annual ryegrass (Lolium rigidum). 2. Chlorsulfuron resistance involves a wheat like detoxification system. Plant Physiol., 1991, 95, 1036-1043.
68. Anonymous, A Benefit/Risk Assessment for the Introduction of Herbicide Tolerance Crops in Iowa, Iowa State University, Ames, 1991, 15 pp.

BIOHERBICIDES: THEIR ROLE IN TOMORROW'S AGRICULTURE

M.P. Greaves and M.D. MacQueen
Department of Agricultural Sciences, University of Bristol, AFRC Institute of Arable Crops Research, Long Ashton Research Station, Bristol, UK.

ABSTRACT

The current status of bioherbicide development is briefly reviewed and factors constraining development are considered. The problems of strain selection, inoculum production, bioherbicide formulation, pesticide interactions and genetic manipulation and variability are addressed and indications given of the progress towards resolving some of the problems. It is concluded that bioherbicides have considerable potential to play a significant role in tomorrow's agriculture, not only in niche markets but also against serious weeds in major crops. This potential will be realised only if the appropriate multidisciplinary, fundamental, biological and biochemical research is established and maintained with adequate and stable funding from both the public and private sectors.

INTRODUCTION

The biological control of weeds exploits the ability of selected organisms to reduce the population of an undesirable plant species to below the level at which it causes economic or environmental damage. This may be achieved by different strategies. The classical, or inoculative, strategy relies on release of small quantities of the control organism and its ability to establish, reproduce and spread throughout the area infested by the target weed. The aim is, generally, to keep the target weed population at an acceptable low level rather than to eradicate it. The bioherbicide strategy, in contrast, aims to eradicate the target by inundating it with the control organism. As this strategy is suitable for commercial development, it is desirable that the inoculum does not persist from one season to the next. Classical biocontrol, on the other hand, requires that the introduced control agent persists for at least several seasons and, thus, has little or no commercial potential.

Most attempts to control weeds using the classical strategy have used insects as the control agent, although other organisms, such as nematodes [1], are not precluded. In the last twenty years, attention has been given increasingly to pathogenic fungi, especially obligate pathogens. Two recent reviews [2], [3] give excellent, detailed and complementary overviews.

The classical strategy is an ecological approach to weed control where success may be constrained by many factors, such as genetic variability, spatial distribution of the target population and unsuitable environmental conditions for establishment and spread. In the bioherbicide, or inundative, strategy there are many opportunities for overcoming these constraints through a technological approach to the problems.

Bioherbicides are host-specific organisms, usually indigenous in areas infested by their weed targets, and are generally inoculated inundatively by spraying an appropriate formulation of the organism at a susceptible weed growth stage, in the same way as chemical herbicides. The inundative nature of the inoculation requires that large quantities of the inoculum can be produced easily. For this reason, bioherbicide development has so far focused on micro-organisms, particularly non-obligate pathogenic fungi. Ideally, inoculation is done only once in the season, with the aim of quickly reducing the weed's population to below the economic damage threshold. The biological control agent then dies out, generating the need to re-inoculate in following seasons if the weed is again troublesome. In this sense, the strategy has much potential for use in agriculture and, thus, for commercial development. The present emphasis on reducing our dependence on chemicals in agriculture does much to strengthen the potential offered by bioherbicides.

The Past and Current Status of Bioherbicides
The potential of bioherbicides to give a practical means of controlling weeds was established by the introduction of two commercial products some ten years ago. DEVINE was the first registered and consists of a chlamydospore suspension of a

pathotype of Phytophthora palmivora, native to Florida, which is used to control stranglervine (Morrenia odorata), an introduced weed of Florida citrus groves. This bioherbicide was quickly followed by COLLEGO, a wettable powder formulation of dried spores of Colletotrichum gloeosporioides f.sp. aeschynomene, which effectively controls northern jointvetch (Aeschynomene virginica) a leguminous weed of soybean and rice crops in the Southern states of the USA. Subsequently, two further commercial 'products' were announced in the USA, CASST (a preparation of Alternaria cassiae to control sicklepod, Cassia obtusifolia) and BIOMAL (Coll. gloeosporioides f.sp. malvae for the control of round-leaved mallow, Malva pusilla). Unfortunately, neither 'product' has yet appeared on the market. Other preparations are in practical use, although not as commercial products. A preparation of Coll. gloeosporioides f.sp. cuscutae, known as LUBOA II, is used to control dodders (Cuscuta sp.) in China, and Coll. gloeosporioides f.sp. clidemiae (C.g.c) is used to control Koster's curse (Clidemia hirta) in Hawaiian forests. In this latter case, the inoculum is distributed by hikers. One other agent in practical use demonstrates the potential for microbial control of weeds. Cephalosporium diospyri, a wilt-causing pathogen, has been used in Oklahoma, USA, for several years to control weedy persimmon (Diospyros virginiana) in rangelands. The fungus is wound-inoculated into the target.

Most recently, and unusually, a preparation of an obligate pathogen, the rust Puccinia canaliculata is to be sold under the name DR. BIOSEDGE as a bioherbicide against yellow nutsedge (Cyperus esculentus). The fungus will be harvested in bulk from infected crops of yellow nutsedge and used as an inundative inoculum.

In addition to the bioherbicides mentioned above, several others have undergone extensive testing prior to, or during, commercial development (Table 1).

Charudattan [4] summarises current bioherbicide research projects that are 'in the public domain'. His data show that projects are under way in 16 countries and 44 locations, 18 of the latter being in the USA. Some 69 weed targets (33 annual and 33 perennial), of which 41 are broad-leaved herbs, are being studied as targets for 41 genera of pathogenic fungi. These pathogens principally cause destructive diseases such as anthracnoses, wilts and foliar spots, a reflection of the requirement that bioherbicides must have the ability to kill, or at least severely damage, the target plant.

TABLE 1
Bioherbicides in late stages of development (from 4)

Pathogen	Target Weed	
Cercospora rodmanii	Water hyacinth	Eichhornia crassipes
Coll. coccodes	Velvetleaf	Abutilon theophrasti
Bipolaris sorghicola	Johnson grass	Sorghum halepense
Chondrostereum purpureum	Black cherry	Prunus serotina
Coll. orbiculare	Spiny cocklebur	Xanthium spinosum
Coll. gloeosporioides f.sp. jussiaeae	Winged water primrose	Jussiaea decurrens
Coll. malvarum	Prickly sida	Sida spinosa
Fusarium solani f.sp. cucurbitae	Texas gourd	Cucurbita texana
Phomopsis convolvulus	Field bindweed	Convolvulus arvensis
Sclerotina sclerotiorum	Broad spectrum control of broadleaf weeds	

Clearly, the early success with DEVINE and COLLEGO created a considerable activity in bioherbicide research, but this has not yet been translated into the appearance on the market of new products. It is particularly noticeable that the two agents which have received commercial publicity as 'about to be marketed', CASST and BIOMAL, have yet to appear. The obvious inference is that the early successes were, for some reason, unusual and the majority of pathogenic fungi are less suitable as bioherbicides. Alternatively, it is possible that the majority of pathogens are more susceptible to environmental constraints than COLLEGO and require formulation to ensure reliable performance in the field (see later). This certainly seems to be the case with CASST, and possibly with BIOMAL. Examination of the literature, now brought together in two complementary volumes [5,6], shows clearly the complexities of developing a pathogen as a successful bioherbicide. Indeed, development of COLLEGO serves as a model, in this context, in that it took twelve years of patient, exhaustive research to bring it to the market and, even now, further research is attempting to improve it.

Factors Constraining Bioherbicide Development
A major constraint on bioherbicide research is imposed via the need to develop the agent commercially. Even where public sector funding is available to find and establish the feasibility of candidate mycoherbicides, is not easy to convince commercial sponsors

to fund the development of the pathogen to product status. Many reasons are given for this, all of them commercial in nature rather than refusal to accept the concept that bioherbicides are practical possibilities. Most often, the restriction to niche markets, imposed by bioherbicides' inherent high level of host specificity, is cited. Farmers are not perceived as being willing to use a product to control only one weed out of several that infest their crops. This perception may well be more a result of the way the question is asked rather than a true reflection of farmers' attitudes. Certainly, it is not uncommon for a herbicide to be specifically applied to control one weed species which survives treatment with a broad spectrum chemical.

One factor which seems to influence company attitudes is the absence, so far, of a commercial product effective against a major weed in one of the world's main crops. DEVINE and COLLEGO are both parochial in terms of the geographical area in which they are suitable for use and of the weeds against which they are targeted. Regrettably, all the research projects highlighted earlier and the bioherbicides shown in Table 1 are similarly parochial. Bioherbicide research is, therefore, in a Catch 22 situation where essential financial support for development will not be readily forthcoming from industry until a bioherbicide with proven field potential is identified. Potential cannot be proved without proper financial support.

At the scientific level, constraints on bioherbicide development are more tangible. While fungal diseases of weeds are common, they have received relatively little research attention. Plant pathology has emphasised, for good reason, the diseases of crops. The lack of knowledge of weed diseases is now a noticeable constraint on mycoherbicide development in that considerable time and resource must be committed to developing a detailed understanding of the candidate pathogen's basic biology and ecology, before serious attempts can be made to develop any potential as a mycoherbicide. This is not work that attracts commercial funding and is also not supported by the public sector purse. Greater emphasis needs to be placed on exploration to find and characterise new weed pathogens.

Having isolated a range of fungi pathogenic to the target weed, selection is the next important stage. At first sight, this appears to be a simple operation, with the most destructive strain being the best candidate bioherbicide. However, selection for efficacy is a complex process, fraught with difficulties and opportunities to be misled, and there is a justifiable need to develop a commonly acceptable scheme to assure proper selection, evaluation and reporting of efficacy [7]. Charudattan [7] points out that "...there is a danger of under- and over-estimating the potential of candidates, resulting in costly losses in time and effort and lack of progress in this field.". Regrettably, this is all too evident from any literature survey. The problem is often compounded by poor selection of the target weed. It seems pointless to work towards bioherbicides for weeds which are of minor economic importance or are easily controlled by cheap, safe and effective herbicides, yet that is what happens quite commonly. Charudattan [7] has suggested schemes for assessing and rating efficacy of bioherbicide candidates which offer good potential for improving the speed and effectiveness of bioherbicide development. The selection of a candidate bioherbicide and establishment of its efficacy inevitably produce some essential knowledge of its biology and, to a lesser extent, ecology. This knowledge identifies further factors which may be constraints to development. Perhaps the most important of these factors is the ability to produce abundant, durable inoculum in relatively simple culture conditions. At present, submerged fermentation is the only method used to produce the commercial bioherbicides DEVINE and COLLEGO, and it seems likely to be the method of choice for new products in the immediate future. This inevitably limits the number of fungi from which durable inoculum (spores) can be produced in commercial quantities. Other techniques have been used and cited in patent applications. Surface culture on liquid or solid media followed by homogenization of mycelium or harvest of spores to give sprayable suspensions, has been used for Cercospora rodmanii as a control agent for water hyacinth [8]. Submerged fermentation to produce mycelium, followed by harvesting and treatment of the mycelial biomass to induce sporulation has been used with fungi that do not sporulate readily in liquid culture [9]. In general, although these techniques can be used easily to provide sufficient, effective inoculum for laboratory and small-scale field experiments, a more economical process is required to allow bioherbicides to reach the market. The complexities of the production of inoculum and the associated economics are clearly described by Stowell [10]. While acknowledging that alternative production technologies may have a place for bioherbicides aimed at small (niche) markets, Stowell concludes that advances in submerged fermentation and downstream processing are essential in order to speed the development and release of new bioherbicides.

Just as chemical herbicides must be formulated to ensure optimal efficacy, so must bioherbicides. A useful review of this subject is given by Boyette et al. [11]. Bioherbicides can rarely, if ever, be applied as simple aqueous suspensions which result in uneven inoculum distribution on the target weed, unacceptable amounts of run-off and, perhaps most important, rapid desiccation of the inoculum. Fungal pathogen spores, the preferred inoculum, require periods of exposure to free water or high relative humidity in order to germinate, grow and infect their target. These periods are commonly longer than twelve hours. In practice, in most temperate regions of the world, dew periods which provide the necessary free water are shorter than twelve hours and may not occur every night. Thus, bioherbicides applied as aqueous suspensions, which rapidly dry, may not, even if the leaf is subsequently wetted by dew, be exposed to enough free water to ensure successful infection, and so allow the pathogen to continue growth without further dependence on free-water external to the leaf. This is the major constraint to efficacy of all bioherbicides worked on to date, and much attention is now being given to formulation to reduce bioherbicide dependency on dew. The complexity of this work is increased by the need to use surfactants to ensure even distribution of inoculum, especially on plants with waxy cuticles. These also tend to spread the inoculum droplets into thins sheets, so increasing their drying rate.

At present, much formulation research is focused on the use of emulsions and invert (water-in-oil) emulsions as inoculum carriers. Invert emulsions, in particular, are claimed to reduce evaporation of the water phase significantly and allow spore germination and infection of the target to proceed without check due to desiccation. The early invert emulsions had a thick consistency and needed to be applied with specialised sprayers [12]. More recent developments of low viscosity invert emulsions allows conventional hydraulic sprayers to be used successfully (Connick, 1991, pers. comm.). Typical data for such a low viscosity formulation shows that its water content drops from 49% to 35% during 48 hours at 21° and a relative humidity of 65%. Such developments, together with new oil-in-water emulsions being developed, promise to eliminate desiccation of inoculum as the major constraint on bioherbicide efficacy in the field. This may open the door for several bioherbicides, currently suffering from variable performance in the field, to be marketed in the not-too-distant future.

Many bioherbicides under development show maximum effect on their targets only if applied at a very young (cotyledon) growth stage or if they infect specific parts of the plant heavily. For example, our own work with cleavers, Galium aparine, shows that infection at or below the cotyledonary node is essential to guarantee plant death. If this site is not infected, lateral shoots will be produced at that node and, even if the main shoot is completely killed, the plant will survive. Similarly, itch grass, Rottboellia cochinchinensis, will grow away from very heavy infections as the meristem is protected from infection within the leaf sheath (Martin and Ellison, 1991, pers. comm.). In such cases, great benefit can be obtained by formulating the bioherbicide with a low dose of conventional herbicide. In the case of itch grass, some of the new sulfonyl urea herbicides appear to be highly effective at doses as low as 5% of the recommended field rate. The effect of the herbicide appears to be one of checking the growth of the weed and, at the same time, synergising the infectivity of the pathogen. The net result is complete kill of the target. Such synergistic action has been reported to increase significantly the effect of bioherbicides [7, 11].

A further aspect of herbicide or pesticide interactions with bioherbicides has quite different connotations. As living organisms, bioherbicides are, clearly, likely to be sensitive to many pesticides, especially fungicides, used in crop management programmes. Thus, spray equipment must be entirely clear of pesticide residues before use with bioherbicides, and appropriate spray windows must be defined when bioherbicides will not be affected adversely by pesticides applied either before or after them. Research in this area of pesticide-bioherbicide interaction is in its infancy but already promises to aid significantly the development of reliable bioherbicides [13].

The problems of pesticide effects on pathogen inoculum and disease spread were addressed during the development of COLLEGO. Benomyl applications in rice or soybean disease control programmes usually have to follow COLLEGO applications by at least three weeks to avoid interference with bioherbicide activity. New strains of the pathogen in COLLEGO have been produced in which benomyl resistance has been induced [14]. These mutant strains appear to be genetically stable and are currently being developed for field use.

The use of mutagens and subsequent selection of strains with desirable traits is very hit-and-miss. Much greater opportunities for improving bioherbicides might arise from the application of the powerful techniques developed by molecular biologists [15, 16]. Considerable advances have been made in the ability to manipulate fungi genetically [17], in a very sophisticated fashion. However, fundamental knowledge of fungal pathogenesis is lacking and it is unlikely that rational design of pathogens to improve their ability to control weeds will be achieved in the foreseeable future. The promise offered by molecular biology will be further constrained by the imposition of regulatory standards, at least in many parts of the world. At present many genetically manipulated pathogens would certainly not obtain permission for release in the field, but the position with unmanipulated bioherbicides is less stringent. Permission to release large inocula of weed pathogens in areas where they are indigenous is generally obtained without unreasonable bureaucratic restraints. Indeed, this is a major factor that might give bioherbicides a cost advantage over chemical herbicides at present.

While genetic variants of pathogens obtained by manipulation may offer further opportunities for bioherbicide improvement, inherent genetic variation in the target weed can be a problem. Successful development of a bioherbicide is highly dependent on the genetic character of both the pathogen and its target weed, and variation in the latter can produce unacceptable variability in susceptibility to the bioherbicide. Variability in the pathogen can affect important characteristics such as sporulation, virulence, host range and tolerance to environmental factors [18]. The increasing problem of development of herbicide resistance in weeds has been seen as evidence that bioherbicide resistance will also develop. However, herbicide resistance usually develops with chemicals with target-specific modes of action [2], whereas plant pathogens used as bioherbicides usually have a broad spectrum mechanism of action, which appears less likely to lead to resistance. However, resistance may well develop in time in some weed species. Bioherbicides, perhaps, offer some advantage over chemicals in that they are genetically variable and, so, new virulent genotypes can readily be selected for use.

The Future Prospects for Bioherbicides

The role of bioherbicides in tomorrow's agriculture depends on many factors. Some of the major constraints have been touched on in the preceding section. Limitations of space prevent attention being given to other aspects such as shelf-life of bioherbicides, use of bacteria, nematodes and viruses as bioherbicides, problems arising from limited spectrum activity and the possibilities offered by pre-emergence soil treatments. These are all dealt with in recent reviews [6, 19] which present broad and detailed coverage of bioherbicide development.

Although emphasis has been placed on the constraints against bioherbicides, it is not intended to give the impression that bioherbicides have no role in tomorrow's agriculture. On the contrary, by indicating the significant progress that is being made in overcoming the constraints, it is intended to strengthen the conclusion that products will appear in the foreseeable future and make significant contributions to agricultural productivity and environmental safety. The present political and public emphasis on reducing chemical inputs in agriculture can only strengthen that conclusion.

In the short term, it is evident that bioherbicides are most likely to make their mark in niche markets. Already, in Canada and the USA, so-called 'lawn-care' products are being highlighted as having strong potential. This is not as restrictive as it might appear, as the term includes bioherbicides for use in forestry and amenity grassland such as golf courses as well as the urban lawn. At the same time, cost restrictions on the development and/or registration of chemical herbicides for small acreage crops offers opportunities for bioherbicides. In the long term, we expect, with confidence, that bioherbicides for use against some of the world's worst weeds in the major crops will appear on the market. Indeed, we have some hopes that, despite current funding difficulties, our own bioherbicides for use against cleavers, Galium aparine, in cereals and itch-grass, Rottboellia cochinchinensis, in maize, small grains and sugar-cane will find a market. Despite the early commercial 'successes' and the present long list of potential bioherbicides in early stages of development, it is quite clear that the full and proper role of bioherbicides in agriculture will not be achieved unless more research is established. This must be integrated, multidisciplinary research involving pathologists, weed scientists, formulation chemists, fermentation technologists, molecular biologists and others, as appropriate, at different stages of the research. More particularly, the research must be adequately funded by both public and private sectors. At present, in Europe especially, it is not, and that is the biggest constraint of all.

REFERENCES

1. Parker, P.E., Nematodes as Biological Control Agents of Weeds. In Microbial Control of Weeds, ed. D.O. TeBeest, Chapman and Hall, London, 1991, 58-68.

2. Watson, A.K., The Classical Approach with Plant Pathogens. In Microbial Control of Weeds, ed. D.O. TeBeest, Chapman and Hall, London, 1991, 3-23.

3. Bruckart, W.L. and Hasan, S., Options with Plant Pathogens Intended for Classical Control of Range and Pasture Weeds. In Microbial Control of Weeds, ed. D.O. TeBeest, Chapman and Hall, London, 1991, 69-79.

4. Charudattan, R., The Mycoherbicide Approach with Plant Pathogens. In Microbial Control of Weeds, ed. D.O. TeBeest, Chapman and Hall, London, 1991, 24-57.

5. Charudattan, R. and Walker, H.L., Biological Control of Weeds with Plant Pathogens, Wiley, London, 1982, 293.

6. TeBeest, D.O., Microbial Control of Weeds. Chapman and Hall, London, 1991, 284.

7. Charudattan, R., Assessment of efficacy of mycoherbicide candidates. Proc. VII Int. Symp. on Biological Control of Weeds, 1989, 455-464.

8. Conway, K.E., Freeman, T.E. and Charudattan, R., Methods and Compositions for Controlling Water Hyacinth. U.S. Patent No. 4,097,261, 1978.

9. Walker, H.L. and Riley, J.A., Evaluation of Alternaria cassiae for the biocontrol of sicklepod (Cassia obtusifolia). Weed Science, 1982, 30, 651-654.

10. Stowell, L.J., Submerged Fermentation of Biological Herbicides. In Microbial Control of Weeds, ed. D.O. TeBeest, Chapman and Hall, London, 1991, 225-261.

11. Boyette, C.D., Quimby, P.C., Connick, W.J., Daigle, D.J. and Fulgam, F.E., Progress in the Production, Formulation and Application of Mycoherbicides. In Microbial Control of Weeds, ed. D.O. TeBeest, Chapman and Hall, London, 1991, 209-222.

12. Quimby, P.C., Fulgham, F.E., Boyette, C.D. and Connick, W.J., An invert emulsion replaces dew in biocontrol of sicklepod - a preliminary study. In Pesticide Formulations and Application Systems, Vol. 8, eds. D.V. Horde and G.B. Beestman, ASTM-STP 980, American Society for Testing and Materials, Philadelphia, 1988, 264-270.

13. Altman, J., Neate, S. and Rovira, A.D., Herbicide-Pathogen Interactions and Mycoherbicides as Alternative Strategies for Weed Control. In Microbes and Microbial Products as Herbicides, ed. R.E. Hoagland, American Chemical

Society, Washington, D.C., 1990, 240-259.

14. TeBeest, D.O., Induction of tolerance to benomyl in Colletotrichum gloeosporioides f.sp. aeschynomene by ethyl methanesulfonate. Phytopathology, 1984, 74, 864.

15. Greaves, M.P., Bailey, J.A. and Hargreaves, J.A., Mycoherbicides: opportunities for genetic manipulation. Pesticide Science, 1984, 26, 93-101.

16. Bailey, J.A., Improvement of mycoherbicides by genetic manipulation. Aspects of Applied Biology 24, The Exploitation of Micro-organisms in Applied Biology, 1990, 33-38.

17. Kistler, H.C., Genetic Manipulation of Plant Pathogenic Fungi. In Microbial Control of Weeds, ed. D.O. TeBeest, Chapman and Hall, London, 1991, 152-170.

18. Weidemann, G.J. and TeBeest, D.O., Genetic Variability of Fungal Pathogens and Their Weed Hosts. In Microbes and Microbial Products as Herbicides, ed. R.E. Hoagland, American Chemical Society, Washington, D.C., 1990, 176-183.

19. Rhodes, D.J., Formulation requirements for biological control agents. Aspects of Applied Biology 24, The Exploitation of Micro-organisms in Applied Biology, 1990, 145-153.

GENETIC ENGINEERING OF PREDATORS AND PARASITOIDS FOR PESTICIDE RESISTANCE

MARJORIE A. HOY
Department of Entomology, University of California
Berkeley, California 94720 USA

ABSTRACT

Genetic selection of phytoseiid predators for pesticide resistance has been shown to be a practical and cost effective tactic for the biological control of spider mites. Field tests have been conducted with several manipulated phytoseiid species and some are being used in integrated pest management programs in agriculture. Development of resistant strains of parasitoids and insect predators currently lags behind efforts with predatory mites, but several laboratory-selected insect natural enemies are being evaluated for incorporation into integrated pest management programs. The use of mutagenesis and recombinant DNA (rDNA) techniques could improve the efficiency of genetic improvement projects. Critical research needs include identifying and cloning useful resistance genes, developing methods for maintaining fitness of the manipulated strains, learning how to manage and maintain released strains, and developing improved methods for inserting resistance genes into the germline of beneficial arthropods. Protocols for evaluating risks associated with the release of arthropod natural enemies that have been manipulated with rDNA methods need to be developed well in advance so that excessive delays in evaluating efficacy and fitness in the field can be avoided.

INTRODUCTION

Genetic manipulation of arthropod natural enemies has been proposed as a method of enhancing the biological control of arthropod pests for over 70 years (1). If genetic improvement projects are to be successful in pest management programs, extensive ecological, biological, and genetic information is required. This paper reviews the current status of developing pesticide resistant strains of arthropod natural enemies as a practical solution to pest management problems and discusses potential directions for future research.

First, some definitions may be useful. Biological control employs pathogens, parasitoids, and predators to control arthropod pests and weeds and typically involves one or more of three basic tactics singly or in combination--classical, augmentation, and conservation (2). Genetic improvement involves directed, purposeful alterations to the genome of a natural enemy to enhance its efficacy as a biological control agent (3, 4). Such genetic alterations theoretically can be achieved through artificial selection, heterotic effects, or, potentially, the use of recombinant DNA (rDNA) techniques (5, 6). Genetically-manipulated natural enemies could be used in classical biological control programs and augmentative releases. The identification and use of naturally-occurring biotypes and the maintenance of genetic quality is not genetic improvement, although it is crucial to maintain quality if the natural enemy is to be effective in pest management programs. Maintaining quality has always been considered to be a constraint on genetic improvement as a tactic in biological control (7).

Genetic improvement projects can be considered to have three phases if the improved natural enemy strain is actually used in a pest management program. In Phase I, the project is conceived and the problem is identified. During Phase II, the improved strain is developed and evaluated in the laboratory. During Phase III, the newly-developed strain is evaluated in the field, implementation is achieved, and the success or failure of the strain is documented and economic analyses are conducted (4, 5, 8). Only a few genetic improvement projects have progressed through all three phases. We need a great deal of fundamental research before genetic manipulation of arthropod natural enemies can become a consistent and fully accepted component of biological control.

A BRIEF HISTORY OF GENETIC IMPROVEMENT

Mally (1) probably was the first to discuss the selection and breeding of a variety of ecological and behavioral attributes beneficial insects. DeBach (3), another early proponent of genetic improvement, suggested that parasitic insects have a narrow range of adaptations to environmental conditions which limits their success in classical biological control. DeBach defined critical components of genetic improvement programs: 1) determine characters needing improvement, 2) identify or provide for adequate genetic variability, 3) select appropriately, and 4) maintain integrity of the new strain in the field. He pointed out that laboratory manipulation may not only duplicate and greatly speed up events which occur rarely in the field but, because genetic variability can in some cases be increased considerably beyond that which might ever be obtained under natural conditions, laboratory manipulation can even produce results that might never occur under field conditions.

Sailer (9) suggested that interspecific hybridization might provide useful genes for genetic improvement and he thereby anticipated the possibilities of rDNA techniques when he speculated that "...there is no reason why a wide variety of useful characters cannot be similarly moved from one species to another and combined to form strains of insect parasites or plant pollinators that are superiorly adapted to the environments where they are needed".

PESTICIDE-RESISTANT NATURAL ENEMIES

Several reviews of genetic improvement projects with the goal of developing pesticide-resistant natural enemy strains are available (4, 5, 8, 10-12). Selection of parasitoids for resistance to pesticides was recognized as a potent method for enhancing biological control in agriculture early on. Thus, during the 1950s, *Macrocentrus ancylivorous* was selected for resistance to DDT, achieving a 12-fold increase in tolerance in 19 generations (13, 14). No information is provided to suggest that field releases were made, however. In the 1960s, the boll weevil parasite *Bracon mellitor* was selected for resistance to DDT and toxaphene (15). These authors also concluded that this parasite has a mechanism for developing resistance to insecticides, a conclusion at variance with the widely held belief that arthropod natural enemies lack genetic and physiological mechanisms for developing resistance to pesticides. Unfortunately, they did not confirm that the parasite survived in pesticide-treated cotton fields.

Despite extensive field selection of arthropod natural enemies in orchards, glasshouses, and fields during the 1950s and 1960s, only a relatively few species were shown to have developed resistance to pesticides before the 1970s and 1980s (Table 1). Whether this is due to the slow responses of arthropod natural enemies to field selection, relative lack of genetic variability, relative lack of detoxifying enzymes, or failure to adequately monitor natural enemies for resistance is not known (11, 12).

Predatory mites in the family Phytoseiidae proved to be models for genetic selection for pesticide resistances during the 1970's and 1980's (10). Extensive research during the 1960's and 1970's had documented that phytoseiids could be highly effective predators of spider mites, particularly in glasshouses, orchards, and vineyards (16, 17). As shown in Table 1, a number of phytoseiid species, including *Amblyseius andersoni, A. fallacis, A. longispinosis, A. newsami, A. sojaensis, Euseius hibisci, Metaseiulus occidentalis, Phytoseiulus persimilis, Typhlodromus caudiglans,* and *T. pyri* developed resistances, most often to parathion and azinphosmethyl. DDT resistance was probably induced in several phytoseiid species, although this was not well documented. Carbaryl and sulfur resistances were selected less often.

TABLE 1

Some pesticide-resistant arthropod natural enemies developed through field selection.

Species	Pesticide(s)	Citation number
Amblyseius andersoni	azinphosmethyl	18, 19
Amblyseius fallacis	azinphosmethyl	20, 21, 22, 23,
	parathion, dimethoate, propargite	24
	several OPs	25
	carbaryl	21, 26
	DDT	27
Amblyseius longispinosis	methidathion, EPN, mecarbum, phenthoate, methomyl, carbaryl	28, 29
	permethrin	30
Amblyseius newsami	carbaryl, lime sulfur, DDT	31
Amblyseius sojaensis	methidathion, carbaryl	32
Aphidoletes aphidimyza	azinphosmethyl	33
Aphytis africanus	methidathion	34
Aphytis holoxanthus	malathion	35
Aphytis melinus	parathion, methidathion, dimethoate, malathion	36
	OPs & carbaryl	37
Chiracanthium mildei	malathion	38
Chrysoperla carnea	fenvalerate, permethrin, phosmet	39
	DDT, parathion, azinphosmethyl, carbaryl, pyrethroids, phosmet	40
Coleomegilla maculata	DDT	41
Comperiella bifasciata	methidathion	34
Diglyphus begini	oxamyl, methomyl, fenvalerate, permethrin	42
Euseius hibisci	parathion	43
	dimethoate & other OPs	44
Metaseiulus occidentalis	parathion	45, 46
	azinphosmethyl	47
	azinphosmethyl, diazinon	48
	sulfur	49
	DDT ?	S. C. Hoyt, pers. commun.
Phytoseiulus persimilis	parathion & other OPs	50, 51
Typhlodromus caudiglans	DDT	52
Typhlodromus pyri	azinphosmethyl	53, 54, 23, 55
	azinphosmethyl, parathion, carbaryl	56
	parathion & propoxur	57
	parathion	58
	azinphosmethyl, carbaryl, parathion	59

The integrated mite management program developed for Washington apple orchards has served as a useful model for using phytoseiids in pest management programs around the world (47). *Metaseiulus occidentalis* developed resistance to azinphosmethyl after approximately eight years of field selection. Azinphosmethyl, used to control codling moth, could subsequently be used without causing spider mite outbreaks. This program documented that an arthropod natural enemy had sufficient genetic variability to develop a useful level of resistance and that the resistant strain could provide effective biological control of spider mites.

Schulten and van de Klasthorst demonstrated that a predatory mite could be artificially-selected for usable levels of resistance in the laboratory when they selected *Phytoseiulus persimilis* for resistance to parathion (50). The laboratory-selected strain was not utilized in glasshouses because a glasshouse-selected strain with a higher level of resistance was found. This example demonstrates the importance of effective sampling before beginning a genetic improvement program. Another significant step was achieved when the azinphosmethyl-resistant strain of *M. occidentalis* from Washington was released into southern California apple orchards where it established and survived pesticide applications (60); now we knew it was possible to transplant phytoseiid biotypes into new agricultural environments. Subsequently, laboratory-selected and wild strains of *M. occidentalis, A. fallacis, P. persimilis*, and *T. pyri* have been established as effective predators of spider mites in apples, pears, peaches, strawberries, and glasshouse vegetable crops around the world (for a recent review see 11).

GENETIC IMPROVEMENT OF *METASEIULUS OCCIDENTALIS*

In 1976 *M. occidentalis* became the target of a genetic improvement project with the explicit goal of determining whether this biological control agent could be selected for a usable level of pesticide resistance in the laboratory, established in orchards or vineyards, survive field rates of pesticides, overwinter, control spider mites, and maintain a stable level of resistance (10, 61).

Initial surveys were conducted to determine whether populations were resistant to carbaryl and permethrin, but detectable levels of resistance were not found. Laboratory selection subsequently yielded strains resistant to both pesticides and laboratory, greenhouse, and small plot evaluations were conducted with both strains (62-64). The permethrin-OP resistant strain established in apple and pear orchards in California, Washington, and Oregon, survived low rates of permethrin, and persisted. However, this strain was not used because permethrin has not become a component of any pest management program in these crops. After the carbaryl-OP resistant strain was evaluated in the field, it was crossed with a sulfur-OP resistant strain to obtain a carbaryl-OP-sulfur resistant (COS) strain, which has been commercially mass reared and released into California almond orchards (65, 66).

In all evaluations conducted with the COS strain, it has performed as well as, or better than, the wild strains with which it has been compared and has become part of an a highly cost-effective integrated mite management program in almond orchards (61, 66, 67). Adoption of the integrated mite management program has resulted in reduced pesticide applications to control spider mites and savings to growers of $24 to $44 per acre due to reduced costs for acaricides. Adoption of the program has resulted in savings to the almond industry of more than $21 million, primarily due to reduced pesticide costs (66). The benefits of reduced pesticide applications to the environment were not estimated. Because the COS strain is being used in other crops, and has been released in South Africa, Chile, China, Canada, and Australia, the actual economic benefits are likely to be in excess of the above figure. Thus, genetic manipulation can be an economically sound investment in agricultural research.

The COS strain of *M. occidentalis* was selected for nondiapause and evaluated in small plot trials on glasshouse roses where it performed well and reduced the numbers of acaricides applied (65, 68). Despite the fact that two traits, pesticide resistance and nondiapause, were selected, the strain was shown to be an effective biological control agent.

Why have phytoseiids responded to selection for pesticide resistances in the laboratory and field? The answer is that we don't know whether the answer lies in their genetic system, ecology, behavior, or physiology, or a combination of all these factors. Of approximately 1000 species of phytoseiids, at least 12, including *Amblyseius andersoni, A. fallacis, A. longispinosis, A. newsami, A. nicholsi , A. pseudolongispinosus, A. sojaensis, Euseius hibisci, Metaseiulus occidentalis, Phytoseiulus persimilis, Typhlodromus caudiglans , and T. pyri,* have been documented to have acquired resistances to pesticides (Tables 1, 2).

It appears that once a phytoseiid species becomes resistant to one pesticide, resistances to other pesticides can develop through laboratory or field selection. For example, *M. occidentalis* is resistant to OPs, carbaryl, pyrethroids, DDT, and abamectin; *T. pyri* is now resistant to OPs, carbaryl, and synthetic pyrethroids; and *A. fallacis* has developed resistances to OPs, carbaryl, pyrethroids, and DDT. *Amblyseius longispinosus* is a recent example of a phytoseiid developing resistances to OPs and carbamates (28, 29). Once resistances developed through field selection, the resistant strains of *A. longispinosus* became highly effective predators of the Kanzawa spider mite, *Tetranychus kanzawai,* in Japanese tea plantations (28, 29). Laboratory selection was also conducted with methidathion and the resulting strain was apparently fit (29). Widespread biological control of *T. kanzawai* was attributed to the pesticide-resistant populations of *A. longispinosus* in many prefectures in Japan, except when synthetic pyrethroids were applied (29). By 1990, Mochizuki (30) reported that *A. longispinosus* collected from Shizuoka prefecture in Japan had developed resistance to permethrin, with the most tolerant strain 34-fold more resistant than the most susceptible strain. It will be interesting to learn whether this resistance is polygenic or monogenic and whether it is a harbinger of field-selected pyrethroid resistance in other phytoseiid species.

RESISTANT INSECT PARASITOIDS AND PREDATORS

The facility with which insect parasitoids or predators can become resistant to pesticides remains controversial (11, 12). Relatively few insect predators or parasitoids have been documented to have become resistant to pesticides (Tables 1, 2), especially when compared to the 400+ pest arthropods that have become resistant (69).

Early examples of laboratory selection of parasitoids for resistance include a modest 4.4-fold increase in tolerance to DDT after nine months of selecting *Macrocentrus ancylivorus,* a parasite of the Oriental fruit moth (*Grapholitha molesta*) (13). Robertson (14) continued the selection with DDT and increased the resistance to 12-fold over that of the base colony, but when DDT selection was terminated, the strain lost its resistance within 13 generations. Adams and Cross (15) found that *Bracon mellitor,* a parasite of the boll weevil, *Anthonomus grandis,* responded to laboratory selection with DDT, carbaryl, and methyl parathion. Resistances increased about four-fold within five generations and a population selected with a mixture of DDT and toxaphene yielded a resistance level 8-fold greater than that of the original population. Abdelrahman (70) found that *Aphytis melinus* colonies selected with malathion responded with a 4-fold increase in resistance. Delorme et al. (71) selected *Encarsia formosa* for parathion and deltamethrin resistance, but were unable to obtain more than a 0.5 to 4.2-fold increase in tolerance. None of these resistant strains were evaluated in the field to determine whether they were sufficiently resistant and fit to be effective natural enemies. Thus, they were considered by many to be a laboratory novelty of no practical value.

Examples of pesticide resistance developed through field selection in insect parasitoids or predators are rare before the 1970s (Table 1). Croft (11) cited resistance in eight predators (*Geocoris pallens,* the spider *Chiracathium mildei, Aphidoletes aphidimyza, Chrysoperla carnea, Coleomegilla maculata, Stethorus punctum, Stethorus punctillum, Labidura riparia*), but in several cases the resistances are not fully documented because comparisons between susceptible and resistant strains were not made. Croft (11) listed eleven parasitoid species in which pesticide resistance has been found, including *Aphytis africanus, Aphytis coheni, Aphytis holoxanthus, Aphytis melinus, Bracon mellitor, Comperiella bifasciata, Encarsia formosa, Macrocentrus ancylivorus, Pholetesor ornigis, Trichogramma evanescens,* and *Trioxys pallidus.* However, some of these resistance levels were probably not sufficiently high to be useful in integrated pest management programs. Thus, Strawn (36) found substantial variability to organophosphorus insecticides in California populations of *Aphytis melinus,* but concluded that none were sufficiently resistant to be useful in a citrus pest management program. Rosenheim and Hoy (37) examined California populations of *A.*

melinus about 10 years later and also found variability in responses to carbaryl and OPs but concluded, that none were sufficiently high to be useful in citrus pest management.

The questions raised by these apparent failures to achieve high levels of resistance to pesticides in large numbers of insect parasitoids and predators have been discussed for more 30 years, yet remain unresolved. Possible reasons include one or more of the following: parasitoids (and insect predators to a lesser extent) lack the ability to detoxify pesticides; parasitoids, especially hymenopteran parasitoids, lack variability upon which selection can operate; insect parasitoids and predators lack mechanisms for high levels of resistance; and, trophic level and ecological differences between arthropod natural enemies and pest arthropods may explain the greater likelihood for pest arthropods to become resistant to pesticides.

Rathman et al. (42) demonstrated high levels of resistance to methomyl, oxamyl, permethrin, and fenvalerate in a Hawaiian population of the *Liriomyza* leafminer parasitoid *Diglyphus begini*. Resistance ratios, compared to a California population, were 20, 21, 17, and 13, respectively. One population from a heavily sprayed tomato greenhouse had LC_{50} values for permethrin and fenvalerate 10 and 29 times greater than the field rate, respectively. This is a rare example of a parasitoid with a high level of resistance that was developed through field selection.

TABLE 2

Arthropod natural enemies successfully selected in the laboratory for resistance to pesticides. Colonies with resistances underlined were evaluated in the field.

Species	Pesticide	Citation number
Amblyseius fallacis	carbaryl	21
	pyrethroids	72, 73
A longispinosus	methidathion	29
A. nicholsi	phosmet	74, 75, 76, 77
A. pseudolongispinosus	dimethoate	78
Aphytis holoxanthus	azinphosmethyl	35, 79
A. melinus	carbaryl	80
Bracon mellitor	DDT & toxaphene	15
Chrysoperla carnea	carbaryl	81
Encarsia formosa	parathion, deltamethrin	71
Macrocentrus ancylivorus	DDT	13, 14
Metaseiulus occidentalis	permethrin	64
	carbaryl, dimethoate	62, 63, 82
	carbaryl, nondiapause	65, 68
	propoxur	83
	carbaryl-OP-sulfur	65
	abamectin	84
	pyrethroids	85
Phytoseiulus persimilis	parathion	50
	other OPs	51
	deltamethrin	86
	pyrazophos, pirimor, pirimiphosmethyl	87
	methidathion	88, 89, 90
Trichogramma japonicum	fenvalerate, decis	91
Trioxys pallidus	azinphosmethyl	92, 93, 94, 95
Typhlodromus pyri	cypermethrin, deltamethrin, fenvalerate	96, 97, 98, 99, 100

CURRENT RESEARCH

Other phytoseiid species have been successfully selected in the laboratory for resistance to pesticides, including *Amblyseius fallacis*, *A. longispinosis*, *A. pseudolongispinosus*, *Typhlodromus pyri*, *A. nicholsi*, and *Phytoseiulus persimilis* (Table 2). Fitness and efficacy of the strains and methods for implementing several strains are being investigated, including *T. pyri*, and *A. nicholsi*.

Typhlodromus pyri, has been a central element of New Zealand's integrated mite management program in apple orchards, where it controls the European red mite, *Panonychus ulmi*. Initially, low levels of resistance to OPs were found in *T. pyri* (53), but after additional field selection resistance levels increased (54). Because of fears about the potential loss of azinphosmethyl to the integrated control program through resistance in codling moth or leafrollers, Markwick (96, 97) began selecting *T. pyri* in the laboratory with synthetic pyrethroids. Field selections were also conducted during in between 1978 and 1982 in an experimental orchard, but initially appeared to fail. However, by 1985, predators from this block were found with useful levels of cypermethrin resistance (98, 99). The laboratory selection also produced a selection response, and field trials were carried out with the two resistant strains during the 1985-86 field season. Several years of field evaluation indicate the resistant strains of *T. pyri* establish and provide excellent mite control after the first field season. During the first year, European red mite populations are not adequately controlled by the predators. Because the new strain disperses slowly, control is enhanced if predators are distributed throughout the release block. Resistance is relatively stable, although an application of pyrethroids every year is optimal for maintaining high levels of resistance. The insecticides and resistant predators will be sold as a package with technical advice on compatible pesticide programs and other management tactics. If pyrethroid-resistant *T. pyri* are successfully implemented in the 8000 hectares of New Zealand apple and pear orchards, savings are estimated to be at least NZ$ 3 million (100).

Amblyseius nicholsi is an effective predator of the citrus red mite *Panonychus citri* in Chinese citrus orchards near Guangdong Province. Native populations are susceptible to phosmet, which is used to control citrus psylla, *Diaphorina citri*. (74). A genetic selection project yielded a resistant strain with a 19-fold increase in resistance, determined by a single semidominant gene. The phosmet-resistant strain was subsequently selected for resistance to Dimehypol (75). Laboratory evaluation of the phosmet-resistant strain indicated that oviposition rate, hatch rate, survival of immatures, and sex ratio of the resistant strain is similar to the susceptible. However, resistant females live longer and have a higher predation rate than susceptible females. Interestingly, the resistant strain is smaller than that of the susceptible strain, which was perceived to be advantageous under conditions when prey densities are low (76). Field tests have indicated the resistant strain can survive phosmet applications and control citrus red mites (77). The resistant strain overwintered and retained its high level of resistance, and thus appears to be promising for implementation.

Laboratory selection for pesticide resistance has been considered extremely high risk research because of the perception that insect parasitoids and predators lack genetic variability and mechanisms for resistance to pesticides. While the predator *Chrysoperla carnea* and the parasitoids *Trioxys pallidus* and *Aphytis melinus* recently have been selected successfully in the laboratory (Table 2), this perception probably is still appropriate (12).

The project with *T. pallidus* illustrates some of the issues. Surveys indicated that colonies of this parasitoid varied in their responses to azinphosmethyl, indicating that natural selection for resistance was occurring although no evidence was found that any strain had a usable level of resistance (12, 92). Laboratory selection yielded an azinphosmethyl-resistant strain of *T. pallidus* and the resistant strain was released into commercial walnut orchards in California in June 1988 (94). The strain established in four of the five release sites, survived on azinphosmethyl or methidathion residues, parasitized walnut aphids, and persisted in the orchards throughout the growing season. At two sites the resistant strain dispersed to nearby nonrelease blocks. The resistant strain was released in several sites again in 1989 and in 1991. Resistant parasites overwintered successfully in several sites (95), but fully effective control of walnut aphids was not always achieved. Because this azinphosmethyl-resistant strain is cross resistant to methidathion, chlorpyrifos, diazinon, phosalone, and endosulfan, establishment was thought to be enhanced since application of any of these pesticides, all commonly used in walnut IPM, should select for the resistant strain (93). This parasite can disperse rapidly over great distances, which could enhance the likelihood that it will establish and spread.

Factors that delay or reduce the possibility of establishing the azinphosmethyl-resistant strain of *T. pallidus* include: 1) not all walnut orchards are treated with pesticides, 2) nor can all pesticide applications provide complete coverage of the foliage. Thus, selection for the resistant strain may be less than fully effective. 3) Releases could be very costly because this host-specific parasitoid is difficult and expensive to rear (95, 101) and very large numbers may have to be released if the more than 180,000 acres of walnuts are to be colonized. 4) The release rates required to replace one (susceptible) biotype with another are unknown (101). Considering how readily resistant pest populations seem to replace susceptible populations, it would seem to be a relatively easy task, but a model for establishing this new biotype in the Central Valley of California was lacking until recently.

Caprio et al. (101) developed a population genetics model to optimize release strategies with this strain. The role of initial frequencies of the resistant strain in the release block, population size within the block, selection intensity, refuge size (amount of orchard not treated with pesticides), relative fitness of the resistant and susceptible parasites, and dispersal within and between orchards were varied to determine how important these factors might be. Implementation was simulated as a two-stage procedure. Stage I simulations examined the outcome of releasing the resistant strain into a single 80-acre orchard. Stage II simulations examined the dispersal of the resistant strain from the 80-acre site and its establishment in a matrix of 80-acre orchards in an 8000-acre region.

The model indicated that establishment within an 8000-acre region was promoted by migration of the parasites between orchards, low survival of the wild strain when treated with pesticides, and minimal refuge from insecticide applications. Establishment was defined as the resistant strain making up 50% of the individuals in at least 90% of the orchards.

Several factors had a significant negative impact on the rate of establishment of the resistant strain of *T. pallidus* in the model simulations (101). Under current cultural practices in California walnut orchards, establishment of the resistant strain in the 8000-acre region was estimated to require at least five years and possibly up to ten years. The long time required to replace the susceptible strain with the resistant strain was unexpected given that the resistant strain survived significantly better after an azinphosmethyl treatment than the susceptible strain in the simulations. The model simulations suggest that refuge size and total parasite population density are important factors influencing establishment, because survival by susceptible parasites in even small untreated refuges can translate into large absolute numbers. For example, 0.1% survival of a susceptible parasite population consisting of 10,000,000 individuals still leaves 10,000 following selection with azinphosmethyl. With a refuge size of 10%, one million susceptible parasites would remain following selection with a pesticide. Another important factor is the high reproductive rate of the parasite, which allows susceptible parasite populations to recover rapidly after pesticide applications. Finally, the model indicates that use of pesticides other than azinphosmethyl for codling moth control will delay establishment of the resistant strain because susceptible parasite populations recover more rapidly after alternative pesticides than they do after azinphosmethyl treatments. Although alternative pesticides select for the resistant strain, the residues are toxic for a shorter interval and result in reduced selection intensity.

FUTURE RESEARCH

Both fundamental and applied research is needed if genetic improvement programs are to become a common tactic in biological control. Artificial selection on naturally-occurring variability has been successful with a variety of arthropod natural enemies (Table 2). It is not clear, however, whether variability is adequate to allow successful selections for all target species. Table 3 lists some failures to select beneficial species for pesticide resistance in the laboratory. These examples are probably incomplete because of the failure to report negative data and the likelihood that some failures are buried in published reports about related topics.

TABLE 3

Some failures to obtain a laboratory selection response for pesticide resistance in arthropod natural enemies. No doubt this is an abbreviated list because negative results are less often reported.

Species	Pesticide	Citation number
Amblyseius longispinosus	methomyl	29
Aphytis holoxanthus	malathion	35
Aphytis melinus	methidathion	80
Chrysoperla carnea	diazinon	81
Phytoseiulus persimilis	fenvalerate	96

If natural variation is insufficient to allow genetic manipulations to occur, should mutagenesis be used in genetic improvement projects to provide variability? Mutagenesis has been used extensively to obtain desirable genes in *Drosophila*. Mutagenesis by irradiation or chemicals has serious drawbacks; mutations are random, deleterious changes can occur in nontarget genes, and very large numbers of natural enemies must be screened to obtain a potentially useful mutation. Since some natural enemies are difficult to rear in the thousands, mutagenesis may have a limited applicability. For species that are reared on artificial diets or inexpensive hosts, mutagenesis could prove useful (12).

The possibility of developing transgenic natural enemies should be explored (12). Pesticide resistance genes from both prokaryotes and eukaryotes have been identified and cloned. If these genes can be inserted into arthropod natural enemies, be expressed appropriately and stably, rDNA techniques offer the possibility of revolutionizing genetic improvement of arthropod natural enemies. Recombinant DNA techniques could make genetic improvement projects more efficient than artificial selection because once a useful gene is cloned it could be used to transform many beneficial species (5, 6, 8, 102). A few genes have been cloned that could be used in the genetic manipulation of arthropod natural enemies. One such gene is the parathion hydrolase gene isolated from *Pseudomonas diminuta* (103). This prokaryotic gene has been successfully introduced into *Drosophila melanogaster* via P element-mediated transformation where it was expressed, albeit at a low level (104). It was also expressed in the fall armyworm after it was introduced with a virus vector (105). ffrench-Constant et al. (106, 107) cloned a cyclodiene resistance gene from *Drosophila melanogaster*. While this gene is of limited value for agricultural pest management, it could serve as a marker gene. Other genes from *Drosophila* could be used as probes to identify homologous genes in other insect species (6) and these cloned genes could even be modified by *in vitro* mutation.

P-element mediated transformation of *Drosophila* has provided a valuable method for investigating a wide array of questions in fundamental developmental genetics but it is unclear whether the P element can be used as a vector outside the genus *Drosophila* (108, 109). Additional methods for transforming insects will probably be necessary. While transposable elements are probably present in most arthropod species, it seems unlikely that it will be cost effective to develop specific transposable element vectors for each beneficial insect species under consideration. Furthermore, because stability of the transformed natural enemy strain will be crucial if permission is to be obtained for its release into the environment, the use of transposable element vectors should probably be avoided (8).

Research should probably focus on other methods of introducing exogenous DNA, such as transformation of insect cells in culture and their transfer back into developing embryos, which has been achieved with *Drosophila* embryos. Other options may include developing minichromosomes. Eukaryotic chromosome telomeres have been cloned, and it may be possible to develop minichromosomes with a centromere, telomeres and a series of cloned genes. Insertion and stable transmission of minichromosomes to progeny in a variety of beneficial species could provide a stable transformation system.

Viruses offer promise as vectors but such viral vectors should not be capable of spreading to other (particularly pest) arthropod species. Transformation by microinjecting DNA directly into cells has been generally successful with an array of plants and animals; research on its use in beneficial insects should be a high priority because it offers the promise of providing a rapid, cost-effective transformation method that is not dependent upon extensive formal genetic information on the target species. Once transformation can be achieved efficiently, inserted genes must be expressed appropriately (in both time and proper tissues), be stable, and transmitted to progeny. Information on gene regulation and expression must be developed for arthropod natural enemies.

Once genetically engineered beneficial species are obtained through transformation or artificial selection, they must be evaluated to determine whether they perform well in agricultural cropping systems. Efforts to predict field efficacy have focussed on life table analyses, mode of inheritance of the target trait, and evaluation of the stability of the target trait in the manipulated strains. Ultimately, however, the fitness and efficacy of the improved strain must be evaluated under the field conditions in which it is to operate (8).

Protocols for evaluating the risks associated with releasing parasitoids and predators that have been manipulated with rDNA techniques do not currently exist. Such protocols should be developed now--before improved strains are ready for release. Guidelines for release of natural enemies manipulated with rDNA methods will likely include the following questions or principles: 1) Evaluate each strain on a case by case basis. 2) Determine whether the host range/preference of the manipulated parasitoid or predator strain has been altered. (Host/prey specificity is important for efficacy and safety; thus potential changes should be evaluated through laboratory or greenhouse tests.) 3) Determine whether the geographic range of the manipulated predator/parasitoid has been altered. (Temperature/relative humidity tolerances and diapause attributes often restrict geographic distribution and any changes should be demonstrable with growth chamber and laboratory tests.) 4) Determine whether the new strain is stable. (Laboratory tests should determine whether the trait is transmitted faithfully to progeny; whether the inserted genes are maintained in their original insertion site; whether the inserted gene can be transmitted to pest species.)

Whether a parasitoid or predator has been improved through laboratory selection, rDNA techniques, or hybridization, small plot trials will be necessary to evaluate the fitness of the improved strain under field conditions. Such field trials are difficult to design in some cases because we lack critical information on the behavior of beneficial species. For example, the field releases of the azinphosmethyl-resistant strain of *Trioxys pallidus* were complicated by our lack of knowledge about its dispersal. Improved techniques for marking and recovering released insects are highly desirable. Options for improved markers include identifying restriction fragment length polymorphisms (RFLPs) for specific biotypes, developing DNA markers using random primers for the polymerase chain reaction (PCR), or inserting segments of unique noncoding DNA into the genome of specific colonies of beneficial organisms.

Once the strain has proven efficacious and fit in small plot trials, large plot trials should be conducted. This will require efficient and effective mass rearing and monitoring methods so that efficacy, fitness, persistence, and dispersal can be demonstrated in commercial agricultural ecosystems (8).

The implementation of genetically manipulated biological control agents (or any other natural enemy) requires adaptive research and effective educational programs. Such research may require as much or more funding as was required to develop the improved strain (66). Furthermore, once the strain is implemented in the cropping system, research may be required periodically in order to maintain the improved natural enemy in the system: agricultural systems unavoidably change when new pests invade, crop varieties are altered, and pesticides are added or eliminated from the IPM program. Once a genetically-improved natural enemy has been implemented, it is crucial that the efficacy and cost effectiveness of the program be analyzed so that we can begin to generalize about genetic improvement as a tactic in the biological control of arthropod pests.

CONCLUSIONS

Genetic improvement of arthropod natural enemies has proved to be practical and cost effective when the improved strain retains its fitness and methods for implementation are developed. Currently, traits determined by major genes, such as pesticide resistance or nondiapause, are most appropriate attributes for manipulation because we currently lack methods for manipulating and stabilizing traits that are determined by complex genetic systems.

Genetic improvement can be a useful tactic when the natural enemy is known to be a potentially effective biological control agent except for its susceptibility to pesticides, we can locate variability upon which selection can be directed, and the released strain can be maintained in reproductive isolation through the use of pesticide selectivity or geographic isolation. In the next few years, advances in molecular genetics of arthropods should provide fertile new opportunities for genetic improvement of arthropod natural enemies. However, risk assessment issues will have to be resolved before any such transgenic natural enemies can be released into the environment for evaluation or utilization in integrated pest management programs.

ACKNOWLEDGEMENTS

I thank the organizers of this conference for inviting me to participate and for their support. Research in this paper was supported in part by Regional Research Project W-84 and the California Agricultural Experiment Station.

REFERENCES

1.   Mally, C. W., On the selection and breeding of desirable strains of beneficial insects. S. Afr. J. Sci., 1916, 13, 369-385.

2.   Garcia, R., Caltagirone, L. E., Gutierrez, A. P., Comments on a redefinition of biological control. BioScience, 1988, 38, 692-694.

3.   DeBach, P., Selective breeding to improve adaptations of parasitic insects. Proc. X Intern. Cong. Entomol., 1958, 4, 759-768.

4.   Hoy, M. A., Use of genetic improvement in biological control. Agric. Ecosys. Environ., 1986, 15, 109-119.

5.   Beckendorf, S. K. and Hoy, M. A., Genetic improvement of arthropod natural enemies through selection, hybridization or genetic engineering techniques. In Biological Control in Agricultural IPM Systems, eds., M. A. Hoy and D. C. Herzog, Academic Press, N.Y., 1985, pp. 167-187.

6.   Mouches, C., Cloning and characterization of genes responsible for resistance in mosquitoes to insecticides: future prospects of genetic engineering of beneficial arthropods. Proc. Intn. Symp. Fruit Flies, Hania, Crete, 1986, p. 16.

7.   Bartlett, A. C., Guidelines for genetic diversity in laboratory colony establishment and maintenance. In Handbook of Insect Rearing, eds., P. Singh and R. Moore, 1985, Elsevier Science Publ., Amsterdam, pp.7-17.

8.   Hoy, M. A., Genetic improvement of arthropod natural enemies: becoming a conventional tactic ? In New Directions in Biological Control, UCLA Symp.

318

Molecular & Cellular Biology, New Series, 1990, **112**, eds., R. Baker & P.Dunn, Alan R. Liss, N.Y., pp. 405-417.

9. Sailer, R. I., Possibilities for genetic improvement of beneficial insects. In Germ Plasm Resources, 1961, American Assoc. Advance.Science, Washington, D.C., pp. 295- 303.

10. Hoy, M. A., Recent advances in genetics and genetic improvement of the Phytoseiidae. Ann. Rev. Entomol., 1985, **30**, 345-370.

11. Croft, B. A., Arthropod Biological Control Agents and Pesticides, Wiley-Interscience, New York, 1990, 723 pp.

12. Hoy, M. A., Pesticide resistance in arthropod natural enemies: variability and selection responses. In Pesticide Resistance in Arthropods, eds., R. T. Roush and B.E. Tabashnik, Chapman and Hall, New York, 1990, pp. 203-236.

13. Pielou, D. P. and Glasser, R. F., Selection for DDT resistance in a beneficial insect parasite. Science, 1952, **115**, 117-118.

14. Robertson, J. G., Changes in resistance to DDT in *Macrocentrus ancylivorus* . Can. J. Zool., 1957, **35**, 629-33.

15. Adams, C. H. and Cross, W. H., Insecticide resistance in *Bracon mellitor*, a parasite of the boll weevil. J. Econ. Entomol., 1967, **60**, 1016-20.

16. McMurtry, J. A., Huffaker, C. B., and van de Vrie, M., I.Tetranychid enemies: their biological characters and the impact of spray practices. Hilgardia, 1970, **40**, 331-390.

17. Huffaker, C. B., van de Vrie, M. and McMurtry, J. A., II. Tetranychid populations and their possible control by predators: an evaluation. Hilgardia, 1970, **40**, 391-458.

18. Gambaro, P. I., Selezione di popolazioni di Acari predatori resistenti ad alcuni insetticidi fosforati-organici. Inform. Fitopatol. [In Italian], 1975, **25**, 21-25.

19. Caccia, R., Baillod, M., E. Guignard, and Kreiter, S., Introduction d'une souche de *Amblyseius andersoni* Chant (Acari, Phytoseiidae) resistant a l'azinphos, dans la lutte contre les acariens phytophages en viticulture, Rev. Suisse Vitic. Arboric. Hortic., 1985, **17**, 285-90.

20. Motoyama, N., Rock, G. C., Dauterman, W. C., Organophosphorus resistance in an apple orchard population of *Typhlodromus (Amblyseius) fallacis*. J. Econ. Entomol., 1970, **63**, 1439-42.

21. Croft, B. A. and Meyer, R. H., Carbamate and organophosphorus resistance patterns in populations of *Amblyseius fallacis*. Environ. Entomol., 1973, **2**, 691-5.

22. Strickler, K. and Croft, B. A., Variation in permethrin and azinphosmethyl resistance in populations of *Amblyseius fallacis* (Acarina: Phytoseiidae). Environ. Entomol., 1981, **10**, 233-36.

23. Watve, C. M. and Lienk, S. E., Toxicity of carbaryl and six organophosphorus insecticides to *Amblyseius fallacis* and *Typhlodromus pyri* from New York apple orchards. Environ. Entomol. 1976, **5**, 368-70.

24.   Rock, G. C. and Yeargan, D. R., Relative toxicity of pesticides to organophosphorus-resistant orchard populations of *Neoseiulus fallacis* and its prey. J. Econ. Entomol. 1971, **64**, 350-52.

25.   Croft, B. A., Brown, A. W. A. and Hoying, S. A., Organophosphorus-resistance and its inheritance in the predaceous mite *Amblyseius fallacis*. J. Econ. Entomol. 1976, **69**, 64-68.

26.   Croft, B. A. and Hoying, S. A., Carbaryl resistance in native and released populations of *Amblyseius fallacis*. Environ. Entomol., 1975, **4**, 895-98.

27.   Smith, F.S., Henneberry, T. J. and Boswell, A. L., The pesticide tolerance of *Typhlodromus fallacis* (Garman) and *Phytoseiulus persimilis* A. H. with some observations on the predator efficiency of *P. persimilis*. J. Econ. Entomol. 1963, **56**, 274-8.

28.   Hamamura, T., Studies on the biological control of Kanzawa spider mite, *Tetranychus kanzawai* Kishida by the chemical resistant predacious mite, *Amblyseius longispinosus* (Evans) in tea fields (Acarina: Tetranychidae, Phytoseiidae). Bull. Nat. Res. Inst. Tea [In Japanese], 1986, **21**, 121-201.

29.   Hanamura, T., Biological control of the Kanzawa spider mite, *Tetranychus kanzawai* Kishida, in tea fields by the predacious mite, *Amblyseius longispinosus* (Evans), which is resistant to chemicals (Acarina: Tetranychidae, Phytoseiidae). J.A.R.Q. 1987, **21**, 109-116.

30.   Mochizuki, M., A strain of the predatory mite *Amblyseius longispinosus* (Evans) resistant to permethrin, developing in the tea plantation of Shizuoka Prefecture (Acarina: Phytoseiidae). Japn. J. Appl. Ent. Zool.,[In Japanese], 1990, **34**, 171-74.

31.   Anon., Studies on the integrated control of the citrus red mite with the predaceous mite as a principal controlling agent. Acta Entomologica Sinica [In Chinese] 1978, **21**, 260-270.

32.   Inoue, K., Osakabe, M. and Ashihara, W., Identification of pesticide resistant phytoseiid mite populations in citrus orchards, and on grapevines in glasshouses and vinyl-houses (Acarina: Phytoseiidae) [In Japanese]. Jpn. J. Appl. Ent. Zool. 1987, **31**, 398-403.

33.   Adams, J. and Prokopy, R., Apple aphid control through natural enemies. Massachusetts Fruit Notes, 1977, **64**(6), 6-10.

34.   Schoonees, J. and Giliomee, J. H., The toxicity of methidathion to parasitoids of red scale, *Aonidiella aurantii* (Hemiptera: Diaspididae). J. Entomol. Soc. South Africa, 1982, **45**, 261-73.

35.   Havron, A. and Rosen, D., Selection for pesticide resistance in *Aphytis*. Proc. Sixth Intern. Citrus Congress, Tel Aviv, Israel, R. Goren and K. Mendel (Eds.), Balaban Publ., Philadelphia/Rehovot, 1988. pp. 1187-93.

36.   Strawn, A. J., Differences in response to four organophosphates in the laboratory of strains of *Aphytis melinus* and *Comperiella bifasciata* from citrus groves with different pesticide histories, M.S. thesis, 1978, Univ. California, Riverside.

37.   Rosenheim, J. A. and Hoy, M. A., Intraspecific variation in levels of pesticide resistance in field populations of a parasitoid, *Aphytis melinus* (Hymenoptera:

Aphelinidae): the role of past selection pressures. J. Econ. Entomol., 1986, **81**, 1161-73.

38. Mansour, F., A malathion-tolerant strain of the spider *Chiracanthium mildei* and its response to chlorpyrifos. Phytoparasitica, 1984, **12(3-4)**, 163-66.

39. Grafton-Cardwell, E. E. and Hoy, M. A., Intraspecific variability in response to pesticides in the common green lacewing, *Chrysoperla carnea* (Stephens) (Neuroptera: Chrysopidae). Hilgardia, 1985, **53(6)**, 1-32.

40. Pree, D. J., Archibold, D.E. and Morrison, R. K. Resistance to insecticides in the common green lacewing *Chrysoperla carnea* (Neuroptera: Chrysopidae) in southern Ontario. J. Econ. Entomol. 1989, **82**, 29-34.

41. Atallah, Y. H. and Newsom, L. D., Ecological and nutritional studies on *Coleomegilla maculata* DeGeer (Coleoptera: Coccinellidae). III. The effect of DDT, toxaphene and endrin on the reproductive and survival potentials. J. Econ. Entomol., 1966, **59**, 1181-87.

42. Rathman, R. J., Johnson, M. W., Rosenheim, J. A., and Tabashnik, B. E., Carbamate and pyrethroid resistance in the leafminer parasitoid *Diglyphus begini* (Hymenoptera: Eulophidae). J. Econ. Entomol., 1990, **83**, 2153-58.

43. Kennett, C. E., Resistance to parathion in the phytoseiid mite *Amblyseius hibisci*. J. Econ. Entomol., 1970, **63**, 1999-2000.

44. Tanigoshi, L. K. and Congdon, B. D., Laboratory toxicity of commonly-used pesticides in California citriculture to *Euseius hibisci* (Chant) (Acarina: Phytoseiidae). J. Econ. Entomol. 1983, **76**, 247-50.

45. Huffaker, C. B. and Kennett, C. E., Differential tolerance to parathion of two *Typhlodromus* predatory on cyclamen mite. J. Econ. Entomol., 1953, **46**, 707-708.

46. Morgan, C. V. G. and Anderson, N. H., Notes on parathion-resistant strains of two phytophagous mites and a predacious mite in British Columbia. Can. Entomol., 1958, **90**, 92-97.

47. Hoyt, S. C., Integrated chemical control of insects and biological control of mites on apple in Washington. J. Econ. Entomol., 1969, **62**, 74-86.

48. Hoy, M. A. and Knop, N. F., Studies on pesticide resistance in the phytoseiid *Metaseiulus occidentalis* in California. In J.G. Rodriguez, ed., Recent Advances in Acarology, 1979, **I**, Academic Press, N. Y., pp. 89-94.

49. Hoy, M. A. and Standow, K. A., Inheritance of resistance to sulfur in the spider mite predator *Metaseiulus occidentalis*. Entomol. Exp. Appl. 1982, **31**, 316-23.

50. Schulten, G. G. M., and van de Klashorst, G., Genetics of resistance to parathion and demeton-s-methyl in *Phytoseiulus persimilis* A-H (Phytoseiidae).Proc. IV Intn. Cong. Acarol., Saalfelden, Austria, 1974, pp. 519-24.

51. Schulten, G. G. M., van de Klashorst, G. and Russell, V. M., Resistance of *Phytoseiulus persimilis* A. H. (Acari: Phytoseiidae) to some insecticides. Zeitschrft. angew. Entomol. 1976, **80**, 337-41.

52. Putman, W. L. and Herne, D.C., The role of predators and other biotic factors in regulating the population density of phytophagous mites in Ontario peach orchards. Can. Entomol., 1966, **98**, 808-20.

53. Hoyt, S. C., Resistance to azinphosmethyl of *Typhlodromus pyri* (Acarina: Phytoseiidae) from New Zealand. N. Z. J. Sci., 1972, **15**, 16-21.

54. Penman, D. R., Ferro, D. N. and Wearing, C. H., Integrated control of apple pests in New Zealand. VII. Azinphosmethyl resistance in strains of *Typhlodromus pyri* from Nelson. N. Z. J. Exp. Agric., 1976, **4**, 377-80.

55. Baillod, M. and Guignard, E., Resistance de *Typhlodromus pyri* Scheuten a lazinphos et lutte biologique contre les acariens phytophages en arboriculture. Rev. Suisse Vitic. Arboric. Hortic., 1984, **16**, 155-60.

56. Kapetanakis, E. G. and Cranham, J. E., Laboratory evaluation of resistance to pesticides in the phytoseiid predator *Typhlodromus pyri* from English apple orchards. Ann. Appl. Biol., 1983, **103**, 389-400.

57. van de Baan, H. E., Kuijpers, L. A. M., Overmeer, W. P. J. and Oppenoorth, F. J., Organophosphorus and carbamate resistance in the predacious mite *Typhlodromus pyri* due to insensitive acetylcholinesterase. Exper. Appl. Acarol., 1985, 1, 3-10.

58. Overmeer, W. P. and van Zon, A. Q., Resistance to parathion in the predaceous mite *Typhlodromus pyri* Scheuten. Meded. Fac. Landbouwwet. Rijksuniv. Gent., 1983, **48**, 237-51.

59. Hadam, J. J., Aliniazee, M., and Croft, B. A., Phytoseiid mites (Parasitiformes: Phytoseiidae) of major crops in Willamette Valley, Oregon, and pesticide resistance in *Typhlodromus pyri* Scheuten. Environ. Entomol., 1986, **15**, 1255-63.

60. Croft, B. A. and Barnes, M. M., Comparative studies on four strains of *Typhlodromus occidentalis*: III. Evaluations of releases of insecticide resistant strains into an apple orchard ecosystem. J. Econ. Entomol., 1971, **64**, 845- 50.

61. Hoy, M. A., Integrated mite management for California almond orchards. In Spider Mites. Their Biology, Natural Enemies and Control, W. Helle and M. W. Sabelis, eds., 1985, **1B**, 299-310.

62. Roush, R. T. and Hoy, M. A., Genetic improvement of *Metaseiulus occidentalis*: selection with methomyl, dimethoate, and carbaryl and genetic analysis of carbaryl resistance. J. Econ. Entomol., 1981, **74**, 138-141.

63. Roush, R.T. and Hoy, M. A., Laboratory, glasshouse, and field studies of artificially selected carbaryl resistance in *Metaseiulus occidentalis*. J. Econ. Entomol., 1981, **74**, 142-47.

64. Hoy, M. A. and Knop, N. F., Selection for an genetic analysis of permethrin resistance in *Metaseiulus occidentalis*: genetic improvement of a biological control agent. Entomol. Exp. Appl., 1981, **30**, 10-18.

65. Hoy, M.A., Genetic improvement of a biological control agent: multiple pesticide resistance and nondiapause in *Metaseiulus occidentalis*. Acarolology VI, 1984, **2**, 673-79.

66.   Headley, J. C. and Hoy, M. A., Benefit/cost analysis of an integrated mite
        management program for almonds. J. Econ. Entomol., 1987, **80**, 555-59.

67.   Bruce-Oliver, S. J. and Hoy, M. A., The effect of prey stage on the life table attributes
        of a genetically-manipulated COS strain of *Metaseiulus occidentalis* (Nesbitt) (Acari:
        Phytoseiidae). Exp. Appl. Acarol., 1990, **9**, 201-217.

68.   Field, R. P.and Hoy, M. A., Evaluation of genetically improved strains of *Metaseiulus
        occidentalis* (Nesbitt) for integrated control of spider mites on roses in greenhouses.
        Hilgardia, 1985, **54(2)**, 1-32.

69.   Roush, R. T. and Tabashnik, B. E., eds., Pesticide Resistance in Arthropods, 1990,
        Chapman and Hall, New York, 303 pp.

70.   Abdelrahman, I., Toxicity of malathion to the natural enemies of California red scale,
        Aonidiella aurantii (Mask.) (Hemiptera: Diaspididae). Australian J. Agric. Res., 1973,
        24, 119-33.

71.   Delorme, R., Angot, A. and Auge, D., Variations de sensibilite d'*Encarsia formosa*
        Gahan (Hym: Aphelinidae) soumis a des pressions de selection insecticide: approches
        biologique et biochemique. Agronomie, 1984, **4**, 305-09.

72.   Strickler, K.and Croft, B. A., Selection for permethrin resistance in the predatory mite
        Amblyseius fallacis. Entomol. Exp. Appl., 1982, **31**, 339-45.

73.   Whalon, M. E., Croft, B. A. and Mowry, T. M., Introduction and survival of
        susceptible and pyrethroid-resistant strains of *Amblyseius fallacis* in a Michigan
        apple orchard. Environ. Entomol., 1982, **11**, 1096-99.

74.   Huang, M. D., Ziong, J. J. and Du, T.Y., The selection for and genetical analysis of
        phosmet resistance in *Amblyseius nicholsi*. Acta Entomol.Sinica ,1987, **30**, 133-39.

75.   Du, T. Y., and Xiong, L. J., The selection for and genetical analyses of phosmet-
        dimehypol resistance in *Amblyseius nicholsi*. [In Chinese], In: Studies on the
        Integrated Management of Citrus Insect Pests, Huang Ming-du, ed., 1989, Academic
        Book and Periodical Press, Beijing, 56-62.

76.   Du, T. Y., Xiong, L. J., and Huang, M. D., Observation on bionomics of phosmet-
        resistant strain in *Amblyseius nicholsi* Ehara et Lee. Natural Enemies of Insects, [In
        Chinese], 1987, **9 (3)**, 173-176.

77.   Xiong, J. J., Du, T. Y. and Huang, M. D., Field studies in citrus orchards of
        phosmet-resistant strain of *Amblyseius nicholsi* Ehara et Lee, Natural Enemies of
        Insects, [In Chinese], 1989, **11**, 50-55.

78.   Ke, L. S., Yang, Y.Y., and Xin, J. L., Selection for and genetic analysis of
        dimethoate resistance in *Amblyseius pseudolongispinosus*. Acta Entomol.Sinica [In
        Chinese], 1990, **33**, 393-97.

79.   Havron, A. and Rosen, D., Selection for pesticide resistance in *Aphytis*. In:
        Proceedings of the Sixth International Citrus Congress, Goren, R. and Mendel, K.,
        ed., 1988, Balaban Publishers, Philadelphia/Rehovot, 1187-1193.

80.   Rosenheim, J. A. and Hoy, M. A., Genetic improvement of a parasitoid biological
        control agent: artificial selection for insecticide resistance in *Aphytis melinus* . J. Econ.
        Entomol., 1988, **81**, 1539-50.

81. Grafton-Cardwell, E. E. and Hoy, M. A., Genetic improvement of common green lacewing, *Chrysoperla carnea* (Neuroptera: Chrysopidae): selection for carbaryl resistance. Environ. Entomol., 1986, **15**, 1130-36.

82. Roush, R. T., Peacock, W. L., Flaherty, D. L. and Hoy, M. A., Dimethoate-resistant spider mite predator survives field tests. California Agriculture, 1980, **34(5)**, 12-13.

83. Roush, R. T. and Plapp, F.W., Biochemical genetics of resistance to aryl carbamate insecticides in the predaceous mite, *Metaseiulus occidentalis*. J. Econ. Entomol., 1982, **75**, 304-307.

84. Hoy, M. A., and Ouyang, Y. L., Selection of the western predatory mite, *Metaseiulus occidentalis* (Acari: Phytoseiidae), for resistance to abamectin. J. Econ. Entomol., 1989, **82**, 35-40.

85. Petrushov, A. Z., Genetic and biochemical mechanisms of Ambush resistance in *Metaseiulus occidentalis* (Acarina: Phytoseiidae). [Abstract] VIII Intern. Congress of Acarology, Ceske Budejovice, Czechoslovakia, Aug. 6-11, 1990.

86. Avella, M., Fournier, D., Pralavorio, M. and Berge, J.,B., Selection pour la resistance a la deltamethrine d'une souche de *Phytoseiulus persimilis* Athias-Henriot. Agronomie, 1985, **5**, 177-80.

87. Konig, V. K. and Hassan, S. A., Resistenz und Kreuzresistenz der Raubmilbe *Phytoseiulus persimilis* (Athias-Henriot) gegenüber organischen Phosphorsäureestern. Zeitsch. angew. Entomol. 1986, **101**, 206-15.

88. Fournier, D., Pralavorio, M., Trottin-Caudal, Y., Coulon, J., Malezieux, S., and Berge, J. B., Selection artificielle pour la resistance au methidathion chez *Phytoseiulus persimilis* AH.Entomophaga, 1987, **32**, 209-19.

89. Fournier, D., Pralavorio, M., Coulon, J., Berge, J. B., Fitness comparison in *Phytoseiulus persimilis* strains resistant and susceptible to methidathion. Exp. Appl. Acarol., 1987, **5**, 55-64.

90. Fournier, D., Pralavorio, M., Cuany, A., and Berge, J. B., Genetic analysis of methidathion resistance in *Phytoseiulus persimilis* (Acari: Phytoseiidae). J. Econ. Entomol., 1988, **4**, 1008-13.

91. Xu, X., Li, K.H., Li, Y.F., Moon, Z.,and Li, L.Y., Culture of resistant strain of *Trichogramma japonicum* to pesticides. Nat. Enemies of Insects, 1986, **8(3)**, 150-54.

92. Hoy, M. A. and Cave, F. E., Guthion-resistant strain of walnut aphid parasite. Calif. Agric., 1988, **42(4)**, 4-5.

93. Hoy, M. A.and Cave, F. E., Toxicity of pesticides used in walnuts to a wild and laboratory-selected azinphosmethyl-resistant strain of *Trioxys pallidus*. J. Econ. Entomol., 1989, **82**, 1585-92.

94. Hoy, M. A., Cave, F.E., Beede, R., Grant, R., Krueger, W., Olson, W., Spollen, W., Barnett, W. and Hendricks, L. C., Release, dispersal, and recovery of a laboratory-selected azinphosmethyl-resistant strain of the walnut aphid parasite *Trioxys pallidus*. J. Econ. Entomol.1990, **83**, 89-96.

95. Hoy, M. A., Cave, F. E. and Caprio, M. A., Guthion-resistant parasite evaluated for implementation. California Agriculture, 1991, in press.

96.   Markwick, N. P., Detecting variability and selecting for pesticide resistance in two species of phytoseiid mites. Entomophaga, 1986, **31**, 225-36.

97.   Markwick, N. P. Induced resistance in beneficials with particular reference to *Typhlodromus pyri* Scheuten [Acarina: Phytoseiidae]. Proc. CSIRO/DSIR Workshop, Insecticide Resistance Management, Canberra, July 1986, pp. 91-98.

98.   Markwick, N. P., Wearing, C. H., and Shaw, P. W., Pyrethroid insecticides for apple pest control: I. Development of pyrethroid-resistant predatory mites. 43rd N. Z. Weed and Pest Control Conference, August 1990, 296-300.

99.   Suckling, D. M., Walker, J. T. S., Shaw, P. W., Markwick, N. P., and Wearing, C. H., Management of resistance in horticultural pests and beneficial species in New Zealand. Pestic. Sci. 1988, **23**, 157-164.

100. Walker, J. T. S., Markwick, N. P., Wearing, C. H., Shaw, P. W. and White, V., Pyrethroid insecticides for apple pest control: II. Field evaluation of mite and insect control. 43rd N.Z. Weed & Pest Control Conf., August 1990, 301-305.

101. Caprio, M. A., Hoy, M. A. and Tabashnik, B. E., A model for implementing a genetically-improved strain of the parasitoid *Trioxys pallidus* Haliday (Hymenoptera: Aphidiidae). American Entomologist, 1991, in press.

102. Steiner, W. W. M. and Teig, D. A., *Microplitis croceipes* (Cresson): genetic characterization and developing insecticide resistant biotypes, Southwestern Entomologist, 1989, **Suppl.12**, 71-87.

103. Serdar, C. M., Murdock, D. C. and Rohde, M. F., Parathion hydrolase gene from *Pseudomonas diminuta* Mg: subcloning, complete nucleotide sequence, and expression of the nature portion of the enzyme in *Escherichia coli*. Biotechnology, 1989, **7**, 1151-55.

104. Phillips, J. P., Xin, J.H., Kirby, K., Milne, C. P., Krell, P., and Wild, J. R., Transfer and expression of an organophosphate insecticide-degrading gene from *Pseudomonas* in *Drosophila melanogaster*. Proc. Natl. Acad. Sci. USA, 1990, **87**, 8155-59.

105. Dumas, D. P., Wild, J. R. and Raushel, F. M., Expression of *Pseudomonas* phosphotriesterase activity in the fall armyworm confers resistance to insecticides. Experientia, 1990, **46**, 729-31.

106. ffrench-Constant, R. H., Roush, R. T., and MacIntyre, R. J., Isolation, characterization and progress in cloning of cyclodiene insecticide resistance in *Drosophila melanogaster*. In Molecular Insect Science, H. H. Hagedorn et al., Eds., Plenum Press, N. Y., 1990, 41-48.

107. ffrench-Constant, R. H., Mortlock, D. P., Shaffer, C. D., MacIntyre, R. J. and Roush, R. T., Molecular cloning and transformation of cyclodiene resistance in *Drosophila*: an invertebrate GABA$_A$ receptor locus. Proc. National Acad. Sci. U.S.A., in press.

108. Handler, A. M. and O'Brochta, D. A., Prospects for gene transformation in insects. Annu. Rev. Entomol., 1991, **36**, 159-83.

109. Walker, V. K., Gene transfer in insects, Adv. Cell Culture, 1989, **7**, 87-124.

POTENTIAL NEW SITES FOR FUNGICIDES

JOHN J. BISAHA AND ROBERT S. LIVINGSTON
E. I. du Pont de Nemours and Company
Agricultural Products Stine-Haskell Research Center
Newark, DE 19714 USA

ABSTRACT

One method of combating pesticide resistant fungi is to develop areas of chemistry active at a novel site in the pathogen. This can be accomplished in two ways: 1) large-scale screening of a wide variety of different structural classes of chemistry, identification of a class of chemistry active on a specific pathogen and optimization of that activity by an analoging program; or 2) selection of a specific process believed to be essential to the pathogen, and through biochemical knowledge at the molecular, enzyme, cellular and plant level develop active areas of chemistry to inhibit that process. In reality, the discovery process may incorporate both mechanisms to varying degrees. Due to the somewhat serendipitous aspect of the first method of discovery, this presentation will focus on the latter. Specific topics will deal with target site selection and verification, establishment of an appropriate assay, synthesis concerns and correlation of assay/greenhouse results. In addition to the general process, specific examples will be presented from our group's work on the sterol biosynthesis pathway.

INTRODUCTION

The development of resistance to agricultural chemicals in a population can be due to any one of a number of different mechanisms such as modification of target enzyme, metabolism or uptake/export systems. One strategy of dealing with the problem of resistance in fungi or other organisms is to develop a chemical material with a new mode of action. In this paper we will address the process of discovery of new modes of action followed by several examples from our labs in which we utilize these principles for the discovery of novel fungicides.

BIOCHEMICAL APPROACH TO NEW TARGET SITES
(GENERAL CONSIDERATIONS)

The process of discovery of a new mode of action can be undertaken in two ways, a chemistry driven approach or a biochemistry driven approach. Traditionally, the chemistry driven approach has been the primary method of developing new classes of agrochemicals. This approach involves the large-scale synthesis of a wide variety of chemicals of different structural classes, and the evaluation of these compounds for fungicidal activity in a pathogen/host plant system. Typically, mode of action studies are not initiated until the most active analog enters the product development stage. Clearly this method works, having given us the wide variety of agrochemicals we possess today, and will continue to produce new areas in the future. However, discovery of a new mode of action by this process is somewhat serendipitous and may not be the most efficient way to develop resistance breaking agrichemicals. In contrast, the primary topic of this paper will be the biochemically driven approach and its utilization in the discovery process. This process may also be termed target site directed. In this approach, the order of steps is somewhat different. At the outset, a specific process known to be important to the pathogen is identified and through the interaction of biochemists and synthesis chemists, areas of chemistry that inhibit that process are identified.

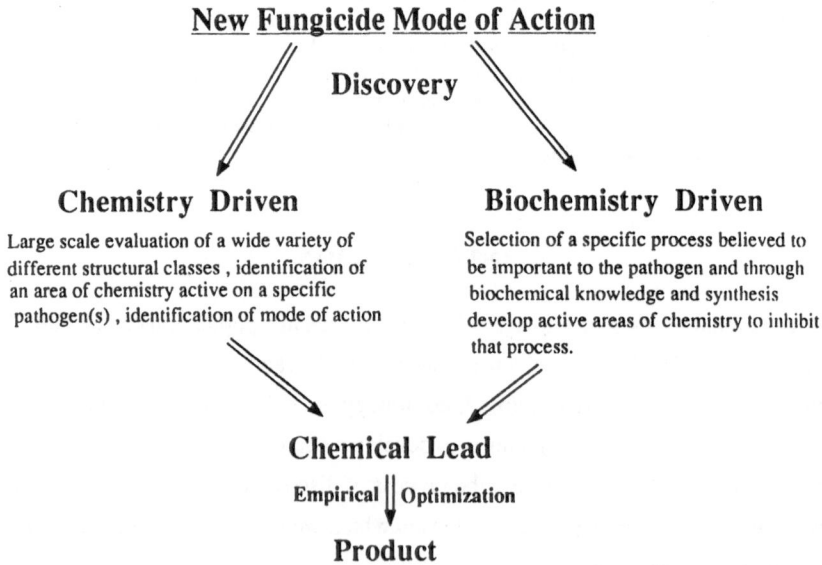

Figure 1. A comparison of two approaches to discovery of new fungicide modes of action.

Both approaches are designed to identify a chemical lead. But an important distinction needs to be made. In the chemical approach, any compound, having shown disease control activity, is a fungicide lead and, through the power of empirical optimization, has the potential to be developed into a marketable product. In the target driven approach, the lead is a chemical that inhibits the targeted process and may or may not in itself be a fungicide lead. Further development may be required to overcome such problems as delivery to the active site, fungal or plant metabolism, photodegradation, etc. We will return to this point shortly.

The biochemistry driven approach can further be divided into three phases: target selection, target validation and chemical optimization. The starting point is target selection; which process are we going to attempt to inhibit? Much discussion can occur concerning target site selection, ranging from trying to inhibit every enzyme to a full-scale program of gene disruption to identify important steps; several major points are summarized here.

The market is the major driving force in the fungicide business. What is needed at the grower level? With the onset of resistance to several classes of fungicides, the market will be ripe for development of effective agrochemicals with new modes of action. Do we attempt to target a process that would give us broad efficacy, such as inhibition of microtubule formation as with benomyl, or do we target processes found only in a specific pathogen, such as wheat rust? These decisions need to be made at the initial stages of a program.

A second factor that enters into target selection is knowledge of the biochemistry of fungal systems. Understanding of the germination, penetration, proliferation and sporulation events can assist in selecting a target for disruption of the pathogen life cycle. It is the necessity of this knowledge base that should encourage organizations to support programs in fungal biochemistry, even if they are not immediately coupled to product development.

Fungal genetics can also serve as a useful selection tool and could indicate whether or not a lesion in a biosynthetic pathway would render an organism nonpathogenic. Additional factors could be included in target site selection and it remains for each organization to decide exactly how to proceed. At this point, we turn to the next phase of the discovery process, that of target site validation.

Once a potential target has been selected, research will concentrate on validation of the target. Would inhibition of that step effectively limit fungal growth and development of plant disease? Several criteria that can be used to identify a "good" target include: the utility of known compounds on selected pathways, clear biological evidence that a selected process is crucial to organism viability and correlation of target inhibition with control of plant disease. With these points in mind, how do we try to uncover new areas of chemistry to attack these targets? A flexible assay is needed to answer the initial question of whether or not a specific compound has any effect on the selected target. The assay may be at the target level (for instance, isolated enzyme, receptor or nucleic acid binding) or at the cellular level. Results from these assays will provide different information, such as factors that affect the ability of the chemical to reach the target, and it is important to recognize this when interpreting assay results. Ideally, an assay should use the pathogen of interest, but in some cases this may not be possible. It then becomes important to identify a selected organism with biochemical processes similar to the desired target organism.

In organic synthesis, the primary question is not how to synthesize a compound but what to synthesize. In a program directed against an enzymatic process, potential inhibitors can be designed as substrate or transition state analogs taking advantage of the natural binding characteristics of the enzyme pocket. These methods offer a useful starting point for an active synthesis program. While initial lead compounds may posess quite complex structures, use of structure activity relationships, identification and trimming down of the less important domains of a compound, or incorporation of substrate isosteres can lead to new areas of chemical design. Similar chemical starting points can be gained from identification of target inhibition by known compounds, whether derived from natural or synthetic sources. The chemical design process will be a cyclical one in which assay results direct further synthetic effort.

Design of an inhibitor directed by a target assay will give rise to a lead compound for that target but to establish a fungicide lead, the compound will have to exhibit plant disease control. In order to bridge this gap, it is necessary to correlate assay results with the greenhouse or field results. The chance of finding an excellent *in vitro* inhibitor which is also an excellent fungicide is small, and the parameters present in the plant system are far too numerous, and currently not well enough understood, to allow design of a fungicide from a strictly molecular level.

While an assay can help to answer the question of whether a compound acts by a putative mode of action, it also drives the initial phases of a program. Plant disease control results will drive the latter phases. It is the application of information derived from both sources that is the key to the biochemically driven approach to discovery. If plant disease control results reflect target activity and proper controls have been included to ensure that the observed biochemical effect is the primary mode of action for disease control, then that area of chemistry can proceed to the optimization phase. In the event that the assay and greenhouse results do not agree, and the commitment still exists for the target to be developed, the next step is to ask why. Pharmacokinetics can perhaps be used to try to answer this question, and experiments conducted, often using a radiolabeled compound, to determine such factors as leaf permeability, compound mobility, and eventual fate of the compound in the plant and fungus. Results of these studies will then feed back into the chemical design process where synthetic efforts to overcome the problem can be utilized, perhaps by modification of partition coefficient or acidity, or blocking a site of metabolism.

    A representation of the validation sequence is shown in Figure 2. The process is a circular one where biological data, from both the assay (*in vitro*) and plant disease control (*in vivo*) direct the synthesis program and indicate the hurdles that need to be overcome in terms of pharmacokinetics for a given class of compounds. The product of this validation phase will be one or more lead areas that will be carried on to the optimization phase of the program.

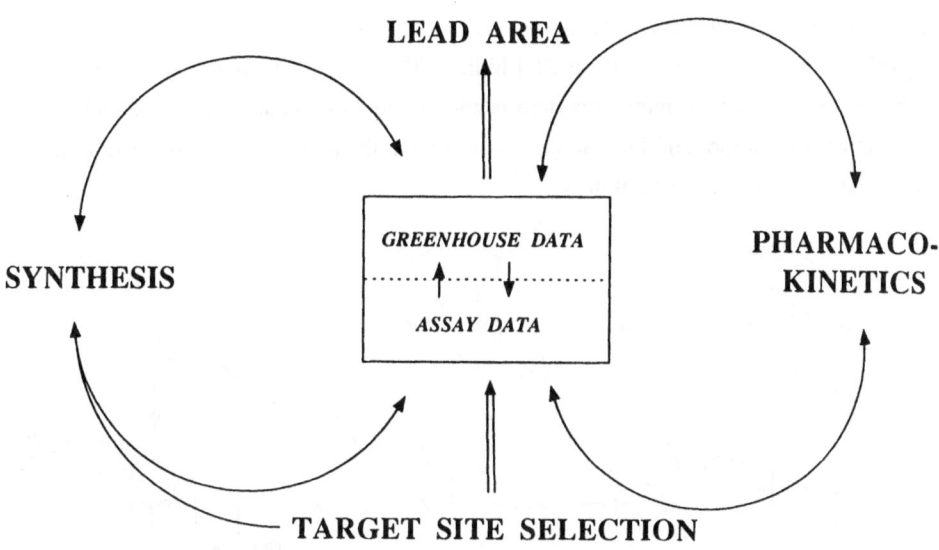

Figure 2. Process of lead area discovery in the biochemistry driven approach.

    At this point, leads with the novel mode of action generated in the prior stage will feed into a program of empirical optimization of disease control to determine their potential as fungicide candidates. Using quantitative structure activity relationships to select the varying substituents for a series of lead analogs, synthesis will be carried out followed by evaluation on the plant/pathogen of interest. During this evaluation phase, a fungicide lead may fail not only due to poor disease control but also for environmental, toxicity or economic reasons. If the validation of the new mode of action has been carried out properly, but a particular lead area does not generate a commercial product, the target may still be valid. Utilization of the biochemical knowledge gained in the validation phase to further chemical design of second generation lead areas or using the target assay to screen compound files for new lead structures, offer possible ways to proceed.

NEW TARGETS IN STEROL BIOSYNTHESIS

Sterol biosynthesis is a proven essential pathway in fungi. Two major classes of fungicides that inhibit the pathway are the triazoles, which inhibit 14-demethylase (1a), and the morpholines, which show inhibition of both the 14-reductase and the 8-isomerase (1b). A major market exists encompassing disease control of powdery mildews, rusts, eyespot and Septoria which in cereals alone totals $1.1 billion. The biosynthetic pathway from acetate to the final sterol contains numerous steps in addition to those mentioned above, and inhibition of one or more of these processes clearly has the potential of leading to a new class of sterol biosynthesis inhibitors.

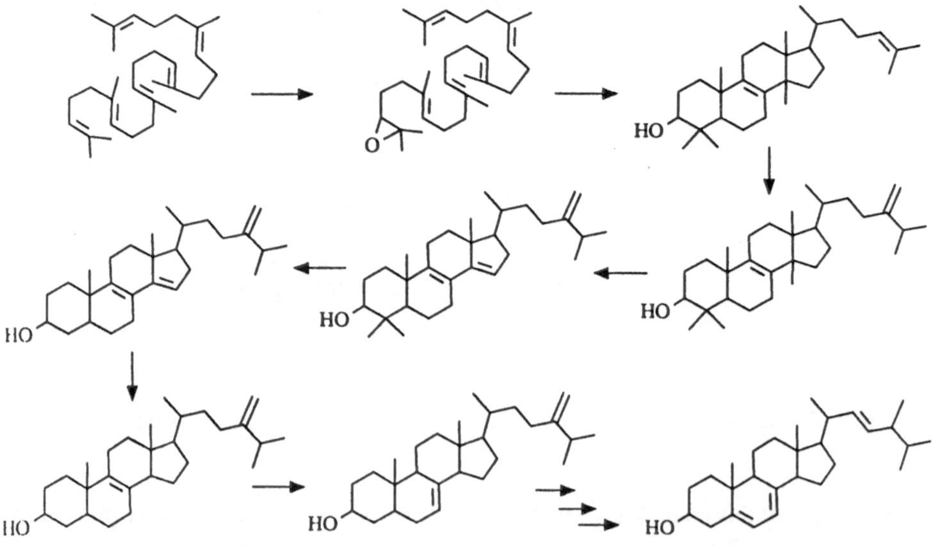

Figure 3. A general pathway of sterol intermediates resulting from acetate incorporation in *Ustiago maydis*.

Figure 3 shows a cascade of sterols from squalene to ergosterol. The exact order of these steps may be different for various fungi but inhibition of one of these steps, such as the 14-demethylase, would lead to a buildup of the precursors, such as 14-α methyl fecosterol, obtusifoliol, eburicol or lanosterol. Treatment of an actively growing fungal culture in liquid medium, containing radiolabeled acetate and an inhibitor, would result in the accumulation of radiolabeled sterol intermediates. Identification of those sterols would indicate at which step the pathway is blocked. The general protocol of the assay (2) is outlined on Figure 4.

Sterol Biosynthesis Assay

--   2 ml cells

--   10 ul inhibitor stock solution

--   50 ul ^{14}C acetate

--   incubate 30 °C, 90 min

--   kill cells with EtOH / KOH

--   100 °C for 60 min to extract / saponify sterols

--   extract sterols with pet ether

--   dry, dissolve in EtOH

--   analyze sterols by C8 HPLC with UV / radioactive detectors

Figure 4. Protocol for the sterol biosynthesis assay.

Figure 5 shows HPLC traces obtained from the assay using various classes of inhibitors (3). The sterols are identified by their retention times relative to cholesterol. In several cases where we have observed a new radioactive peak, the sterol was isolated and its structure determined by NMR and mass spectral analysis. Currently we can identify inhibition at ten discrete points in the pathway. Since the assay utilizes living cells, inhibition indicates that the test compound passes into the cell and is not immediately metabolized or excreted. In conjunction with the sterol assay, a growth assay is conducted under similar conditions. Good correlation between target inhibition and growth inhibition indicates that we are on the right track for a lead area.

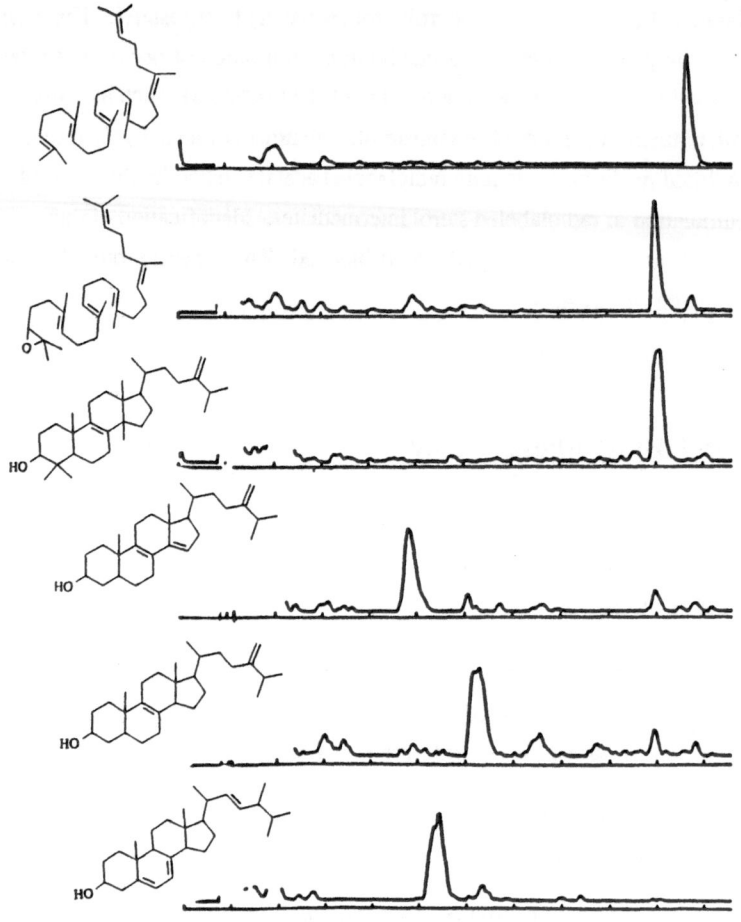

Figure 5. HPLC separation of radiolabeled sterols resulting from inhibition of various sites in the *Ustilago maydis* pathway: (a)squalene epoxidase, (b) squalene oxide cyclase, (c) 14-α demethylase, (d) 14-reductase, (e) 8-isomerase, (f) no inhibition of pathway.

a) Squalene Epoxidase

Squalene epoxidase catalyzes the epoxidation of squalene at the terminal olefin, and is the last enzyme in the pathway that has a non-steroidal product. Inhibition of this step would then result in a complete lack of sterols in the fungal membrane. Coupled with the inability of externally supplied sterols to reduce the toxicity of traditional SBI fungicides, inhibition of this enzyme was proposed as a target for an agricultural fungicide. The allylamine class of antimycotics such as terbinafine, discovered by Sandoz in 1974 (4), have as their primary mode of action epoxidase inhibition (5). Another class of chemistry that shows epoxidase inhibition is exemplified by tolnaftate (6). In terms of resistance, mutants of *Cercospora beticola* resistant to triazole fungicides were not resistant to the allylamines (7) and *Ustilago maydis* mutants isolated after UV irradiation which were resistant to the allylamines were not resistant to the triazoles (3).

Squalene Epoxidase

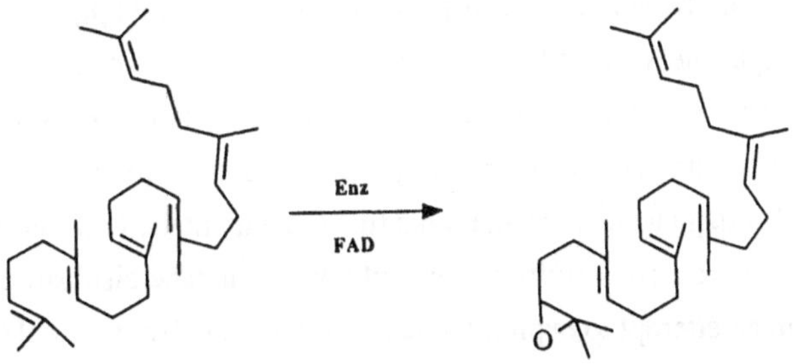

o Early step in the pathway

o Known mode of action for the antimycotics terbinafine and tolnaftate

o No cross resistance of the allylamines with triazoles

o Slow turnover of the enzyme

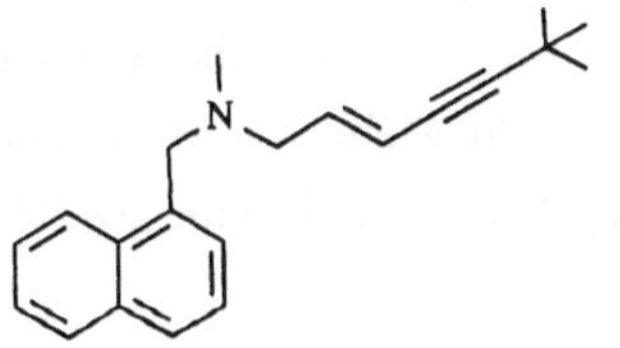

Terbinafine

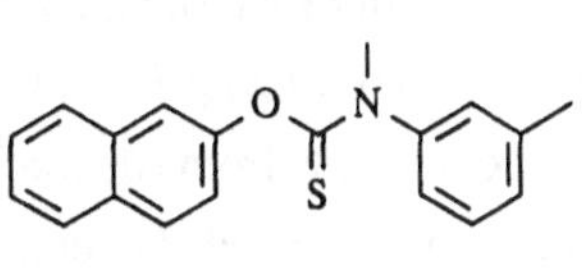

Tolnaftate

Figure 6. Squalene epoxidase as a fungicide target.

Due to the inability to isolate sufficient amounts of the enzyme from fungal sources, the enzyme was obtained from a mammalian source. Squalene epoxidase was detergent solubilized from rat liver microsomes and purified to 50% homogeneity. Homogeneous cytochrome P-450 reductase was isolated from rat liver microsomes or baker's yeast. Initial rate studies suggested a sequential kinetic mechanism where random addition of reaction components to enzyme forms a quaternary complex (Enz/FAD/P-450/Reductase/Squalene). The rate limiting step was determined, using a series of FAD analogs, to be the transfer of electrons from P-450 reductase to squalene epoxidase. The epoxidase cofactor, FAD, obtains electrons from P-450 FMN not P-450 FAD. The high redox potential of P-450 FMN may contribute to the slow turnover of squalene epoxidase (0.5/min). Kinetic data suggest the allylamines bind most tightly to the enzyme/FAD/reductase complex and are competitive with substrate squalene (8).

In the synthesis effort for this project several approaches were tried. Modified squalene analogs that incorporated a latent electrophilic group such as an acetylene or an allene at the epoxidation site were prepared. Here we had hoped that epoxidation would lead to a potent alkylating agent that would serve to deactivate the enzyme. Aza-squalenes with an allylic functionality at the terminus similar to the allylamines were also examined. Several flavin/squalene bisubstrate analogs were also prepared in an attempt to exploit the cofactor binding affinity. In these areas, very little or no epoxidase inhibition was observed in the sterol biosynthesis assay. Finally, we looked at modification of the terbinafine structure, varying both the aromatic portion and the side chain. Many of these compounds did show significant enzyme inhibition and several of the better analogs are shown on Figure 7.

The compounds U9483, 38449 and E9463 had enzyme inhibition I_{50} values in the same range as the allylamines terbinafine and naftifine, but in general were less effective in the *in vitro* growth assay. While U9483 and 38449 were active at a single site in the pathway, E9463 also inhibited the 8-isomerase. In the *in vivo* plant disease screens, these and other compounds generated by the program gave only minimal control of wheat powdery mildew and botrytis on grapes.

After a substantial amount of work, we concluded that the current classes of chemistry did not offer an effective fungicide lead, and the inability to isolate sufficient quantities of enzyme from economically important pathogens limited the possible mechanistic studies needed to continue.

Although this program did not result in squalene epoxidase inhibitors significantly more active than the original allylamines, the flexibility of the assay allowed verification of other sites in the pathway inhibited by test compounds.

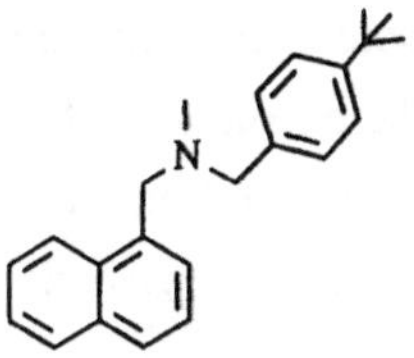

U9483 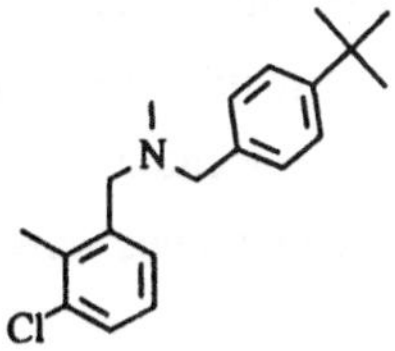   38449 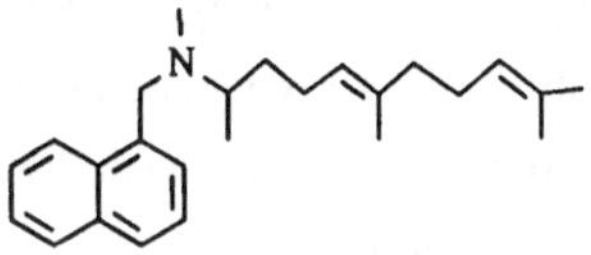   E9463

Sterol Biosynthesis Assay (U. maydis)

	I_{50}(erg)	I_{50}(growth)	Site
U9483	0.008 ug/ml	0.089 ug/ml	epox
38449	0.018	0.810	epox
E9463	0.033	0.780	epox/isom
terbinafine	0.013	0.044	epox
naftifine	0.005	>2	epox

Plant disease control screens - weak activity on wheat powdery   mildew and botrytis

Figure 7. Selected examples of sterol biosynthesis inhibitors. I_{50} values refer to the concentration of inhibitor at which incorporation of acetate into ergosterol or growth of *U. maydis* in liquid culture is reduced by one half.

b) 8-Isomerase

One site of inhibition which we have subsequently examined is the 8-isomerase, which is inhibited, in conjunction with the 14-reductase, by the morpholine class of fungicides. To determine whether the 8-isomerase was a good target in itself, and if design of a good inhibitor of this enzyme would lead to a useful fungicide, synthesis focused mainly on aza squalene and aza sterol partial structures designed to mimic the putative carbocationic transition state of the isomerization. A number of molecules prepared in this program showed excellent inhibition of the isomerase with I_{50} values comparable to the commercial morpholines. Although designed as isomerase inhibitors, many of these structures also inhibited the 14-reductase at slightly higher concentrations. However, where isomerase was the sole block in the pathway, the test compounds were not active in the *in vivo* plant disease control assay.

8 - Isomerase

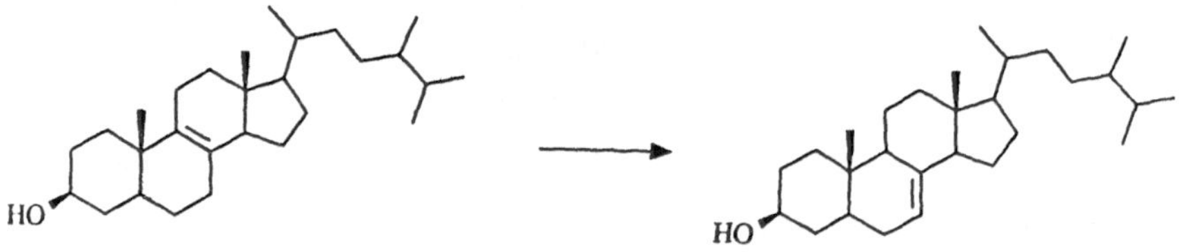

○ Inhibited by the morpholine class in addition to the 14-reductase

○ Inhibitors designed to mimic the carbocationic TS
   - aza squalene and aza sterol partial structures

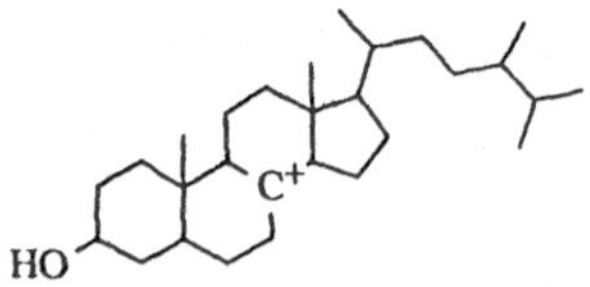

○ Whole cell assay -- I_{50} - 0.001 ug/ml (comparable to triazoles)

○ Isomerase inhibition closely coupled to 14-reductase activity

○ Isomerase inhibition alone -- little effect on growth or whole plant assays

Figure 8. 8-Isomerase as a fungicide target.

In another approach to test the necessity of the isomerization, or whether buildup of Δ8 sterols caused any deleterious effects, we have studied a *Neurospora crassa* mutant, erg1, which has been reported to have a lesion in the pathway at the isomerase step (9). The erg1 strain was cultured and the major sterols were isolated. Analysis by NMR and mass spectra served to identify a family of Δ8 sterols with the structures and relative amounts shown in Figure 9.

Sterols Isolated from Neurospora crassa mutant erg 1

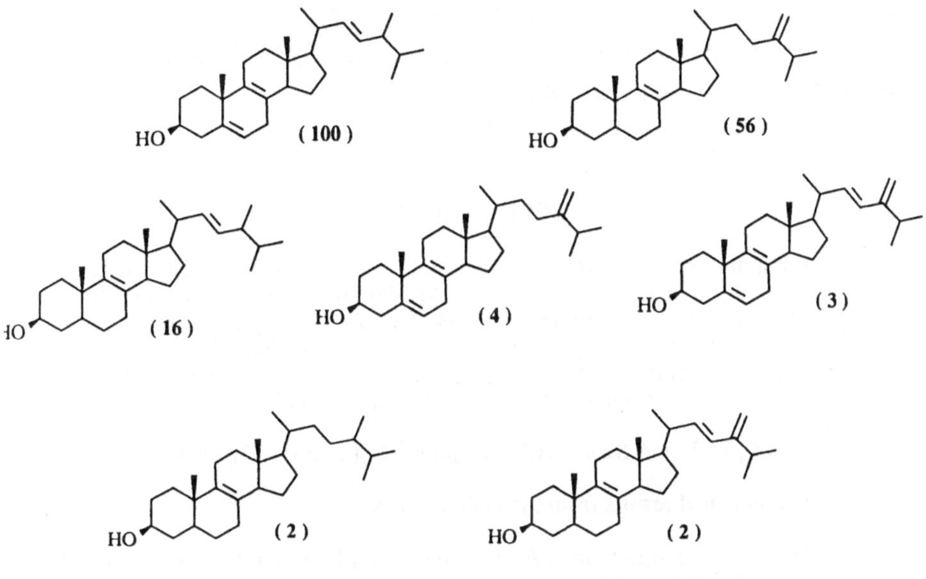

○ erg 1 grows normally with Δ8 sterols

⟹ 8 - 7 isomerase may not be a lethal target

Note: The above sterols were identified by both GC/MS and by NMR after isolation

Figure 9. Sterols isolated from *N. crassa* erg 1. Numbers in parentheses refer to relative peak areas of the sterol fraction in the GC using an FID detector.

A key feature of the erg1 mutant of *N. crassa*, is that it grows nearly as well as wild type. If any ergosterol is produced it is at an undetectable level. These findings indicate that accumulation of Δ8 sterols is not toxic and they appear to substitute well for ergosterol in this non-pathogen membrane system. This in conjunction with the lack of growth inhibition with the better isomerase inhibitors leads us to question the importance of the 8-isomerase as a target (10).

In conclusion, the biochemical driven approach presented here offers an opportunity to discover new modes of action. While clearly not intended to displace other methods of fungicide discovery and development, it is hoped that it will offer a new approach to combating resistance to agricultural fungicides.

ACKNOWLEDGEMENT

We thank the many people at Du Pont who support the biochemical approach to discovery and those involved in the sterol biosynthesis project specifically Dr. J. Steffens, Dr. D. Jordan, Dr. S. Vollmer, and Dr. R. Olson.

REFERENCES

1. For a discussion of fungicide modes of action see: (a) Buchenauer, H., Mechanism of action of triazolyl fungicides and related compounds. In Modern Selective Fungicides, ed. H. Lyr, John Wiley and Sons, New York, NY, 1987, pp 205-232. (b) Kerkenaar, A., Mechanism of action of morpholine fungicides. In Modern Selective Fungicides, ed. H. Lyr, John Wiley and Sons, New York, NY, 1987, pp 159-172.

2. Henry, M.J. and Sisler, H.D., Pesticide Sci., 1981, 12, 98-102.

3. Unpublished results from the author's labs.

4. Stutz, A., Georgopoulos, A., Granitzer, W., Petranyi, G., and Berney, D., J. Med. Chem., 1986, 29, 112-125.

5. Ryder, N.S., Antimicrobial Agents and Chemotherapy, 1985, 27, 252-256.

6. Ryder, N.S., Frank, I. and Du Pont, M., Antimicrobial Agents and Chemotherapy, 1986, 29, 858-860.

7. Henry, M.J. and Trivellas, A.E., Pestic. Biochem. Physiol., 1989, 35, 89-96.

8. Jordan, D.B., Squalene Epoxidase: the elusive flavoenzyme of sterol biosynthesis. In Flavins and Flavoproteins, ed. B. Curti, S. Ronchi and G. Zanetti, Walter de Gruyter, Berlin-New York, 1991, pp 865-868.

9.   The mutant strain *Neurospora crassa* erg 1 was obtained from the Fungal Genetics Stock Center, Department of Microbiology, University of Kansas Medical Center, Kansas City, Kansas 66103. FGSC stock no. 2722, mating type A.

10.   A recent report has indicated that deletion of the 8-isomerase gene in yeast did not prove lethal to the organism. See: Ashman, W.H., Barbuch, R.J., Ulbright, C.E., Jarrett, H.W. and Bard, M., <u>Lipids</u>, 1991, 26, 628-632.

POTENTIAL FOR SYNERGISING HERBICIDES THROUGH MODIFICATION OF METABOLISM

W. JOHN OWEN
DowElanco Ltd.
Letcombe Laboratory, Letcombe Regis, Wantage, Oxon., OX12 9JT, U.K.

ABSTRACT

There are numerous examples in the literature of herbicides whose species selectivity has been primarily attributed to differential metabolism. In addition, resistant weed biotypes (e.g. chlotoluron-resistant blackgrass (*Alopecurus myosuroides*) have emerged for which elevated metabolism appears to form the basis for the acquired resistance. Clearly, metabolism can play a key role in the response of a plant to the particular herbicide and, consequently, an ability to manipulate herbicide detoxification in weed species offers a potential opportunity for synergy. This may lead to a broadening of spectrum to include previously difficult to control species as well as synergistic mixtures for eradication of resistant biotypes. The present paper briefly outlines the key herbicide detoxification reactions responsible for species selectivities, with emphasis on oxidative reactions and glutathione conjugation and reference to the underlining biochemistry. Published literature on the modification of these degradative mechanisms by such mixed function oxidase inhibitors as 1-aminobenzotriazole (ABT), piperonyl butoxide, tetcyclacis and paclobutrazol (plant growth regulators), a number of azole fungicides and the herbicide tridiphane is reviewed. The effect of tridiphane on glutathione *S*-transferase mediated detoxification will also be discussed. Some comparisons between *in vivo* and *in vitro* results are made. Finally, some of the problems associated with the possible implementation and further exploitation of this approach to synergy are discussed, not least of which would be the need for selective synergists to avoid loss of crop selectivity.

INTRODUCTION

Herbicide mixtures are commonly used in agriculture to broaden the spectrum of weed species controlled. In some cases combinations give spectacularly good control at dose rates considerably below those normally utilised in single applications. Such an effect is clearly synergistic, and thus a mixture that provides a greater herbicidal effect than that expected from the sum of each of the components used separately is referred to as a synergistic mixture. Though there are numerous examples of herbicide synergy in the literature it is not the intention to review these here, and the reader is referred to the key publications of Colby[1] and Akubundu *et.al.*[2] and the recent reviews of Caseley & Kemp[3] and Gressel[4] for further information.

The majority of agricultural herbicides are of necessity selectively phytotoxic to a spectrum of unwanted weed species whilst leaving unharmed the crop for which they have been developed. What is achieved through the use of synergists is an expansion of the selectivity thresholds between crop and weeds which may lead to the development of synergistic mixtures for eradication of resistant biotypes as well as to a broadening of efficacy spectrum to include hitherto difficult to control species. Clearly there are a number of mechanisms which may form the basis of such synergistic interactions and in most cases will reflect those which have been determined to account for herbicide selectivity. These include modified uptake, translocation or active-site sensitivity or the deliberate creation of herbicide resistant crops through mutational breeding or gene-transfer techniques. However, the most important crop selectivity mechanism is differential metabolism of the herbicide between crops and weeds. This mechanism accounts for crop selectivities to numerous herbicides including triazines, acetanilides, phenoxy-alkanoic acids, diphenyl ethers, imidazolinones, sulphonylureas and others (see recent reviews by Cole et.al.[5], Hatzios and Penner[6], Brown et.al.[7] and Owen[8]). In addition, resistant weed biotypes such as chlortoluron-resistant blackgrass (*Alopecurus myosuroides*) have emerged for which elevated metabolism appears to form the basis for the acquired resistance.

For these reasons an ability to manipulate herbicide detoxification in weed species offers a potential opportunity for synergy. Because of the importance of oxidative metabolism in detoxification processes emphasis will be placed on developments in this area of plant/herbicide interactions. Consideration is also given to glutathione conjugation which has already been implicated as the mechanism that accounts for the synergy of atrazine by tridiphane. Reports of synergistic effects of azole fungicides in combination with urea herbicides will also be evaluated and discussed in the context of interference with oxidative metabolism.

OXIDATIVE METABOLISM

Cytochrome P450

Oxidative metabolism is a major mode of detoxification of xenobiotics by plants[9], and is commonly referred to as a Phase 1 or activation step. In mammalian systems, subsequent conjugation of the newly formed -OH group with such constituents as glutamic acid, cysteine or glutathione (Phase 2) generates a more hydrophilic moiety that can be excreted in urine or bile. In plants there is no simple route for the elimination of oxygenated products which generally have lower biological activity, but may either become sequestered into structural polymers or glycosylated (Phase 2) as they traverse the tonoplast membrane, resulting in at least temporary compartmentation in the vacuole.

The various types of oxidation reactions encountered in herbicide detoxification include aliphatic and aromatic hydroxylations, O-and N-dealkylations, epoxidation, sulphoxidation and the formation of N-Oxides. O- and N-dealkylations proceed via an initial oxidation event but the frequent instability of the hydroxylated intermediate commonly results in cleavage of the hydroxy-alkyl moiety. It is frequently assumed that all of these Phase 1 reactions are catalysed by the microsomal cytochrome P450 monoxygenase system, but many reactions presumed to be P450-mediated may be catalysed by peroxidases or other oxygenases[10]. In mammals cytochrome P450 is well characterised and exists as a group of isoenzymes with differing substrate specificities, which have protohaeme as a prosthetic group and which bind CO to form a complex with a strong absorption band at 450nm. Cytochrome P450 binds both substrate xenobiotic and O_2 to form a ternary complex within which it catalyses the transfer of one atom of oxygen to oxygenate the xenobiotic, usually forming an -OH group, whilst the second atom of oxygen forms water.

$$NADPH + H^+ + O_2 + RH \longrightarrow ROH + H_2O + NADP^+$$

Oxidised cytochrome P450 can exist in either high or low spin forms and binds substrate, RH, to generate complexes with characteristically altered spectroscopic properties. Substrate binding to the high spin form generates a Type 1 complex characterised by an absorbance maximum at 390nm, whereas binding to the low spin form (Type 2) is associated with an absorbance maximum around 420nm. The catalytic requirement for reducing equivalents provided by reduced pyridine nucleotide is mediated by NADPH: cytochrome P450 reductase and/or by the cytochrome b_5 system in concert with NADH-cytochrome b_5 reductase. Both reductases are membrane-bound flavoproteins and have been purified from plant tissues[11,12].

In green plants, cytochrome P450s are much less well characterised resulting in part from their low abundance in plant tissues coupled with problems of stability under conditions hitherto employed in the isolation of plant microsomal preparations which denatured cytochrome P450 to the inactive P420 form. As will be dicussed below, however, recent improvements in technique are beginning to overcome these difficulties with the result that evidence in support of the belief that plant P450s have properties similar to those of their mammalian counterparts is accumulating.

Role of Microsomal Cytochrome P450s in Plant Metabolism

Some mammalian P450 enzymes are concerned with the normal metabolism of endogenous substrates, such as the oxygenation reactions required in the formation of steroid sex hormones, bile acids, and the active forms of vitamin D. In plants, P450 is implicated in the 4-hydroxylation of trans-cinnamic acid[13], a key step in the synthesis of most coumarins, tannins, lignins and flavanoids, the hydroxylation of geraniol and nerol[14] in the formation of alkaloids, and in the hydroxylation of kaurene[15], an obligate step in the biosynthesis of the giberellin family of plant hormones. P450 is also involved in the hydroxylation of some fatty acids[16], the formation of phytoalexins and the demethylation steps of sterol biosynthesis, notably the obtusifoliol 14 α-methyl demethylase.

The giberellin biosynthetic pathway has long been recognised as a validated target for the development of novel plant growth regulators for use in agriculture. Two recently commercialised compounds of differing chemistries, paclobutrazol and tetcyclacis(Fig.3), to emerge from this approach have been characterised as potent inhibitors of the cytochrome P450-dependent kaurene oxidase[17,18].

Following the identification of the 14 α-demethylation step in ergosterol biosynthesis as the fungal target site of commercial triazole and imidazole fungicides (Fig.1; see Kato[19]), it has been demonstrated that these compounds may also bind to the haeme moiety of the cytochrome P450s in plant and mammalian cells in addition to fungi and yeasts. In the case of these azoles the binding is characterised by a Type 2 spectrum with a maximum absorption around 420-430nm[20]. In plants, triazole fungicides and plant growth regulators have been shown to interact with microsomal P450-dependent obtusifoliol 14 α-methyl demethylase from maize[21]. Not surprisingly, therefore, a constant challenge in the search for efficient fungicides has been their inherent plant growth regulator effects. This property has evidently been exploited in order to develop herbicides[18]. More recently the activity of a novel class of herbicidal imidazole carbonic acid esters has been attributed to inhibition of plant sterol biosynthesis at the obtusifoliol 14 α-methyl demethylase step[22]

The involvement of cytochrome P450 in hydroxylation reactions can be indicated indirectly with the use of inhibitors. Inhibition by carbon monoxide through binding to the haeme prosthetic group can be reversed by exposure to light of 450nm. Similarly 1-amino benzotriazole (ABT) and some of its analogues are very effective inhibitors of many P450s. ABT functions as a suicide substrate of the haeme centre in that oxidation of the 1-amino group results in generation of benzyne which binds covalently to the adjacent nitrogen atoms of the porphyrin ring thereby displacing the iron[23].

Whole plant studies with known P450 inhibitors : A role for a species of cytochrome P450 in the detoxification of several herbicides has been suggested by observed modifications to their oxidative metabolism resulting from co-treatment with ABT. In the case of the substituted phenylurea herbicides chlortoluron (Fig. 1) and isoproturon, which are metabolised in wheat by a combination of ring-methyl hydroxylation and N-demethylation[24], application of ABT via the root system gave rise to strong inhibition of their metabolism and had a synergistic effect[25]. The dominant reaction in the detoxification of both herbicides is hydroxylation of the ring alkyl substituent which was particularly susceptible to ABT treatment. Products of *N*-monodemethylation were not significantly affected. Similar results were obtained[26], in a study with wheat cell suspension cultures, suggesting to the authors that either ABT was more specific in its binding to P450s than previously indicated or that *N*-monodemethylation of

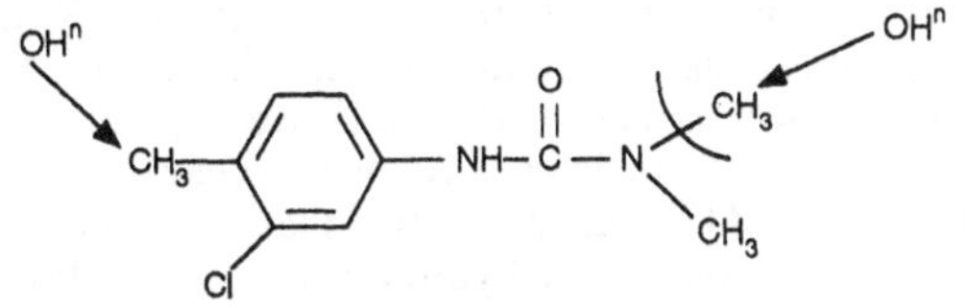

Figure 1. Sites of oxidative metabolism of chlortoluron

phenylureas is not mediated by cytochrome P450. Interestingly, no differential inhibition of the alternative modes of oxidation of chlortoluron was observed in cell suspension cultures of maize or cotton[27]. Tetcyclacis was observed to be a much more potent inhibitor of chlortoluron metabolism in both cell suspension culture studies, as was paclobutrazol in wheat cultures. Neither of these plant growth regulators differentiated between alkyl hydroxylation and *N*-demethylation.

Interesting differential effects of various monoxygenase inhibitors were noted in a study of the metabolism of bentazon in suspension cultures of rice and soybean[28]. Bentazon tolerance in both species has been attributed to ring hydroxylation (Fig. 2) which occurs in both the 6 and 8 positions in soybean but is restricted to the 6 position in rice. Whereas tetcyclacis inhibited metabolism in cell cultures of both species, ABT was effective in rice only. Conversely, the insecticide synergist piperonyl butoxide inhibited the formation of bentazon metabolites in suspension cultures of soybean but not rice. However, because of the complexity of a whole cell system, let alone the whole plant, it is unfortunately not possible to deduce from these studies the extent to which the P450s of the two species differ in sensitivity to these monoxygenase inhibitors, particularly with respect to hydroxylation at either the 6 or 8 positions.

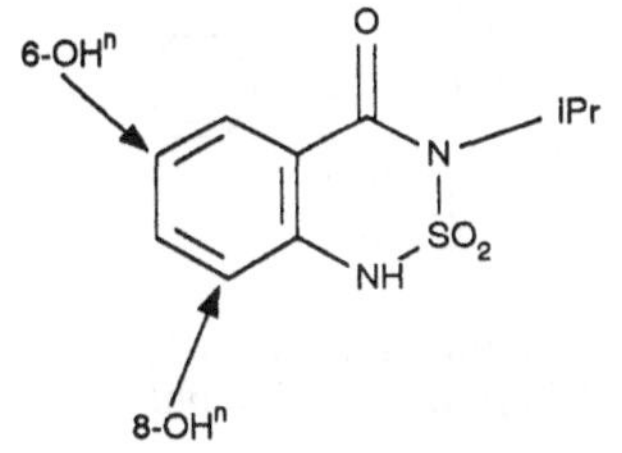

Figure 2. Ring hydroxylation of bentazon

Bentazon hydroxylation by trifoliate leaves of soybean and by wheat has also been reported to be inhibited by tetcyclacis[4,29,30]. There was a differential of 10 and 100-fold between *Chenopodium album* and soybean and *C. album* and wheat, respectively, with respect to the concentration of tetcyclacis required to enhance the activity of bentazon by 50%. Such findings lend support and encouragement to those interested in the concept of designing herbicide synergists capable of selectively targetting oxidative metabolism in weed species. Piperonyl butoxide can be a potent herbicide synergist as indicated by its efficacy in synergising atrazine and terbutryn in maize. However, since oxidative dealkylation plays only a very minor role in the rapid detoxification of atrazine in maize it is likely that in this instance a mechanism other than inhibition of a microsomal monoxygenase is operating.

In a comparative study of chlortoluron metabolism in several tolerant species, Gonneau *et.al.*[31] obtained further evidence to suggest that *N*-monodemethylation of chlortoluron was not inhibited by ABT. Sensitivity of chlortoluron metabolism to ABT varied between species according to the relative importance of the competing modes of oxidative attack. In agreement with Cabanne *et.al.*[25], wheat principally oxidised the ring methyl and this reaction proved to be very sensitive to ABT. By contrast, *Veronica* preferentially *N*-demethylated chlotoluron , a reaction which was relatively insensitive to inhibition. This agreed with the known inability of ABT to synergise chlortoluron on *Veronica* [31]. However, it is inappropriate to conclude from such experiments with whole plants that a particular oxidation step is not mediated by a species of P450. Thus, Reichhart *et.al.*[32] have reported that ABT is able to inactivate cinnamate-4-hydroxylase but is ineffective on laurate hydroxylase although both enzymes are cytochrome-

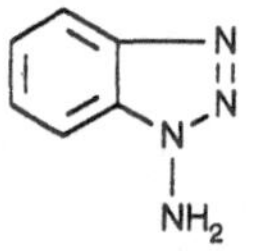

1-aminobenzotriazole (ABT)

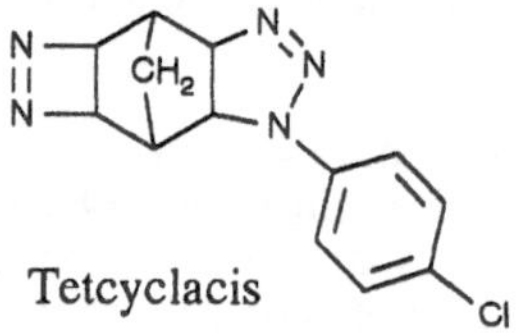

Tetcyclacis

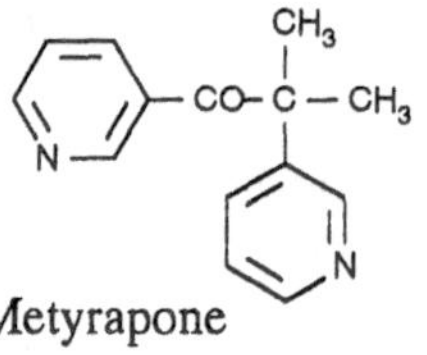

Metyrapone

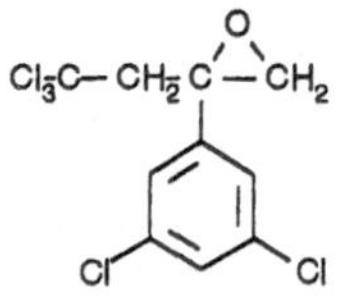

Tridiphane

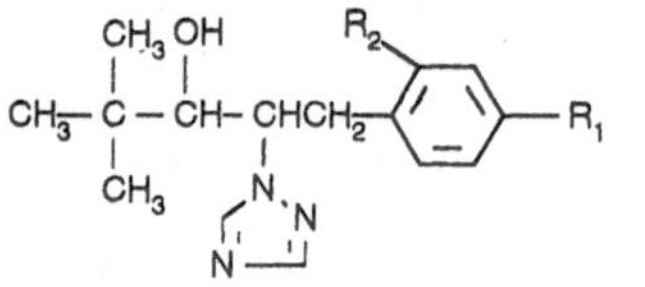

Paclobutrazol, R_1 = Cl; R_2 = H
Diclobutrazol, R_1 = R_2 = Cl

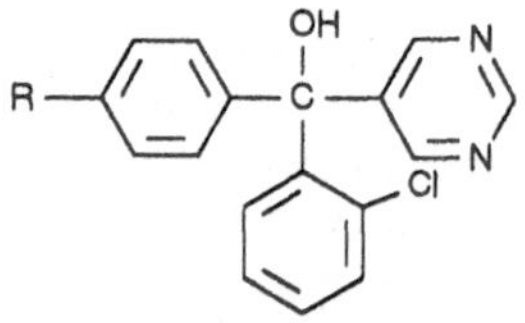

R = Cl, Fenarimol
R = F, Nuarimol

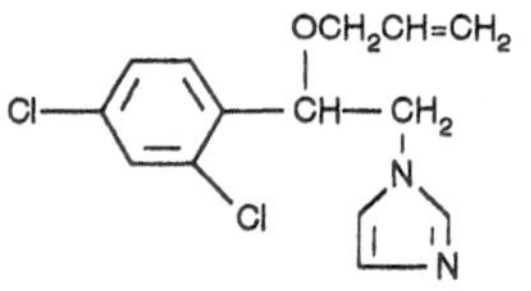

Imazalil

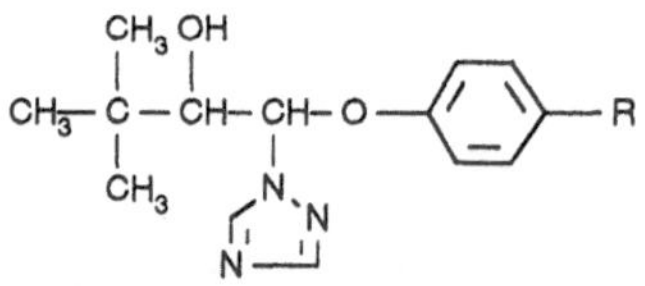

Triadimenol, R = Cl
Bitertanol, R = phenyl

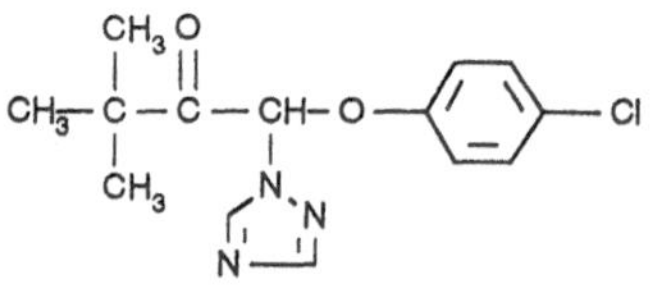

Triadimefon

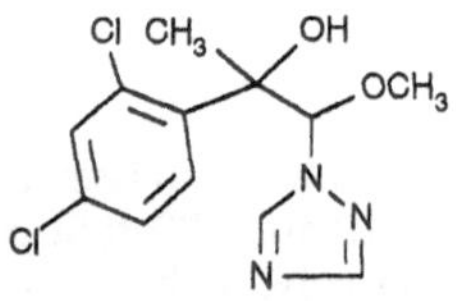

BAS 110

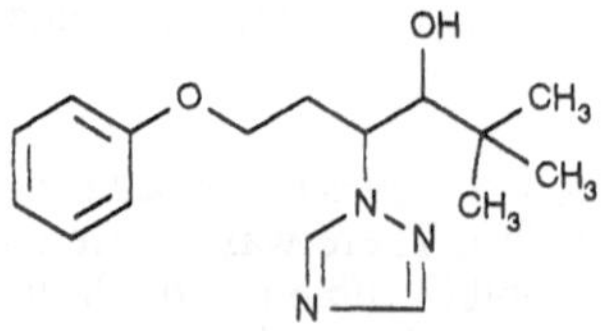

BAS 111

Figure. 3   Structures of cytochrome P450 inhibitors

P450 linked. Furthermore, Fonné[33] has shown that ABT is able to inhibit *in vitro* chlortoluron *N*-demethylase activity of Jerusalem artichoke tuber microsomes , albeit at high concentrations. It has even been demonstrated that ABT can inhibit classes of monoxygenase that contain a prosthetic group other than cytochrome P450[34].

One of the obvious applications of synergists is in the control of herbicide resistant weed biotypes which can emerge as a consequence of repeated applications of the same or a related class of herbicides over an extended period of time. Recent examples are the diclofop-methyl- and chlortoluron/isoproturon-resistant biotypes of *Lolium rigidum* and *Alopecurus myosuroides* (blackgrass), respectively, which have emerged in Australia and the United Kingdom[35,36, see also this volume]. Tolerance to chlortoluron in the case of *A. myosuroides* biotypes was generally of the order of 2-3 fold. However, a population at Peldon in Essex has consistently demonstrated an exceptionally high resistance (16-fold) compared to a typically sensitive population maintained at the Rothamsted Experimental Station.

Studies aimed at understanding the basis of the acquired resistance[3] pointed to an enhanced rate of oxidative metabolism as the cause. This conclusion was initially based on the ability of ABT to synergise both chlortoluron and isoproturon against the Peldon biotype. When added to the nutrient medium at a concentration of 10ppm, the P450 inhibitor produced a herbicide dose response in the Peldon biotype virtually identical to that of the Rothamsted population. Subsequently the same authors demonstrated in a study with C^{14}-labelled chlortoluron that its metabolism was indeed more rapid in the resistant biotype and that this was substantially inhibited by co-treatment with ABT. Further confirmation was provided by parallel studies using plant cell suspension cultures derived from both chlortoluron-resistant and susceptible biotypes of *A. myosuroides*, the rate of detoxification being elevated in the former. Again, ABT effectively inhibited this metabolism without showing any preferential effect against ring-methyl oxidation versus *N*-demethylation.

The Peldon biotype displays cross-resistance to a range of other herbicides from several classes. Knowledge of their metabolic fate in plants suggests that they have a common susceptibility to oxidative degradation and that they may therefore be subject to the same resistance mechanism, though this remains unconfirmed. If this is true, however, the implication would be that the mutation has generated a microsomal P450 enzyme with very broad substrate specificity.

The diclofop-methyl resistant *L. rigidum* appears to be a more severe problem in that there seems to be cross-resistance to an even broader spectrum of herbicides. In this case ABT and piperonyl butoxide only partially suppressed the evolved resistance to diclofop-methyl and chlorsulfuron(see[4]), suggesting that resistance in *L. rigidum* may have another or multiple basis. This conclusion is supported by the recent report that the detoxification of diclofop in etiolated wheat shoots via aromatic ring hydroxylation was sensitive to inhibition by ABT as well as by carbon monoxide and tetcyclacis[37]. Since this is the only known effective detoxification mechanism for diclofop in plants, it seems unlikely that a modified microsomal P450 is involved in the resistance of *L. rigidum* to diclofop. This has now been verified by Shimabukuro and Hoffer[38] who showed that there was no significant difference in diclofop-methyl metabolism between resistant and susceptible *L.rigidum* biotypes.

Moreland *et. al.*[39] have recently demonstrated unequivocal interaction between tridiphane (see Fig.3) and cytochrome P450 isozymes in mammalian (mouse liver) microsomes. P450 binding was confirmed by the appearance of a classical Type 1 binding spectrum. Interference with the plant microsomal P450 system involved in the detoxification of isoproturon in resistant biotypes of *A. myosuroides* would thus appear to offer a plausible explanation for the observed synergy of tridiphane in this system (J.C.Caseley, pers. comm., 1991). Other explanations may be tenable, however, since tridiphane is itself a moderately active herbicide for which a precise biochemical target is yet to be identified.

In view of the known mechanism of action of the pyridine, pyrimidine, triazole and imidazole fungicides as inhibitors of the sterol 14 α-demethylase involved in ergosterol biosynthesis, and the ability of the azoles to bind to at least some plant cytochrome P450s some workers have explored the potential of these fungicides as herbicide synergists. By far the most comprehensive study of this type was associated with the phenylurea-resistant blackgrass problem in order to provide clues as to how it might be addressed, based on a better understanding of the resistance mechanism. In all, some 13 heterocyclic nitrogen compounds were evaluated, many of which had a substantial effect on chlortoluron phytotoxicity [40].

A number of compounds modified herbicidal efficacy because of their own growth inhibiting and phytotoxicity effects. These were characterised by affecting both chlortoluron-resistant and susceptible plants without any apparent interaction with chlortoluron. This was the case with the pyridinol fungicides nuarimol and fenarimol and the imidazole imazalil (Fig.3). On the other hand, other compounds tested were indeed synergistic, enhancing chlortoluron phytotoxicity only in the resistant (Peldon) biotype in a manner analogous to that previously recorded for ABT[3]. There was evidence of synergy as opposed to additive phytotoxicity with the pyridine compound metyrapone (Fig.3), which is known to cause inhibition of certain mammalian cytochrome P450 systems[41], and this is supported by the recent demonstration that this compound was a potent inhibitor of ring hydroxylation of bentazon in soybean leaves[29].

Synergy was clearly evident with the 1,2,4-triazoles, the 3,3-dimethyl-1-(1H-2,4-triazol-1-yl)butan-2-ol derivatives exemplified by diclobutrazol, bitertanol and triadimenol being particularly active. However, the most synergistic compound was the ketone analogue triadimefon(Fig.3). Incorporation of triadimefon into the culture medium at a concentration of 1ppm eradicated resistance in the Peldon population such that its response to chlortoluron became modified to that of the susceptible Rothamsted population. It would be interesting to explore in this context the efficacy of the triazole plant growth regulators BAS 110 and BAS 111 (Fig.3) shown to be more potent than tetcyclacis in inhibiting hydroxylation of bentazon in soybean[29].

Reliable interpretation of the relative synergies of these azoles was complicated by the fact that they were generally tested as their enantiomeric mixtures, the compositions of which varied. These enantiomers can differ greatly in their ability to selectively inhibit the biosynthesis of fungal sterols, plant sterols or gibberellic acids[18] and it is thus highly likely that such differences will exist with respect to herbicide synergism. For this reason the resolved enantiomers should be tested for their relative ability to synergise herbicides detoxified by monoxygenases.

Studies with Isolated Microsomes : The use of whole plant studies in the interpretation of the relative synergistic efficacies of the various P450 inhibitors in relation to their inherent abilities to inhibit microsomal P450s is problematical because of likely differences in rates of uptake, translocation and metabolism between the inhibitors themselves. These difficulties can be circumvented by resorting to the use of isolated plant microsomes to determine the herbicide and inhibitor specificities of P450-linked monoxygenases from different species, particularly crops vs. weeds, so that the potential for selective synergy based on this mechanism can be fully evaluated biochemically.

With the exception of the monuron N-demethylase and 2,4-D p-hydroxylase activities of cotton[42] and cucumber[43], respectively, until very recently attempts to obtain plant microsomal preparations active in the oxidative metabolism of herbicides, or indeed any xenobiotic, have met with very little success. Improvements to extraction procedures such as the inclusion of glycerol, thiol reagents, PVP/XAD, protease inhibitors and, occasionally, lithium carbonate to grinding media and /or the addition of a sephadex G-25 gel filtration step, which presumably removes endogenous inhibitors, have yielded extracts active in the metabolism of several herbicides from a number of sources. These developments are summarised in Table1.

In many cases activity was barely detectable until plant tissue was pre-treated with agents such as phenobarbital, 2,4-D or certain safeners which resulted in induction of P450s thus contributing greatly to these recent successes. This is illustrated by the experimental safener CGA 154281, an N-substituted 1,4-benzoxazine, which stimulated hydroxylation of chlortoluron and primisulfuron by maize microsomal preparations some 15-fold[44,45]. A Type 2 binding spectrum was apparently obtained for tetcyclacis[46] which would confirm its interaction with P450. Mougin et. al.[47] and Zimmerlin & Durst[48] referred to atypical Type 1 spectra following treatment of wheat microsomes with chlortoluron and diclofop, respectively, though a wheat microsome diclofop binding spectrum with the general characteristics of a Type 1 has been published[37], supporting the involvement of P450. Jones

TABLE 1.
Plant microsomal systems active in oxidative herbicide metabolism

Herbicide	Source of microsomes	P450-inducing agent	Inhibitors	Reference
Bentazon	*Sorghum bicolor*	-	CO; piperonyl butoxide; tetcyclacis	Moreland *et. al.*, (1990)
Chlortoluron	*Alopecurus myosuroides*	-	-	Jones & Caseley, (1989)
Bentazon	Maize	Naphthalic anhydride	CO; tetcyclacis	McFadden *et. al.*, (1990)
Chlortoluron	Maize	CGA 154281 (a safener)	CO; tetcyclacis; paclobutrazol; propiconazole	Fonné-Pfister & Kreuz, (1990)
Chlortoluron	Wheat	2,4-D	Tetcyclacis; paclobutrazol; prochloraz; piperonyl butoxide; triadimenol; tridimefon; tridiphane; phenylurea herbicides, notably diuron	Mougin *et. al.*, (1991)
Chlortoluron	*Helianthus tuberosus*	-	-	Fonné, (1985)
Chlortoluron	Wheat	Cyometrinil	CO (ring OH^T only); ABT	Mougin *et. al.*, (1990)
Diclofop	Wheat	Phenobarbital	CO	Zimmerlin & Durst, (1990)
Diclofop	Wheat	-	CO (weak); tetcyclacis; ABT	McFadden *et. al.*, (1989)
Primisulfuron	Maize	CGA 154281	CO; tetcyclacis	Fonné-Pfister *et. al.*, (1990)
Triasulfuron	Wheat	Naphthalic anhydride	CO; piperonyl butoxide; tetcyclacis; paclobutrazol; tridiphane (weak); monoxygenase herbicide substrates	Frear *et. al.*, (1991)

and Caseley[49] also claim to have obtained a Type 1 binding spectrum for chlortoluron with microsomal preparations from resistant *A. myosuroides.*

In most, but not all cases (see[50]), herbicide oxidation was strongly inhibited by carbon monoxide. Interestingly, carbon monoxide inhibited ring-methyl oxidation of chlortoluron but not *N*-demethylation[47]. Such observations should not, however, be taken as confirmation that a P450 is not involved since other plant P450s have demonstrated various degrees of sensitivity to carbon monoxide[16]. ABT, on the other hand, inhibited both ring-methyl oxidation and *N*-demethylation[47] which might not have been expected on the basis of the effects of ABT on chlortoluron metabolism in plants of wheat and *Veronica*, discussed above. An attempt should now be made to determine whether the microsomal *N*-demethylase responsible for chlortoluron metabolism in the latter species is insensitive to ABT.

The plant growth regulator tetcyclacis was invariably found to be a potent inhibitor in the microsomal assays as was paclobutrazol and to a lesser extent the insecticide synergist, piperonyl butoxide (Table1). Fungicidal azoles such as propiconazole, prochloraz, triadimenol and triadimefon were also inhibitory to differing degrees[44,51], which possibly accounts for their varying ability to synergise chlortoluron and isoproturon-resistant blackgrass[40]. With respect to inhibition of chlortoluron *N*-demethylation by wheat microsomes triadimefon, which was the best blackgrass synergist, emerged as the most active triazole tested. None of the fungicides or plant growth regulators differentiated between the various modes of oxidative attack. Whether this reflects a very broad specificity for P450s or the involvement of a single monoxygenase in oxidative herbicide metabolism in the species examined cannot be deduced and requires further investigation.

A range of related substituted phenylurea herbicides were tested for their ability to inhibit chlortoluron metabolism by wheat microsomes. With the curious exception of isoproturon, which is known to be rapidly detoxified in wheat, all analogues demonstrated some activity, with diuron emerging as the most active (approx. 30% inhibition at 500μM). Other classes of herbicides metabolised by oxidation were recently tested as inhibitors of triasulfuron hydroxylation by wheat microsomes[50]. In this latter study the greatest inhibition was shown by chlorsulfuron while the least active herbicide was diclofop (66% and 10% inhibition, respectively, at 100μM). The possibility of some selectivity in the action of tridiphane was indicated. Using wheat microsomes this synergist was without effect on triasulfuron metabolism, but inhibited oxidation of chlortoluron by some 30% at an equivalent concentration[50,51].

SUPPRESSION OF GLUTATHIONE CONJUGATION

Though tridiphane has been implicated above as having potential for the synergy of herbicides based on monoxygenase inhibition, its ability to enhance the phytotoxicity of atrazine via a different mechanism is already well documented. Atrazine is rapidly metabolised in resistant species such as maize and the problematical grass weed giant foxtail (*Setaria faberii*) to a non-toxic glutathione conjugate, a reaction catalysed by a glutathione *S*-transferase (GST) enzyme. Tridiphane selectively synergises atrazine toxicity in grassy weeds such as barnyard grass, crabgrass, fall panicum and green and yellow foxtail in addition to giant foxtail. Synergism is not observed in maize.

A role for metabolism in tridiphane synergism was strongly suggested in a study by McCall *et.al.*[53] who found that detoxification of atrazine by glutathione conjugation was blocked in young giant foxtail plants treated with a post-emergence spray of tridiphane and atrazine. No such suppression of atrazine metabolism is observed in maize. A full explanation was provided in a very elegant study by Lamoureux and Rusness[54] who demonstrated that tridiphane was itself a glutathione *S*-transferase substrate in both maize and giant foxtail and that its glutathione conjugate was 4-fold more potent in inhibiting the giant foxtail GST enzyme. A further contributory factor was that in contrast to maize the glutathione conjugate was further metabolised only very slowly in foxtail. The need for metabolic activation of tridiphane also allows for considerable atrazine detoxification to occur during an initial lag period[4]. It is very plausible, therefore, that this combination of factors explains why tridiphane selectively synergises atrazine in giant foxtail.

However, the synergistic properties of tridiphane are not unique to atrazine since synergy has also been recorded with EPTC (a thiocarbamate) and alachlor (an α-chloroacetamide) in maize (see[4]) and cyanazine (an atrazine analogue) in blackgrass (J.C. Caseley, pers. comm., 1991). Since all these herbicides are known to be metabolised in plants by glutathione conjugation (see [8]) these effects were not particularly surprising, though the mechanisms were not investigated. A number of other classes of herbicide are known substrates for glutathione S-transferases from various species. These include diphenylethers, sulphonylureas, and the sulphoxides of methylthio-triazines and triazinones (e.g. metribuzin, see [8]). The studies on tridiphane have paved the way for investigating the properties of GST enzymes catalysing these detoxifications in key weed species to facilitate the biorational identification of additional synergists. The excellent correlation between enzyme and whole plant studies obtained with tridiphane in maize and giant foxtail suggests that such an approach might well prove fruitful.

CONCLUSION

The studies cited above clearly indicate the potential for exploiting chemistries which modify herbicide metabolism as synergists. Apart from a few notable exceptions, however, these studies have been made using either plants of crop species or enzyme preparations obtained from them. Knowing what we do about enzyme specificity there is now an obvious need to focus effort on important weeds, to learn more about their herbicide detoxification enzymology. This would enhance the likelihood of designing synergists with the desirable properties rather than compounds which reduce crop tolerance!

There may well be opportunities for discovering new uses for existing pesticides in synergistic combinations. This was illustrated by the exploratory studies with azole fungicides as potential herbicide synergists based on monoxygenase inhibition. In this case, however, since azoles and several herbicides detoxified by oxidation are targetted on wheat, compatibility needs to be checked. Apart from these concerns, additional applications of azoles to coincide with the timing of herbicide treatments would be likely to hasten the emergence of fungal pathogens with higher levels of azole resistance. For these reasons the further development of commercial fungicides as synergists is unlikely to be fruitful. However, the evaluation of related non-fungicidal compounds would certainly be appropriate.

Our knowledge of the microsomal monoxygenases involved in plant herbicide metabolism has advanced as a result of improvements in techniques that have yielded active preparations. Nonetheless, much more work still needs to be done to fully characterise these enzyme systems, from the standpoint of both mechanism and herbicide substrate specificities. This is a prerequisite to any exploitation of this emerging technology. Questions as to the specificity of ABT still need to be addressed i.e. whether the recorded differences are real or reflect variation in isolation protocols between laboratories. There is every reason to suppose, however, that such issues will be clarified as further improvements in methods for the isolation and assay of plant cytochrome P450 enzymes involved in herbicide metabolism are achieved.

REFERENCES

1.	Colby, S.R., Calculating synergistic and antagonistic responses of herbicide combinations. Weeds, 1967, **15**, 20-2.

2.	Akubundu, I.O., Sweet, R.D. and Duke, W.B., A method of evaluating herbicide combinations and determining herbicide synergism. Weed Sci., 1975, **23**, 20-5.

3. Kemp, M.S. and Caseley, J.C., Synergists to combat herbicide resistance. In Herbicide resistance in Weeds and Crops, eds. J.C. Caseley, G.W. Cussans and R.K. Atkin, Butterworth-Heinemann Ltd.,Oxford, 1991, pp. 279-92.

4. Gressel, J., Synergising herbicides. Rev.Weed Sci., 1990, **5**, 49-82.

5. Cole, D.J., Edwards, R. and Owen, W.J., The role of metabolism in herbicide selectivity. In Progress in Pesticide Biochemistry and Toxicology, vol.**6**, eds. D.Hutson and T.R. Roberts, Wiley Interscience, Chichester, 1987, pp. 57-104.

6. Hatzios, K.K. and Penner, D., Metabolism of Herbicides in Higher Plants, Burgess, Minneapolis, 1981.

7. Brown, H.M., Fuesler, T.P., Ray, T.B. and Strachan, S.D., Role of Plant Metabolism in Crop Selectivity of Herbicides. In Pesticide Chemistry : Advances in International Research, Development, and Legislation, ed. H. Frehse, VCH Verlagsgesellschaft mbH, Weinheim, 1991, pp. 257-66.

8. Owen, W.J., Herbicide metabolism as a basis for selectivity. In Target Sites for Herbicide Action, ed. R.C. Kirkwood, Plenum Publishing Corp.,New York, 1991, pp. 285-314

9. Cole, D.J. Oxidation of Xenobiotics in Plants. In Progress in Pesticide Biochemistry and Toxicology, vol.**3**, eds. D.H.Hutson and T.R. Roberts, Wiley Interscience, Chichester, pp. 199-254.

10. Lamoureux, G.L. and Frear, D.S., Pesticide metabolism in higher plants : *in vitro* enzyme studies. In Xenobiotic metabolism - *in vitro* methods, Am. Chem. Soc. Symp. Series, vol.**97**, eds. G.D. Paulson, D.S.Frear and E.P.Marks, Am. Chem. Soc., Washington D.C., pp. 72-128.

11. Benveniste, I., Gabriac, B. and Durst, F., Purification and characterisation of NADPH cytochrome P450 reductase from higher plant microsomes. Biochem. J., 1986, **235**, 365-73.

12. Jollie, D.R., Sligar, S.G. and Schuler, W., Purification and characterisation of microsomal cytochrome b_5 reductase from *Pisum sativum*. Plant Physiol., 1987, **85**, 457-62.

13. Potts, J.R.M., Weklych, R. and Conn, E.E., The 4- hydroxylation of cinnamic acid by sorghum microsomes and the requirement for cytochrome P450. J. Biol. Chem., 1974, **249**, 5019-26.

14. Madhyastha, K.M. and Coscia, C.J., Detergent solubilisation of NADPH cytochrome c(P450) reductase from the higher plant *Catharanthus roseus*. J. Biol. Chem., 1979, **254**, 2419-27.

15 Hanson, E.P. and West, C.A., Properties of the system for mixed function oxidation of kaurene and kaurene derivatives in the microsomes of the immature seed of *Marah macrocarpus*. Plant Physiol., 1976, **58**, 429-34.

16. Soliday, C.L. and Kolattukudy, P.E., Midchain hydroxylation of 16-hydroxypalmitic acid by the endoplasmic reticulum fraction from germinating *Vicia faba*. Arch. Biochem. Biophys., 1978, **188**, 338-47.

17. Sugaranam, B., Diastereoisomers and enantiomers of paclobutrazol : their preparation and biological activity. Pestic. Sci., 1984, **15**, 296-302.

18 Burden, R.S., Carter, G.A., Clark, T., Cook, D.T., Croker, S.J., Deas, A.H.B., Hedden, P., James, C.S. and Lenton, J.R., Comparative activity of the enantiomers of triadimenol and paclobutrazol as inhibitors of fungal growth and plant sterol and gibberellin biosynthesis. Pestic. Sci., 1987, **21**, 253-67.

19. Kato, T., Sterol biosynthesis in fungi, a target for broad spectrum fungicides. In Chemistry of Plant Protection, eds. G. Haug and H. Hoffman, Springer, Berlin, 1986, pp. 1-24.

20. Vanden Bossche, H., Marichal, P., Gorrens, J., Bellens, D., Verhooven, H., Coene, M.L., Lauwers, W. and Janssen, P.A.J., Interaction of azole derivatives with cytochrome P450 isozymes in yeast, fungi, plants and mammalian cells. Pestic. Sci., 1987, **21**, 289-306.

21. Taton, M., Ullmann, P., Benveniste, P. and Rahier, A., Interaction of triazole fungicides and plant growth regulators with microsomal cytochrome P450- dependent obtusifoliol 14 α-methyl demethylase. Pestic. Biochem. Physiol., 1988, **30**, 178-89.

22 Streit, L., Moreau, M., Gaudin, J., Ebert, E. and Vanden Bossche, H., A novel imidazole carboxylic acid ester is a herbicide inhibiting 14 α-methyl demethylation in plant sterol biosynthesis. Pestic. Biochem. Physiol., 1991, **40**, 162-8.

23. Ortiz de Montellano, P.R. and Mathews, J.M., Autocatalytic alkylation of the cytochrome P450 prosthetic haem group by 1-aminobenzotriazole. Biochem. J., 1981, **195**, 761-4.

24. Ryan, P.J., Gross, D., Owen, W.J. and Laanio, T.L., The metabolism of chlortoluron, diuron and CGA 43057 in tolerant and susceptible plants. Pestic. Biochem. Physiol., 1981, **16**, 213-21.

25. Cabanne, F., Huby, D., Gaillardon, P., Scalla, R. and Durst, F., Effect of the cytochrome P450 inactivator 1- aminobenzotriazole on the metabolism of chlortoluron and isoproturon in wheat. Pestic. Biochem. Physiol., 1987, **28**, 371-80.

26. Canivenc, M-C., Cagnac, B., Cabanne, F. and Scalla, R., Induced changes of chlortoluron metabolism in wheat cell suspension cultures. Plant Physiol. Biochem., 1989, **27**, 193-201.

27. Cole, D.J. and Owen, W.J., Influence of monoxygenase inhibitors on the metabolism of the herbicides chlortoluron and metolachlor in cell suspension cultures. Plant Science, 1987, **50**, 13-20.

28. Sterling, T.M. and Balke, N.E., Bentazon uptake and metabolism by cultured plant cells in the presence of monoxygenase inhibitors and cinnamic acid. Pestic. Biochem. Physiol., 1990, **38**, 66-75.

29. Leah, J.M., Worrall, T.C. and Cobb, A.H., A study of Bentazon uptake and metabolism in the presence and absence of cytochrome P450 and acetyl-coenzyme A carboxylase inhibitors. Pestic. Biochem. Physiol., 1991, **39**, 232-9.

30. Fritsch, H., Rademacher, W. and Retzlaff, G., Inhibition of plant growth, gibberellin biosynthesis and cinnamate-4-monoxygenase by selected growth regulators. Proc Int. Bot. Congr., Berlin, 1987, Abstr. p.2-113b.

31. Gonneau, M., Pasquette, B., Cabanne, F. and Scalla, R., Metabolism of chlortoluron in tolerant species : Possible role of cytochrome P450 monoxygenases. Weed Res., 1988, **28**, 19-25.

32. Reichhart, D., Simon, A., Durst, F., Mathews, J.M. and Ortiz de Montellano, P.R., Autocatalytic inactivation of plant cytochrome P450 enzymes : selective inactivation of cinnamic acid 4-hydroxylase from *Helianthus tuberosus* by 1-aminobenzotriazole. Arch. Biochem. Biophys., 1982, **216**, 522-9.

33. Fonné, R., Intervention du cytochrome P450 des vegetaux superieurs dans l'oxydation de composes exogenes : l'aminopyrine et le chlortoluron. These de Doctorat, Universite Louis Pasteur, Strasbourg, 1985.

34. Blee, E. and Durst, F., Hydroperoxide-dependent sulfoxidation catalysed by soybean microsomes. Arch. Biochem., 1987, **254**, 43-52.

35. Powles, S.B. and Liljegren, D., A grass weed (*Lolium rigidum*) biotype displaying cross-resistance to herbicides with different modes of action : first studies on the mechanism of cross-resistance. Weed Sci. Soc. Am. Abstr. 1988, p.187.

36. Moss, S.R., Herbicide resistance in blackgrass (*Alopecurus myosuroides*). Proc. of the British Crop Protection Conference - Weeds, 1987, **3**, 879-86.

37. McFadden, J.J., Frear, D,S. and Mansager, E.R., Aryl hydroxylation of Diclofop by a cytochrome P450 dependent monoxygenase from wheat. Pestic. Biochem. Physiol., 1989, **34**, 92-100.

38. Shimabukuro, R.H. and Hoffer, B. L., Metabolism of Diclofop-methyl in susceptible and resistant biotypes of *Lolium rigidum*. Pestic. Biochem. Physiol., 1991, **39**, 251-60.

39. Moreland, D.E., Novitzky, W.P. and Levi, P.L., Selective inhibition of cytochrome P450 isozymes by the herbicide synergist tridiphane. Pestic. Biochem. Physiol., 1989, **35**, 42-9.

40. Kemp, M.S., Newton, L.V. and Caseley, J.C., Synergistic effects of some P450 oxidase inhibitors on the phytotoxicity of chlortoluron in a resistant population of blackgrass (*Alopecurus myosuroides*). In Factors affecting Herbicidal activity and Selectivity - Proc. of the Eur. Weed Res. Soc. Symp., Wageningen, 1988, pp. 121-6.

41. Liebman, K.C. and Ortiz, E., Metapyrone and modifiers of microsomal drug metabolism. Drug Metab. Dispos., 1973, **1**, 184-9.

42. Frear, D.S., Microsomal *N*-demethylation by a cotton leaf oxidase system of 3-(4-chlorophenyl)1,1-dimethylurea (monuron). Science, 1968, **162**, 674-5.

43. Makeev, A.M., Makoveichuk, A.Yu. and Chkanikov, D.I., Microsomal hydroxylation of 2,4-D in plants. Dokl. Bot. Sci., 1977, **223**, 36-8.

44. Fonné-Pfister, R. and Kreutz, K., Ring-methyl hydroxylation of chlortoluron by an inducible cytochrome P450-dependent enzyme from maize. Phytochemistry, 1990, 29, 2793-6.

45. Fonné-Pfister, R., Gaudin, J., Kreuz, K., Ramsteiner, K. and Ebert, E., Hydroxylation of Primisulfuron by an inducible cytochrome P450-dependent monoxygenase system from maize. Pestic. Biochem. Physiol.,1990, **37**, 165-73.

46. Moreland, D.E., Corbin, F.T., Burton, J.D. and Maness, E.P., Metabolism of Bentazon by excised tissue and a microsomal preparation from grain sorghum seedlings. 7th Int. Congress of Pestic. Chem. Abstr., Int. Union of Pure and Applied Chemistry, Hamburg, 1990, vol.2, p.221.

47. Mougin, C., Cabanne, F., Canivenc, M-C. and Scalla, R., Hydroxylation and N-demethylation of chlortoluron by wheat microsomal enzymes. Plant Sci., 1990, 66, 195-203.

48. Zimmerlin, A. and Durst, F., Xenobiotic metabolism in plants : aryl hydroxylation of diclofop by a cytochrome P450 enzyme from wheat. Phytochemistry, 1990, 29, 1729-32.

49. Jones, O.T.G. and Caseley, J.C., Role of cytochrome P450 in herbicide metabolism. Proc. of the Brighton Crop Protection Conference - Weeds, 1989, vol.3, pp.1175-84.

50   Frear, D.S., Swanson, H.R. and Thalacker, F.W., Induced microsomal oxidation of Diclofop, Triasulfuron, Chlosulfuron and Linuron in wheat. Pestic. Biochem. Physiol., 1991, 41, 274-87.

51. Mougin, C., Polge, N., Scalla, R. and Cabanne, F., Interactions of various agrochemicals with cytochrome P450-dependent monoxygenases of wheat cells. Pestic. Biochem. Physiol., 1991, 40, 1-11.

52. McFadden, J.J., Gronwald, J.W. and Eberlein, C.V., In vitro hydroxylation of Bentazon by microsomes from naphthalic anhydride-treated corn shoots. Biochem. Biophys. Res. Commun., 1990, 168, 206-13.

53   McCall, P.J., Stafford, L.E., Zorner, P.S. and Gavit, P.D., Modeling the foliar behaviour of atrazine with and without crop oil concentrate on giant foxtail and the effect of tridiphane on the model rate constants. J. Agric. Food Chem., 1986, 34, 235-8.

54. Lamoureux, G.L. and Rusness, D.G., Tridiphane [2-(3,5-dichlorophenyl)-2-(2,2,2-trichloroethyl)oxirane] an Atrazine synergist : Enzymatic conversion to a potent glutathione S-transferase inhibitor. Pestic. Biochem. Physiol., 1986, 26, 323-42.

POTENTIAL OF NOVEL CHEMICAL APPROACHES FOR OVERCOMING INSECTICIDE RESISTANCE

JOHN A PICKETT
AFRC Institute of Arable Crops Research
Rothamsted Experimental Station
Harpenden, Herts, AL5 2JQ

ABSTRACT

Novelty in chemical approaches to pest control can provide an important contribution to Insecticide Resistance Management strategies. This novelty can take the form of insecticides with new modes of action, novel molecules not already resisted metabolically, or specifically controlled availability, all of which can be satisfied by purely synthetic chemical leads. Natural products could also be further exploited, but the fact that they are "natural" will not in itself circumvent resistance. Natural product leads fall mainly into three categories:- compounds that regulate internal processes and development of the pest organism, compounds involved in interactions between pests and the crop or other plants, and compounds employed by organisms that attack pest or related species. As well as giving rise to new toxic insecticides, natural product leads include compounds that affect development, or act as behavioural signals, e.g. the semiochemicals. Since these compounds arise from natural biosynthetic processes, they may be exploited by use of molecular biology, an approach not generally available for purely synthetic chemicals. Such approaches include the creation of transgenic crop plants able to generate products conferring defence against pests. Where semiochemicals are used in pest control, the sophisticated analytical methods now available should allow early identification of any related naturally occurring compounds, and thereby help in anticipating potential resistance problems.

INTRODUCTION

History has already amply demonstrated that, for sustained pest control, there is no realistic prospect for methods based solely on the novelty of the control agent. Nonetheless, novelty in chemical approaches can be an important component of Insecticide Resistance Management (IRM) strategies, provided that it is deployed in ways commensurate with combating resistance development [1]. Such novelty can be provided by a novel mode of action to circumvent any existing target site insensitivity, structural types immune from existing metabolic resistance mechanisms, and controlled availability. Although novelty encompasses biorational design and the use of natural product leads, in terms of overcoming resistance, it may just as easily be provided by the conventional approach of directed synthesis that has already given rise to so many useful insecticides. However, biorational design and use of natural product leads can supply new targets and can offer some advantages. When considering constraints on novel chemical control agents, particularly where natural products are concerned, due attention must be given not just to cost but also to safety and selectivity, for often there is a misconception that "natural" is synonymous with "safe".

Directed Synthesis of Novel Chemical Control Agents

Site of action resistance. The main prerequisite is an appropriate bioassay using a library of strains, each carrying an isolated target site insensitivity mechanism. This could be complemented by *in vitro* screening of resistant and susceptible target sites. In both cases, whole-organism tests will eventually be necessary to deal with aspects of penetration and metabolism where other resistance mechanisms may also be manifest.

It should not be assumed that compounds more "finely-tuned" to their biochemical target incur greater resistance factors. A.W. Farnham and B.P.S. Khambay (unpublished data) have demonstrated that the *kdr* (knockdown resistance) allele in houseflies confers approximately 16-fold resistance to all pyrethroids, regardless of their intrinsic toxicity against susceptible insects. However, this picture was less straightforward for houseflies with the *super-kdr* allele. When the pyrethroids were classed according to their alcohol moiety, resistance conferred by *super-kdr* was highest (250-fold) for structures with the α-cyano-3-phenoxybenzyl alcohol, rather lower (60-fold) to the 3-phenoxybenzyl and 5-benzoyl-3-furanyl pyrethroids, and only 12-fold to the pentafluorobenzyl group. Each group of compounds encompassed wide variations in intrinsic toxicity. This work suggested that with an appropriate screen, which in this case involved housefly strains homozygous for target-site insensitivity, it might be possible to design compounds much less affected or even unaffected by contemporary mechanisms. Indeed, although isobutylamides are generally much less active than pyrethroids, Elliott *et al.* [2] reported that *super-kdr* flies show enhanced susceptibility to the synthetic (2*E*,4*E*)-*N*-isobutyl-6-phenylhexa-2,4-dienamide and also to some naturally-occurring isobutylamides compared with a susceptible strain, and this holds for a series of compounds with related structures (A.W. Farnham & B.P.S. Khambay, unpublished data).

The site of action for the pyrethroids and isobutylamides is thought to be the protein comprising the sodium channel involved in axonal transmission of nerve impulses. Although considerable developments have taken place in electrophysiological recording from sodium channels, a natural extension of this work is to clone the sodium channel gene from resistant and susceptible insects to investigate the molecular nature of changes that confer resistance. A number of groups are actively engaged in this work, and already a section of the gene encoding several transmembrane regions together with a substantial portion of the carboxyl terminal sequence has been cloned (M.S. Williamson, unpublished data). Success in this work will, it is hoped, help to rationalise the negative cross-resistance between pyrethroids and isobutylamides, and also permit a more informed search by directed synthesis for compounds that overcome *kdr* resistance. These same approaches are being applied in a number of other areas where target site resistance is involved, for example in modifications of the GABA receptor [3,4].

Metabolic resistance. Overcoming metabolic resistance could involve the synthesis of new types of structures having similar activity to the original lead compounds, or structural modification of any toxophore based on knowledge of the particular metabolic resistance mechanism. Pirimicarb has proved an excellent aphicide, but is now widely resisted metabolically by aphids such as the peach-potato aphid, *Myzus persicae* [5]. However, Rohm and Haas recently launched a compound, RH7988 (structure I), also acting as an inhibitor of acetylcholinesterase, that is chemically sufficiently different to bypass this metabolic mechanism [6].

I

The mechanism by which aphids metabolically resist carbamates and organophosphorus pesticides involves hydrolysis and sequestration by an esterase that is overproduced as a result of gene amplification [7]. Pyrethroids are sequestered rather than hydrolysed by this enzyme and it was therefore perhaps not surprising that analogues without the ester link, such as etofenprox and NRDC 200 [8], exhibited cross-resistance. However, the flexibility of this particular toxophore and the general approach have been illustrated and may yield successes in the future.

A similar approach holds promise for combating widespread field resistance to pyrethroids in the cotton bollworm, *Helicoverpa armigera*, in Australia [9]. Studies with synergists have indicated that oxidative metabolic detoxification, probably via a polysubstrate monooxygenase system, is the major resistance mechanism present, and extensive screening has identified several "resistance-breaking" pyrethroid molecules whose potential for commercial development is being investigated.

Controlled availability. In overcoming persistent pyrethroid resistance, e.g. to permethrin in pig production units, Denholm *et al.* [10] demonstrated the value of using a compound with lower persistence, i.e. bioresmethrin. This may help in some situations but would be of limited value in the field, where persistence in the face of destruction through sunlight is required. However, chemical and physical approaches are available that could give controlled release profiles of greater complexity and could potentially be exploited to manipulate selection pressure and improve IRM strategies. One novel way is to use sunlight to release the active component and a number of protective chemical groups are known where loss is caused by the action of the ultraviolet component of sunlight [11,12]. Other mechanisms, including the action of moisture and biological agents, can be envisaged.

Natural Products

Plant defence compounds. The widely-held view that naturally-produced chemicals are likely to be safer than their synthetic counterparts, and environmentally less damaging as crop protection agents, has no scientific basis. It has recently been demonstrated that products employed naturally by plants as pesticides have no advantage over synthetic pesticides in terms of mammalian toxicology [13,14]. It is important to remember this when advocating the development of insect-resistant plants, whether by genetic engineering or conventional plant breeding.

The tremendous success in developing synthetic pyrethroids from the natural insecticidal lead, pyrethrin I, from the pyrethrum daisy, *Chrysanthemum cinerariifolium* [15], is often cited as another compelling reason why natural products should be investigated more fully. However, it must be appreciated that pyrethrin I was a much better lead than nature would normally offer: it was highly insecticidal, had high selectivity with virtually no toxicity to mammals and its only disadvantage was insufficient stability for field use. This provided one clear objective in developing synthetic analogues.

The isobutylamide class of synthetic insecticides also began as a natural product lead from plants such as black pepper, *Piper nigrum* [2]. In many ways, these compounds are much more typical of insecticidal agents produced by plants in that, as well as being unstable, they also have a more general toxicity. These two aspects combine to make a difficult development programme, but with such relatively simple structures, it is possible to envisage commercial compounds coming from this lead.

Many other lead compounds, although described as having great potential, have too many associated problems for successful development as novel insecticides. The acetogenins from the seeds of plants in the Annonaceae [16] are rather too generally biocidal and the structures too complicated for development, unless a massive synthesis programme were undertaken. Such broad-spectrum activity is consistent with evolutionary expectations, since chemicals active against a wide range of animals, fungal pathogens and even competing plants should be favoured over a highly selective agent. In the Rosaceae, for example, many species protect their seeds by releasing hydrogen cyanide from cyanogenic glycosides on damage [17]. This toxicant could not be contemplated as an insecticide lead in the current circumstances. Even some leads actively being pursued have a high general toxicity, e.g. the avermectins, although compounds already registered in this class are perfectly safe in their use because of the very low field application rates involved. However, if we combine the need for novelty in insecticides with a need for producing more benign agents to satisfy an overcritical public, then the great difficulty in synthesising these multichiral compounds, which at the moment are only commercially accessible by fermentation technology, would necessitate a tremendous effort to produce compounds with substantially lower mammalian toxicity. Azadirachtin, the tetranortriterpenoid from plants in the Meliaceae and particularly the Indian neem tree, *Azadirachta indica* [18,19], is also receiving much attention and is stimulating some extremely exciting synthesis work [20]. In terms of activity, however, it is too weak and the structure is too complex for development of a major new class of insecticides. Nonetheless, the search continues for lead compounds more like pyrethrin I. Indeed, most issues of the journals concerned with the identification of biologically active natural products contain insecticidal compounds worth consideration; for example, rhodojaponin III (structure II), a grayanoid diterpene from *Rhododendron molle* [21], and a novel diterpenoid (structure III) from *Croton linearis* [22], were described in a single recent issue of *Phytochemistry*.

II   III

Compounds that regulate processes within the target organism. Tremendous advances have been made in the study of insect neuropeptides employed as neurotransmitters and neurohormones. Recently, Kataoka *et al.* [23] published the sequence for the 41 amino acid peptide from the tobacco hornworm, *Manduca sexta*, comprising the diuretic hormone and which has activity of possible value in insect control. Much valuable information has already been amassed on the neuropeptide proctolin and the adipokinetic hormone [24,25,26], although their particular activities are probably of little value in pest control. It is not yet clear how any of these materials would be developed for the control of insects, but this area of study must be regarded as having long-term promise. Juvenile hormones have been modified to provide methoprene and kinoprene, although these insecticides are very prone to detoxification by mixed function oxidase. Other synthetic juvenoids such as pyriproxyfen are excluded here as their structures are not directly related to the natural hormones.

Compounds used to attack insects. A wide range of venoms and toxins are employed against insects by predators and parasitoids. Spider and hymenopteran venoms show high activity, but suffer from their high cationic nature preventing cuticular penetration [27,28]. Entomophagous fungi, such as *Metarhizium anisopliae*, produce toxins, the destruxins, which have an interesting insecticidal activity but are also very hydrophilic [29]. Some organisms produce compounds to prevent insect chitin formation, e.g. allosamidin from *Streptomyces* spp. [30], that may lead to the development of a true chitinase inhibitor, unlike the purely synthetic benzoylureas which interfere with chitin deposition. Various insects produce other compounds with simpler structures as toxins and defensive secretions, e.g. a non-protein amino acid dipeptide from the Colorado potato beetle, *Leptinotarsa decemlineata* [31].

Semiochemicals. Semiochemicals, or behaviour-controlling chemicals, are used by insects in interactions within their species (pheromones), or with other organisms (kairomones and allomones) such as host plants, parasitoids or predators. These naturally-produced chemical signals can also provide the novelty required in IRM strategies, and by having non-toxic modes of action, offer advantages in terms of satisfying regulatory criteria. Although the potential of such chemicals has been discussed since the 1960s, a combination of new approaches to their use, better methods for identification, and the realisation that the whole subject of the chemical ecology of the pest must be studied for successful exploitation, may lead to substantially greater use in agriculture [32].

Strategies for Developing Natural Products

Although natural products do not automatically derive toxicological advantage over synthetic chemicals, they confer other advantages in terms of development and exploitation.

Fermentor production. Manufacture of novel natural agents using organisms that perform well in fermentors has long been seen as an approach extending novelty beyond compounds that can be commercially synthesised. The production of useful secondary plant metabolites by tissue culture is often unsuccessful, since undifferentiated cultures lack the necessary organisation for bringing together diverse biosynthetic pathways. With the development of molecular biology, it may be possible to modify fermentation organisms to accomplish synthetic steps expensive to achieve in the chemical reactor. This would be of particular value for chiral compounds. The approach has yet to be clearly demonstrated and suffers considerable criticism from the exponents of chemical synthesis. However, if such an organism could, for example, express the cyclase converting farnesyl pyrophosphate to drimenyl pyrophosphate, the key step in the production of a range of promising insect antifeedants, then this expensive stage in the chiral synthesis of these compounds could be achieved more economically [33]. The recently identified aphid sex pheromones, also showing promise in the field [34], are closely related to compounds from labiate plants in the genus *Nepeta*, and this is greatly facilitating biosynthetic studies which it is hoped will lead to fermentor production of these compounds.

New crop protection strategies. For natural products to be fully exploited as novel agents, they must be used in truly integrated pest management strategies. These might include pest monitoring (to predict accurately if and when control measures should be enacted), and the use of semiochemicals, host plant resistance and trap crops made artificially attractive to lure pests away from the harvestable crop. Slow-acting but highly selective pesticides or biological control agents can then be deployed on the trap crop, thus circumventing the normal requirement for rapid knockdown of pests on the crop to be harvested.

Although some of these methods are already being practised in non-intensive agriculture, often without clear success, they must also be developed for robust and reliable use in intensive systems. An approach that combines making the main crop unacceptable, while aggregating the pest onto a trap crop where it is destroyed by some suitable means, is known as a push-pull or "Stimulo-Deterrent Diversionary Strategy" [35]. The SDDS has been demonstrated in trials within individual plants by placing an antifeedant electrostatically on the upper part of the plant, thus driving pests down onto the lower leaves where an insect growth regulator had been applied [36]. Effectiveness of the antifeedant can be improved by using inhibitors of the kairomones that attract an insect to its host and by using an attractant low down in the canopy to aggregate the pests. Insect growth regulators can be replaced with other agents, ideally fungal pathogens, the spores of which would survive better deep in the canopy away from the desiccating effects of sun and wind. In the field, this combined approach is being investigated for oilseed rape, *Brassica napus* [37].

To understand all aspects of the chemical ecology of a pest and to exploit the different types of semiochemicals effectively, it is necessary to discover what the insect is capable of detecting in its chemical environment. Various methods for obtaining electrophysiological recordings from insect olfactory and gustatory systems have been developed and coupled directly to high resolution capillary column gas chromatography (GC). Single cell recordings from the insect antenna have proved particularly useful. Electrodes are placed into the olfactory sensilla to record from individual sensory cells; these are characterised pharmacologically and the system is then used to monitor the effluent from a high resolution GC [38]. By this means, the sex pheromones for a number of aphid species have recently been identified and their long-range attractiveness to males in the field has now been demonstrated. It was also possible to investigate the activity of attractants from the host plants on which the sexual female aphids aggregate in the autumn [34]. Thus, a control strategy can be envisaged using semiochemicals either to reduce male numbers by trapping, or to lure the males into a trap, from which they escape after contacting a fungal pathogen that would then be transferred to the oviparous female population. In exploring further the role of sex pheromones in aphid chemical ecology, it has been shown that they also act as kairomones in attracting parasitoids searching for hosts in which to lay eggs in order to overwinter [39].

As well as facilitating the identification of compounds which aphids use in locating host plants, GC-electrophysiology also pinpointed volatiles from non-host plants that can be detected by the antenna. For example, the hop aphid, *Phorodon humuli*, and the black bean aphid, *Aphis fabae*, were found to respond to compounds such as 4-pentenyl isothiocyanate [32], a typical component of non-host cruciferous plants. Chemical precursors of these compounds, applied to hops, reduced numbers of *P. humuli* migrating to the crop in the spring (C.A.M. Campbell, unpublished data).

These novel approaches decrease reliance on conventional pesticides but are by no means immune from new resistance risks. Resistance to a pheromone, for instance, may entail insects employing other chemical cues in place of the original pheromonal components. Using the currently available sophisticated analytical methods, combined with electrophysiological techniques, it is possible to identify minor, and as yet inactive, compounds in many pheromone blends. For example, in samples of the hop aphid sex pheromone, although the pheromone itself comprises two particular stereoisomers of nepetalactol [34], there are also three nepetalactones to which the insect does not respond, but which it could subsequently employ as alternative chemical cues. Nonetheless, it should be possible to stay one step ahead as the insect evolves to employ different compounds, provided that registration of the new agents could be dealt with sufficiently flexibly.

Use of natural carrier systems. The successful use of the toxins from entomophagous fungi that attack insects, and those comprising arthropod venoms, is possible because these agents are delivered by a natural carrier system. A number of groups are attempting to exploit such systems, for example in the use of granulosis viruses or baculoviruses to carry peptidic sequences that would act as toxins [40,41] or otherwise cause physiological disruption. Although this has not yet been successful for insect neuropeptides, the approach shows promise.

Expression of pathways by transgenic plants. Already, genes for insecticidal proteins have been successfully transferred into crop plants. The most notable examples are the *Bacillus thuringiensis* (B.t.) endotoxin [42] and the protease inhibitor from the black-eyed bean [43]. Resistance has already occurred to sprayed B.t., and there is considerable concern about the consequences of widespread use of transgenic plants expressing this gene [44].

A great deal of natural defence against insects by plants is based on production of repellent or toxic secondary metabolites and for the future, it is hoped to extend the genetic modification of crop plants to include these as targets. Originally, crop plants were bred by sacrificing defensive secondary metabolism in the interests of yield and nutritional value but now, with the advent of tissue-, and even season-specific, promoters, it may be possible to protect particular parts of the plant more effectively by selective expression of genes. Regardless of the gene or its selective expression, to be effective it must exert an effect on the target insect, and thereby will lay itself open to the development of resistance within the population, as can occur with conventional plant breeding.

It may not be necessary to transfer whole gene sequences, but only to improve the expression within an existing pathway or to add a gene for an enzyme, perhaps even from an animal, that would complete the biosynthesis of a useful secondary metabolite. For example, plants produce farnesyl pyrophosphate, cyclisation products of which could be oxidised by endogenous enzymes to defence compounds if the particular cyclisations necessary were transferred into the plant and expressed in the appropriate parts [33].

With oilseed rape, it has already been noticed that interference with glucosinolate biosynthesis, in order to produce a more suitable crop, can reduce the ability of the plant to deal with pests and diseases, particularly those not adapted to feed or develop on crucifers. Therefore, in this case, it will be necessary to maintain the low glucosinolate requirements of the seed but to increase those metabolites in the vegetative parts of the plant which confer resistance to pests and diseases [33].

CONCLUSIONS

The novelty in chemical methods of pest control required by IRM strategies can still be provided by directed synthesis, with rational approaches applied wherever possible. This meeting is dedicated to the work of Roman Sawicki, who was an ardent exponent of novel approaches, and indeed he and I worked together on the development of the antifeedant (-)-polygodial against virus transmission by aphids [45]. However, Roman was always keen to point out that novel methods would only be of value if employed as components of IRM strategies.

REFERENCES

1. Denholm, I. and Rowland, M.W., Tactics for managing pesticide resistance in arthropods: theory and practice. *Annu. Rev. Entomol.*, 1992, **37**, 91-112.

2. Elliott, M., Farnham, A.W., Janes, N.F., Johnson, D.M., Pulman, D.A. and Sawicki, R.M., Insecticidal amides with selective potency against a resistant (*super-kdr*) strain of houseflies (*Musca domestica* L.). *Agric. Biol. Chem.*, 1986, **50**, 1347-1349.

3. ffrench-Constant, R.H. and Roush, R.T., The cloning and transformation of cyclodiene resistance in *Drosophila melanogaster*: an invertebrate GABA receptor? In *Molecular basis of drug and pesticide action (Neurotox '91)*, Society of Chemical Industry, London, in press.

4. Knipple, D.C., Soderlund, D.M., Doyle, K.E. and Henderson, J.E., Isolation of insect genes coding for voltage-sensitive sodium channels and ligand-gated chloride channels by PCR-based homology probing. In *Molecular basis of drug and pesticide action (Neurotox '91)*, Society of Chemical Industry, London, in press.

5. Devonshire, A.L. and Moores, G.D., A carboxylesterase with broad substrate specificity causes organophosphorus, carbamate and pyrethroid resistance in peach-potato aphids (*Myzus persicae*). *Pestic. Biochem. & Physiol.*, 1982, **18**, 235-46.

6. Jacobson, R.M. and Thriugnanam, M., New selective systemic aphicides. In *Synthesis and Chemistry of Agrochemicals II*, eds. D.R. Baker, J.G. Fenyes, W.K. Moberg, American Chemical Society, 1991, pp. 322-39.

7. Field, L.M. and Devonshire, A.L., this volume.

8. Elliott, M., Lipophilic insect control agents. In *Recent Advances in the Chemistry of Insect Control*, ed. N.F. Janes, Royal Society of Chemistry, London, 1985, pp. 73-102.

9. Forrester, N.W., this volume.

10. Denholm, I., Farnham, A.W., O'Dell, K. and Sawicki, R.M., Factors affecting resistance to insecticides in house-flies, *Musca domestica* L. (Diptera: Muscidae). I. Long-term control with bioresmethrin of flies with strong pyrethroid-resistance potential. *Bull. ent. Res.*, 1983, **73**, 481-9.

11. Pillai, V.N.R., Photoremovable protecting groups in organic synthesis. *Synthesis*, 1980, 1-26.

12. Liu, X., Macaulay, E.D.M. and Pickett, J.A., Propheromones that release pheromonal carbonyl compounds in light. *J. Chem. Ecol.*, 1984, **10**, 809-22.

13. Ames, B.N. and Gold, L.S., Misconceptions on pollution and the causes of cancer. *Angew. Chem. Int. Ed. Engl.*, 1990, **29**, 1197-208.

14. Ames, B.N., Profet, M. and Gold, L.S., Dietary pesticides (99.99% all natural). *Proc. Natl. Acad. Sci. USA*, 1990, **87**, 7777-81.

15. Elliott, M., Progress in the design of insecticides. *Chemistry and Industry*, 1979, 757-68.

16. Alkofahi, A., Rupprecht, J.K., Smith, D.L., Chang, Ch.-J. and McLaughlin, J.L., Goniothalamicin and annonacin: bioactive acetogenins from *Goniothalamus giganteus* (Annonaceae). *Experientia*, 1988, **44**, 83-5.

17. Nahrstedt, A., Recent developments in chemistry, distribution and biology of the cyanogenic glycosides. In *Biologically Active Natural Products*, eds. K. Hostettmann and P.J. Lea, Annual Proceedings of the Phytochemical Society of Europe, Clarendon Press, Oxford, 1987, pp. 213-34.

18. Kraus, W., Bokel, M., Klenk, A. and Pöhnl, H., The structure of azadirachtin and 22,23-dihydro-23β-methoxyazadirachtin. *Tetrahedron Lett.*, 1985, **26**, 6435.

19. Broughton, H.B., Ley, S.V., Slawin, A.M.Z., Williams, D.J. and Morgan, E.D., X-ray crystallographic structure determination of detigloyldihydroazadirachtin and reassignment of the structure of the limonoid insect antifeedant azadirachtin. *J. Chem. Soc., Chem. Commun.*, 1986, 46-7.

20. Ley, S.V., Synthesis of insect antifeedants. In *Pesticide Science and Biotechnology*, eds. R. Greenhalgh and T.R. Roberts, Blackwell Scientific Publications, 1987, 25-34.

21. Klocke, J.A., Hu, M.-Y., Chiu, S.-F. and Kubo, I., Grayanoid diterpene insect antifeedants and insecticides from *Rhododendron molle*. *Phytochemistry*, 1991, **30**, 1797-800.

22. Alexander, I.C., Pascoe, K.O., Manchard, P. and Williams, L.A.D., An insecticidal diterpene from *Croton linearis*. *Phytochemistry*, 1991, **30**, 1801-3.

23. Kataoka, H., Troetschler, R.G., Li, J.P., Kramer, S.J., Carney, R.L. and Schooley, D.A., Isolation and identification of a diuretic hormone from the tobacco hornworm, *Manduca sexta*. *Proc. Natl. Acad. Sci. USA*, 1989, **86**, 2976-80.

24. Grimmelikhuijzen, C.J.P., Graff, D., Groeger, A. and McFarlane, I.D., Neuropeptides in invertebrates. *NATO ASI Series A*, 1987, **141**, 105-32.

25. Holman, G.M., Wright, M.S. and Nachman, R.J., Insect neuropeptides: coming of age. *ISI Atlas of Science: Animal and Plant Sciences*, 1988, 129-36.

26. Goldsworthy, G.J. and Wheeler, C.H., Physiological and structural aspects of adipokinetic hormone function in locusts. *Pestic. Sci.*, 1989, **25**, 85-95.

27. Bruce, M., Bukownik, R., Eldefrawi, A.T., Eldefrawi, M.E., Goodnow, R., Jr., Kallimopoulos, T., Konno, K., Nakanishi, K., Niwa, M. and Usherwood, P.N.R., Structure-activity relationships of analogues of the wasp toxin philanthotoxin: non-competitive antagonists of quisqualate receptors. *Toxicon*, 1990, **28**, 1333-46.

28. Jasys, V.J., Kelbaugh, P.R., Nason, D.M., Phillips, D., Rosnack, K.J., Saccomano, N.A., Stroh, J.G. and Volkmann, R.A., Isolation, structure elucidation, and synthesis of novel hydroxylamine-containing polyamines from the venom of the *Agelenopsis aperta* spider. *J. Am. Chem. Soc.*, 1990, **112**, 6696-704.

29. Roberts, D.W., Toxins of entomopathogenic fungi. In *Microbial control of pests and plant diseases 1970-1980*, ed. H.D. Burges, Academic Press, London/New York, 1981, pp. 441-64.

30. Sakuda, S., Isogai, A., Matsumoto, S., Suzuki, A. and Koseki, K., The structure of allosamidin, a novel insect chitinase inhibitor, produced by *Streptomyces* sp. *Tetrahedron Lett.*, 1986, **27**, 2475-8.

31. Daloze, D., Braekman, J.C. and Pasteels, J.M., A toxic dipeptide from the defense glands of the Colorado beetle. *Science*, 1986, **233**, 221-3.

32. Pickett, J.A., Wadhams, L.J. and Woodcock, C.M., New approaches to the development of semiochemicals for insect control. *Proceedings, Conference on Insect Chemical Ecology, Tábor, Czechoslovakia, August 12-18th, 1990*, in press.

33. Dawson, G.W., Hallahan, D.L., Mudd, A., Patel, M.M., Pickett, J.A., Wadhams, L.J. and Wallsgrove, R.M., Secondary plant metabolites as targets for genetic modification of crop plants for pest resistance. *Pestic. Sci.*, 1989, **27**, 191-201.

34. Campbell, C.A.M., Dawson, G.W., Griffiths, D.C., Pettersson, J., Pickett, J.A., Wadhams, L.J. and Woodcock, C.M., The sex attractant pheromone of the damson-hop aphid *Phorodon humuli* (Homoptera, Aphididae). *J. Chem. Ecol.*, 1990, **16**, 3455-65.

35. Miller, J.R. and Cowles, R.S., Stimulo-deterrent diversion: a concept and its possible application to onion maggot control. *J. Chem. Ecol.*, 1990, **16**, 3197-212.

36. Griffiths, D.C., Maniar, S.P., Merritt, L.A., Mudd, A., Pickett, J.A., Pye, B.J., Smart, L.E. and Wadhams, L.J., Laboratory evaluation of pest management strategies combining antifeedants with insect growth regulator insecticides. *Crop Prot.*, 1991, **10**, 145-51.

37. Pickett, J.A., Towards zero pesticide residues: the biomanagement of pests and diseases of oilseed rape. *AFRC Institute of Arable Crops Research Report for 1989*, pp. 79-82.

38. Wadhams, L.J., The use of coupled gas chromatography: electrophysiological techniques in the identification of insect pheromones. In *Chromatography and Isolation of Insect Hormones and Pheromones*, eds. A.R. McCaffery and I.D. Wilson, Plenum, New York/London, 1990, pp. 289-98.

39. Hardie, J., Nottingham, S.F., Powell, W. and Wadhams, L.J., Synthetic aphid sex pheromone lures female parasitoids. *Entomol. exp. appl.*, 1991, in press.

40. Tomalski, M.D. and Miller, L.K., Insect paralysis by baculovirus-mediated expression of a mite neurotoxin gene. *Nature*, 1991, **352**, 82-85.

41. Stewart, L.M.D., Hirst, M., Ferber, M.L., Merryweather, A.T., Cayley, P.J. and Possee, R.D., Construction of an improved baculovirus insecticide containing an insect-specific toxin gene. *Nature*, 1991, **352**, 85-88.

42. Vaeck, M., Reynaerts, A., Höfte, H., Jansens, S., Beuckeleer, M. De., Dean, C., Zabeau, M., Montagu, M. Van and Leemans, J., Transgenic plants protected from insect attack. *Nature*, 1987, **328**, 33-7.

43. Hilder, V.A., Gatehouse, A.M.R., Sheerman, S.E., Barker, R.F. and Boulter, D., A novel mechanism of insect resistance engineered into tobacco. *Nature*, 1987, **330**, 160-3.

44. Marrone, this volume.

45. Gibson, R.W., Rice, A.D., Pickett, J.A., Smith, M.C. and Sawicki, R.M., The effects of the repellents dodecanoic acid and polygodial on the acquisition of non-, semi- and persistent plant viruses by the aphid *Myzus persicae*. *Ann. appl. Biol.*, 1982, **100**, 55-9.

INDEX OF CONTRIBUTORS